Robotics and automation for improving agriculture

It is widely recognised that agriculture is a significant contributor to global warming and climate change. Agriculture needs to reduce its environmental impact and adapt to current climate change whilst still feeding a growing population, i.e. become more 'climate-smart'. Burleigh Dodds Science Publishing is playing its part in achieving this by bringing together key research on making the production of the world's most important crops and livestock products more sustainable.

Based on extensive research, our publications specifically target the challenge of climate-smart agriculture. In this way we are using 'smart publishing' to help achieve climate-smart agriculture.

Burleigh Dodds Science Publishing is an independent and innovative publisher delivering high quality customer-focused agricultural science content in both print and online formats for the academic and research communities. Our aim is to build a foundation of knowledge on which researchers can build to meet the challenge of climate-smart agriculture.

For more information about Burleigh Dodds Science Publishing simply call us on +44 (0) 1223 839365, email info@bdspublishing.com or alternatively please visit our website at www.bdspublishing.com.

Related titles:

Advances in crop modelling for a sustainable agriculture
Print (ISBN 978-1-78676-240-5); Online (ISBN 978-1-78676-242-9, 978-1-78676-243-6)

Precision agriculture for sustainability
Print (ISBN 978-1-78676-204-7); Online (ISBN 978-1-78676-206-1, 978-1-78676-207-8)

Chapters are available individually from our online bookshop:
https://shop.bdspublishing.com

climate•SMART•publishing
IN AGRICULTURAL SCIENCE

BURLEIGH DODDS SERIES IN AGRICULTURAL SCIENCE
NUMBER 44

Robotics and automation for improving agriculture

Edited by Professor John Billingsley, University of Southern Queensland, Australia

burleigh dodds
SCIENCE PUBLISHING

Published by Burleigh Dodds Science Publishing Limited
82 High Street, Sawston, Cambridge CB22 3HJ, UK
www.bdspublishing.com

Burleigh Dodds Science Publishing, 1518 Walnut Street, Suite 900, Philadelphia, PA 19102-3406, USA

First published 2019 by Burleigh Dodds Science Publishing Limited

Library of Congress Control Number: 2019941174

British Library Cataloguing in Publication Data
A catalogue record for this book is available from the British Library

ISBN 978-1-78676-272-6 (Print)
ISBN 978-1-78676-275-7 (PDF)
ISBN 978-1-78676-274-0 (ePub)
ISSN 2059-6936 (print)
ISSN 2059-6944 (online)

DOI 10.19103/AS.2019.0056

Typeset by Deanta Global Publishing Services, Chennai, India

Contents

Series list

Title	Series number
Achieving sustainable cultivation of maize - Vol 1 From improved varieties to local applications Edited by: Dr Dave Watson, CGIAR Maize Research Program Manager, CIMMYT, Mexico	001
Achieving sustainable cultivation of maize - Vol 2 Cultivation techniques, pest and disease control Edited by: Dr Dave Watson, CGIAR Maize Research Program Manager, CIMMYT, Mexico	002
Achieving sustainable cultivation of rice - Vol 1 Breeding for higher yield and quality Edited by: Prof. Takuji Sasaki, Tokyo University of Agriculture, Japan	003
Achieving sustainable cultivation of rice - Vol 2 Cultivation, pest and disease management Edited by: Prof. Takuji Sasaki, Tokyo University of Agriculture, Japan	004
Achieving sustainable cultivation of wheat - Vol 1 Breeding, quality traits, pests and diseases Edited by: Prof. Peter Langridge, The University of Adelaide, Australia	005
Achieving sustainable cultivation of wheat - Vol 2 Cultivation techniques Edited by: Prof. Peter Langridge, The University of Adelaide, Australia	006
Achieving sustainable cultivation of tomatoes Edited by: Dr Autar Mattoo, USDA-ARS, USA & Prof. Avtar Handa, Purdue University, USA	007
Achieving sustainable production of milk - Vol 1 Milk composition, genetics and breeding Edited by: Dr Nico van Belzen, International Dairy Federation (IDF), Belgium	008
Achieving sustainable production of milk - Vol 2 Safety, quality and sustainability Edited by: Dr Nico van Belzen, International Dairy Federation (IDF), Belgium	009
Achieving sustainable production of milk - Vol 3 Dairy herd management and welfare Edited by: Prof. John Webster, University of Bristol, UK	010
Ensuring safety and quality in the production of beef - Vol 1 Safety Edited by: Prof. Gary Acuff, Texas A&M University, USA & Prof. James Dickson, Iowa State University, USA	011
Ensuring safety and quality in the production of beef - Vol 2 Quality Edited by: Prof. Michael Dikeman, Kansas State University, USA	012
Achieving sustainable production of poultry meat - Vol 1 Safety, quality and sustainability Edited by: Prof. Steven C. Ricke, University of Arkansas, USA	013
Achieving sustainable production of poultry meat - Vol 2 Breeding and nutrition Edited by: Prof. Todd Applegate, University of Georgia, USA	014

Acknowledgements

We wish to acknowledge the following for their help in reviewing particular chapters:

- Chapter 1: Dr Annalisa Milella, ISSIA-CNR, Italy

Introduction

Robotics and automation are having a significant impact on agriculture, one that is accelerating. This book reviews both key advances in their application and the research that is ongoing. It summarises developments in machine vision, navigation, actuation, communication and control technologies. In the second part of the book, ways are discussed to deploy these techniques to save labour, improve precision, speed and efficiency in agricultural operations. The state of the art is reviewed on the use of agricultural robots in applications such as crop spraying, irrigation and weed management. The book also addresses orchard management and harvesting, harvesting of soft fruit and in-field grading of harvested produce. It also reports on the application of robotics in the livestock sector.

Part 1 Technologies

The first part of the book discusses those robotic technologies that are of use in agriculture. Chapter 1 opens the volume by examining machine vision technologies for agricultural robots. After outlining basic concepts of machine vision and processing techniques, the chapter looks at the tasks to which they can be applied, from machine guidance and navigation to pest and disease identification as well as sorting fresh produce.

Moving on from machine vision, Chapter 2 looks at advances in actuation and control in agricultural robots. In recent years, agricultural robots have moved away from being pure sensor-carrying platforms for gathering data in the field into becoming action-delivering platforms providing physical interaction with the environment. The chapter discusses the actuation methods that are most commonly used on intelligent agricultural robots in order to control motion, physical interaction or manipulation. These actuation methods consist of both traditional actuators that have been transformed into precision farming tools, and novel actuators enabled by robotics and autonomous systems. The chapter introduces each actuator type before giving examples in agriculture. It concludes with a case study that looks at the way different actuators are used to automate strawberry production.

The subject of Chapter 3 is advances in communication and control systems in agricultural robots. Unmanned agricultural ground vehicles (UAGVs) have substantial potential to optimize crop yields and increase sustainability. Advances in sensing, communication, and control technologies coupled with Global Navigation Satellite Systems (GNSS) and Geographical Information Systems (GIS) are driving the transition from simple, off-road mechanical machines to machines with intelligence. Controller area networks (CAN) and

GNSS are contributing to the accelerated transition of tractors to become highly automated. These systems need to be safe and robust while operating in sub-optimal conditions compared to other autonomous systems. The variables encountered in the field, such as ground conditions, weather and the sheer size of the operations, highlight just a few of the challenges they face. The chapter shows how developments in communications technologies can help address these challenges. It introduces wireless communication before moving on to consider communication layers, network topology and communications technologies. The chapter considers person-to-machine (P2M) (cellular networks and MIMO broadband radio antenna networks) and machine-to-machine (M2M) (Wi-Fi, Bluetooth, ZigBee, 6LoWPAN and RFID) communication. Finally, the chapter discusses security issues.

Broadening the scope from purely robotic systems, Chapter 4 looks at advances in human-robot collaboration in agricultural robots. These are being developed for many on-farm tasks; though in practice current working agricultural robotics systems are still limited and fully robotized farms are not yet available. The chapter discusses the various types of interaction humans may have with robots and levels of human-robot collaboration. The chapter covers aspects of interface design and human-robot collaborative tasks such as detection, navigation, harvesting and spraying. The chapter concludes that the role of humans in agriculture will not be eliminated by introducing robotic systems and, should more autonomous systems become feasible, humans will still be needed for supervision and collaboration.

The section concludes with a brief overview in Chapter 5 of global positioning systems (GPS) for agriculture. In explaining how the system works, the technical concepts of code and carrier pseudo-ranges are mentioned, to focus on ways in which accuracy can be improved. Base stations and peer-differential systems are making an impact, while new constellations of satellites are promising to improve errors to a few centimetres even for low-cost systems.

Part 2 Applications

The second part of the volume looks at applications of robotic technology in agriculture. Chapter 6 examines the use of agricultural robots in crop spraying. A robotic sprayer can help reduce pesticide application while removing the human operator from the hazardous pesticide environment. The chapter provides an introduction to robotic sprayers and key challenges such as guidance and mapping, target detection and control. It includes a detailed case study of the development of a smart robotic sprayer for use in spraying vineyards, including the development of key components such as an automatic adjustable spraying device. The chapter also describes the development of an operational framework supporting human-robot collaboration.

The subject of Chapter 7 is the use of intelligent/autonomous systems in crop irrigation. Climate change, combined with the need to feed an increasing population with decreasing arable land, requires a radical re-think of the way water is delivered to crops to increase efficiency and minimize wasted water. The chapter examines how robotic and artificial intelligence can be used to improve precision irrigation in vineyards. The chapter pays particular attention to RAPID (Robot Assisted Precision Irrigation Delivery), a novel system currently being developed and tested at the University of California. The chapter presents some of the preliminary results from RAPID testing.

Chapter 8 considers the use of agricultural robots in weed monitoring and control. Weed control is essential for the production of high yielding and high-quality crops, and advances in weed control technology have had a huge impact on agricultural productivity. Any effective weed control technology needs to be both robust and adaptable. Robust weed control technology will successfully control weeds in spite of variability in the field conditions. Adaptable weed control technology has the capacity to change its strategy in the context of evolving weed populations, genetics and climatic conditions. The chapter focuses on key work in the development of robotic weeders, including weed perception systems and weed control mechanisms. The chapter addresses the challenges of robotic weed control, focussing on both perception systems, which can detect and classify weed plants from crop plants, and weed control mechanisms, covering both chemical and mechanical weed control. The chapter provides a case study of an automated weeding system.

Shifting from the field to the orchard, Chapter 9 looks at the use of agricultural robots in orchard management. The use of robotic or automated machines in orchard operations is primarily a response to growing labor shortages and costs. The introduction of robotic technologies is critical for improving yield of high-quality fruit with minimal dependence on seasonal human labor. The chapter provides an overview of robotic technologies for major tree fruit production tasks, including robotic pruning, thinning, spraying, harvesting and fruit transportation.

Chapter 10 turns to advances in automated in-field grading of harvested crops. Mechanical harvesting machines such as canopy and trunk shakers are widely used for the collection of some crops; however, most fruits and vegetables produced for the fresh market have to be collected manually. The chapter reviews the current state of mechanized collection technology, such as the development of harvest-assist platforms, as well as the possibilities of these machines to incorporate artificial vision systems to perform a pre-grading of the product in the field. The main advantages of each system are discussed and the problems encountered in the field are described. The chapter presents a case study on the use of harvest-assist platforms in citrus orchards, describing prototypes that are capable of both inspecting collected fruits and separating them into categories using computer vision.

Turning from fields and orchards to forests, Chapter 11 deals with advances in using robots in forestry operations. Advances in automation will enable forestry operations to be conducted in a more sustainable way. The chapter examines the challenges associated with using robots in forestry operations, focusing on the importance of knowing the state of the machine, where the machine is located, the location of surrounding objects, and how to plan work tasks. The chapter looks at the challenges of moving around in the forest, reaching and handling the trees, converting trees into products, and extracting logs or trees to roadside landings. The chapter also considers remote-controlled operations.

Chapter 12 moves to dairy production, considering advances in robotic milking. In recent years, growth in the number of robotic milking installations on farms has been driven by the need for better labour management and also for improved quality of life for dairy producers. The chapter reviews published research on such robotic milking systems (RMS), considering barn design, feeding management, and udder health in automated systems. The authors' field observations on RMS herds in Minnesota and Wisconsin, USA, are also included. The chapter covers feeding cows in RMS, milk quality and milk production using RMS and the economic considerations of implementing RMS on the farm. The chapter concludes that the trend towards robotic milking is set to continue into the future and the percent of dairy farms around the world using automation for milking their cows will further increase.

The volume's final chapter, Chapter 13, looks at advances in the use of robots in meat processing operations. Meat processing presents a particular challenge to robotics, as it deals with deformable biological products that lack uniformity, which makes automation extremely challenging. The chapter surveys advances in robotic automation of the processing of fish, beef, pork and lamb, as well as poultry, providing a detailed case study of the latter based on the author's own research.

Part 1

Technologies

Chapter 1

An overview of machine vision technologies for agricultural robots and automation

John Billingsley, University of Southern Queensland, Australia

1 Introduction

A mere two-and-a-half decades ago, when our own research concerned machine vision for tractor guidance, the equipment involved a tower-case computer, a black-and-white CRT monitor and a 'Video Blaster' card to grab video stream from a camcorder mounted on the roof of the tractor. Today everything that is needed can be found within a supermarket smartphone that might cost well under a hundred dollars. As an OEM component the price of the camera used in such a phone is measured in cents. When attached to drones, cameras such as this can enable video to be captured, storing hours of image data on a microSD card that is smaller than a little fingernail.

Vision is becoming an essential solution to a vast array of control and measurement tasks, not least on the farm. Any attempt to nail applications down to the equipment and the software of the day runs the danger of becoming obsolete within a very few years. The names of research topics come in and out of fashion. The fundamental principles endure, but are christened with new titles that herald a parade of papers.

Control theory has seen fashions of 'modern', adaptive, predictive, variable structure, neural and fuzzy, to name but a few. The essence of the latter pair is the synthesis of nonlinear functions of multiple variables with back-propagation

http://dx.doi.org/10.19103/AS.2019.0056.03

being used for adaptation. Convolution and correlation underlie the techniques of systems analysis and now 'deep learning' is taking to the stage.

'Deep learning' has already entered the vocabulary of vision research and no doubt there are many more terms to come. But one thing is certain. The greatest impact will continue to be made by advances that are powered by their application in the mass market.

Today every smartphone can not only locate a face in the image, but even assess the quality of the smile of the subject. Surveillance cameras are adept at identifying faces. Here is a technology that can be borrowed and adapted for agricultural use, whether to grade vegetable by shape or identify cattle as individuals.

Some of the more elementary principles of machine vision will be detailed here, but in general this chapter will attempt a broad analysis of the tasks that can be helped by vision and the possibilities for its application.

2 Basic concepts

In general, the main functions of vision are location and recognition. Location allows a vehicle to be guided and fields to be mapped. Recognition allows the thing that you are locating to be identified, such as a tree to be avoided, a ripe fruit to be harvested or a weed to be killed.

In many ways, computer vision has the potential to be greatly superior to the human eye for both functions, though the eye colludes with the brain to achieve some remarkable results of its own. In the human eye, a colour is perceived as the ratios of just three measurements, seen primarily as red, green and blue. Video systems exploit this by measuring and showing just three sets of red, green and blue dots on the screen. But visible light consists of a continuous spectrum of wavelengths which can carry very much more colour information than the human eye can differentiate.

When red light is added to green light, the result is seen as yellow. The light from a sodium street light is also seen as yellow. But in the mixture of red and green light, an object with red and green stripes will show them clearly as colours, while in sodium light they will merely appear as shades of grey.

Common camera sensors peer through a multicoloured film of dots to give an image in three planes of red, green and blue. Though the sensor can detect light a substantial way into the infrared and ultraviolet, these are blocked by a filter in the lens. Monochrome intruder cameras are the exception to this. By adding a specialised colour filter, such a camera can be made sensitive to any combination of wavelengths we choose. By comparing views through two different filters, colour differences can be perceived that the human eye cannot discern. The name given to this is 'multispectral vision' or 'microspectral vision'.

2.1 Discrimination by colour

Discrimination by colour can be implemented in many ways. Of course, in many cases the differences can be seen in the conventional views. If this is not the case, two or more cameras could be used, but it can be more expedient to spin filters of special colours in front of a single camera, as was done in the very early television systems. Simpler still is to split the view of a single camera between stripes of filters and use the progress of the vehicle to move the image of the plant from one colour filter to the next. With autonomous machines that can work at night while the farmer is asleep, lights of special wavelengths can be used instead of filters.

Because illumination is not uniform, it is necessary to consider ratios, rather than absolute brightness. To give an example, a blue object will look brighter than a red one in blue light, but darker than that red one in red light, however bright the lights are. So when seeking to see a colour difference in two sorts of leaf, a spectrophotometer can be used in research to find at least two wavelengths where in one A is more reflective than B, while in the other the reverse is true. Once those wavelengths have been determined, coloured plastic filters can be sufficient to perform the comparison.

2.2 Recognition by shape

The human brain tends to see images in terms of outlines rather than patches of shades. One of the tricks of the eye is to 'invent' lines between areas of slightly differing brightness, as can be seen in Fig. 9. This strongly suggests that outlines are a better basis for object recognition than correlation of the pixels of an image. In graphic computer art there are several ways to represent an image. The original form is the 'bitmap', an array of coloured pixels, as in a .gif, .jpg or .png file. But gaining more prominence is the SVG, scalable vector graphic, that represents a set of outlines of areas each filled with a colour.

By tracing the outline of an object, its shape can be encoded in many fewer bytes. Even fewer bytes are needed to define an 's-psi' plot. This represents the angle of a tangent as we move around the entire circumference. By encoding the angle as an 8-bit byte, giving increments of just under one-and-a-half degrees, a meaningful outline can be defined in just 256 bytes. This s-psi plot will be the same for all objects of the same shape, whatever their size. If the object is tilted, the plot values have a constant added to them. So the s-psi plot represents the shape in a form that is much simpler to use for identification by matching against templates.

3 The tools

Many of the figures in this chapter have been captured from JavaScript applications that can be found at www.essdyn.com/vision. Some of these show a window of code that can be changed in a browser to experiment with new values.

3.1 The camera

Evolving from the 'flying spot scanner' of the early days of television, today's cameras still use the concept of a 'raster'. The view is scanned by a line that runs left to right, moving down the picture from top to bottom to present a frame of data. In digital terms, each spot of the image is a 'pixel', so in a full HD scan of 1080 lines of 1920 pixels they number a couple of million, arriving at some fifty million per second. From this barrage of data our purpose might be to resolve a simple binary question such as 'is it a weed?' or just to obtain a couple of numbers defining the location of a landmark.

The software driver of the camera will grab each frame to the computer memory as a string of bytes, 3 or 4 to a pixel. Though one might think of these bytes as an array laid out as an image, they are in fact just a linear sequence. The pixel at location (50, 100) is held by the bytes starting from location (100 * width + 50) times the number of bytes per pixel. So until the fundamental camera design changes, any software has to start with this data set in order to extract the information that will perform the actual control or identification.

3.2 Image processing

Many software libraries exist with routines for processing image or video information, of which one of the best known is 'Open CV' (2018). Many of the functions in these libraries have the purpose of making the image more appealing to look at, such as 'filters' that identify edges, or sharpen or blur images by adding weighted sums of neighbouring pixel values.

The filter consists of a matrix of weights and can be thought of as a 'patch'. The patch is moved to be centred over each pixel of the image in turn then the sum is taken of each weight times the pixel value beneath it. The result is a new array of pixel values to be displayed.

3.3 Binary images

For shape analysis, images are often subjected to a threshold to simplify their analysis. Consider a very simple binary image, where each pixel is represented as either black or white, denoted by a value of either 0 or 1. It might appear as Fig. 1.

It can be 'smoothed' by adding up the pixel values in a 3 by 3 patch. The totals are shown in Fig. 2.

The new value of the central pixel is then set to 1 if the total exceeds 4, or to 0 otherwise, as shown in Fig. 3.

This has removed the straggling pixels and holes. It can be regarded as the application of the filter patch

Figure 1 Rough binary image.

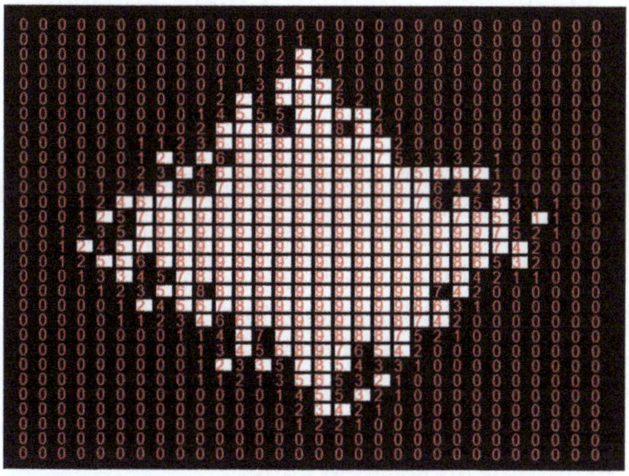

Figure 2 Numbers are shown for calculating smoothing.

Figure 3 Result of smoothing.

1	1	1
1	1	1
1	1	1

and then applying a threshold of 5.

If we now apply the 3 by 3 filter

0	–1	0
–1	4	–1
0	–1	0

to the new image, we obtain the totals shown in Fig. 4.

By setting pixels to 1 only where the total is greater than 0, we obtain an outline as shown in Fig. 5.

The central region where pixels are surrounded North, South, East and West are eliminated.

An alternative filter

–1	–1	–1
–1	8	–1
–1	–1	–1

can be used to identify edges in a slightly different way. If the central '8' is located above a pixel that is white, having value 1, the result will be 8 minus the values of any pixels around it. If all of these are also 1, so that it is surrounded, the total is 0 and the new pixel value is black. Otherwise the answer is non-zero and the new pixel continues to be white. In this way, all interior pixels are eliminated and we are left with just the edge.

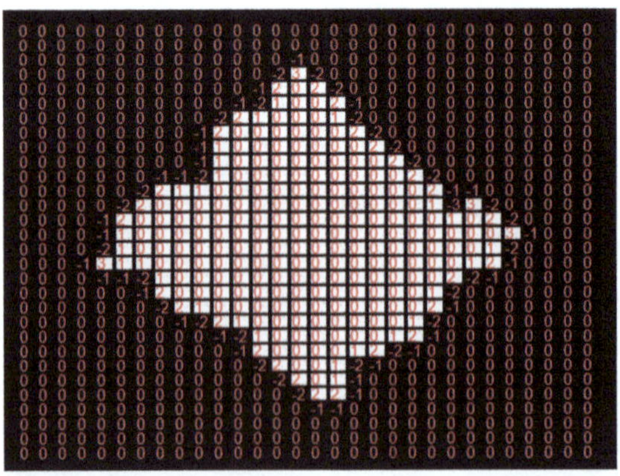

Figure 4 Numbers for detecting the edge.

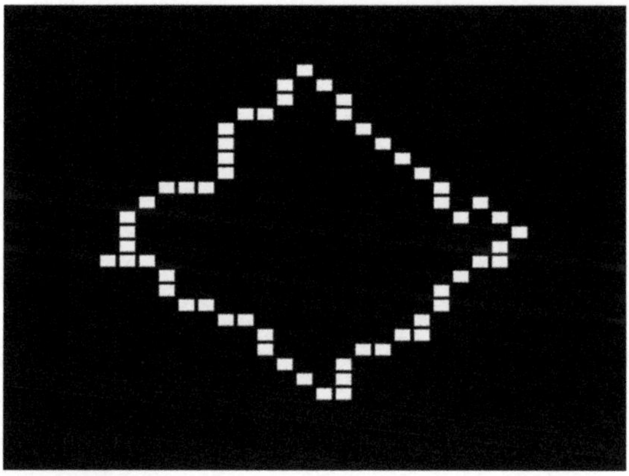

Figure 5 The edge displayed.

3.4 Colours and grey scales

When the pixel values represent shades that do not have to be binary 1s or 0s, many different functions can be performed. For more effective smoothing and removing speckles a larger filter array can be used, 5 by 5 or more, with a resulting pixel value that also represents a shade, usually an integer in the range 0–255 for each of red, green and blue.

The filters can perform blurring or sharpening. Figure 6 shows an image where the background has a gradient and a slightly lighter diamond is superimposed upon it.

Application of the 3 by 3 filter

1	−2	1
−2	4	−2
1	−2	1

where 128 is added to each resulting grey level, gives the result shown in Fig. 7.

The filter is a discrete approximation to 'del-squared' – the result is a sort of second derivative of the original.

Note that the image gradient has been removed and the bordering pixels are unaltered.

There are routines that can find whether a feature exists within an image, which work in a very similar way. A 'patch' is created from the feature being sought by subtracting its mean pixel value from each pixel within it. It thus contains positive and negative values, with a 0 total. When correlated with the image being searched, it is therefore not affected by changes in brightness. Moved around in the same way as a filter, it searches for a product that is above a threshold value in order to register a 'hit'. It is computationally very intensive.

Figure 6 Graded background and diamond.

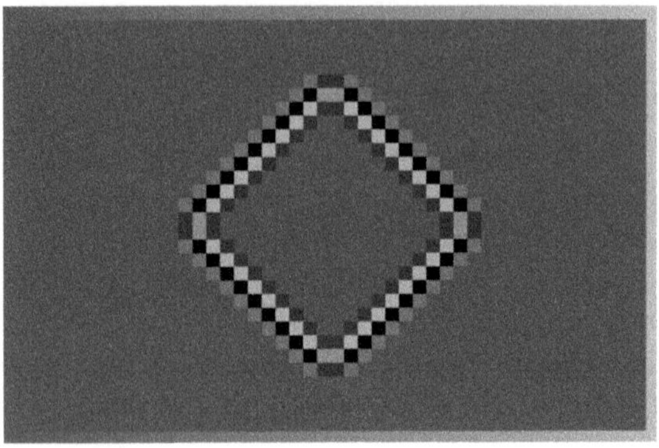

Figure 7 Image sharpened by a 3 by 3 filter.

There are other routines for use with 'visual streaming'. Successive video frames will have images in them that move from frame to frame. Sections of a past image can be correlated with a later one over a range of small displacements, so that the movement between the frames can be measured and thus relative camera velocity can be assessed.

3.5 Other methods

Two methods that my teams have developed are not to be seen in OpenCV.

The first concerns image blurring and sharpening. The technique of convolving the image with an array can be considered as a 'FIR' finite impulse-response filter. A pixel value will only influence the values of new pixels within a local range, depending on the size of the matrix.

In contrast, an analogy can be drawn with the filtering of a signal stream, such as a sound signal. The low-pass filter that consists of a resistor-capacitor network is deemed an 'infinite impulse response' filter. The influence of a sample value dies away exponentially, rather than having a finite limit. This can involve much less computation. A low-pass filter can be applied to a series x_n by just two lines of code per step, assigning to a 'slow' variable z a new value $z + (x_n - z)/k$ then saving each z in turn to a new series y_n. Here k is a parameter that represents the 'time constant' in sample steps.

If we apply the same technique to an image, the value of k will become a 'distance constant' in pixels. This will of course smear the image to the right if it is performed in the normal order. But unlike a real-time sound signal, we can then apply this same technique again from right to left. This corrects the smear and results in zero corresponding phase shift. To complete the process, we apply this down and then up for each column of pixels. The result is a blurred image. When applied to the image of Fig. 6, the result is as shown in Fig. 8.

We are more likely to wish to sharpen an image than to blur it, but if we now subtract a proportion of the blurred image from the original image and add a compensating brightness, the image is sharpened. By our choice of k and the mix proportion, we can approximate to the gradient of the brightness or merely compensate for uneven lighting (Fig. 9).

Figure 8 Image blurred by the IIR filter.

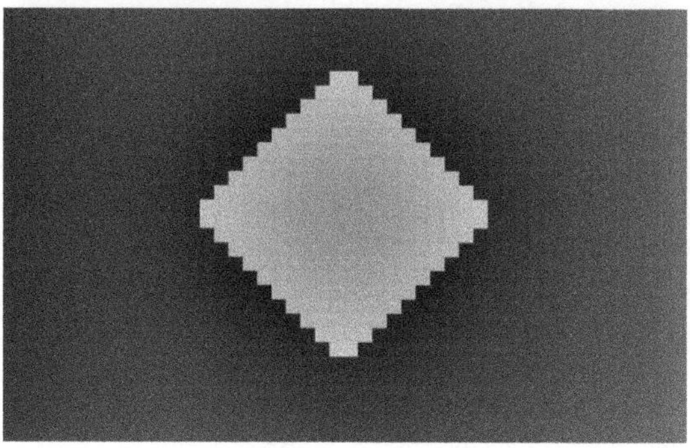

Figure 9 Image sharpened by IIR filtering.

3.6 Image tracing

The other technique is image tracing. Each pixel along the path that is to be traced is tested for being 'bright' or 'dark'. A spiral search is made to find the first boundary point where there is a change. The search then follows the path that resembles a dancing quickstep. If the present pixel is dark, the search steps forward. If the new pixel is also dark, the next step is taken forward again, but at an angle of 45° to the first, until a bright pixel is found. If the present pixel is bright, the search steps to the side, repeatedly stepping with a negative turn if the new pixel is also bright. Figure 10 shows the result of applying this to the 'diamond' image of Fig. 6.

The result of such a tracing is called a 'Freeman Chain' (Hunt, 2018), an ordered set of way points that progressively traces the boundary, rather than just the set of marked points given by the filter method. For identification the boundary data can be refined to give the s-psi plot mentioned earlier.

At each change, the coordinates of a 'way point' can be noted, enabling a graph to be drawn of the boundary. But the direction from one way-point to the next can instead be recorded, using values 0, 32, 64, 96, 128, 160, 192 and 224 for the eight points of the compass. The many points of the path can then be smoothed by taking averages of neighbouring points to reduce the array to just 256 entries. From this 's-psi' representation, a very close representation of the shape can be reconstructed.

In Fig. 11, an example is shown of tracing the image of a steer. Separating an object from its boundary can be difficult, so we have 'cheated' by setting up a blue background sheet and using 'blue or not blue' as the test. The image shows the traced image in the bottom right and the chart of gradients on the left. The red initial values have been interpolated to give the black trace. At this stage, 377 points are considered.

Figure 10 The yellow path shows the traced image.

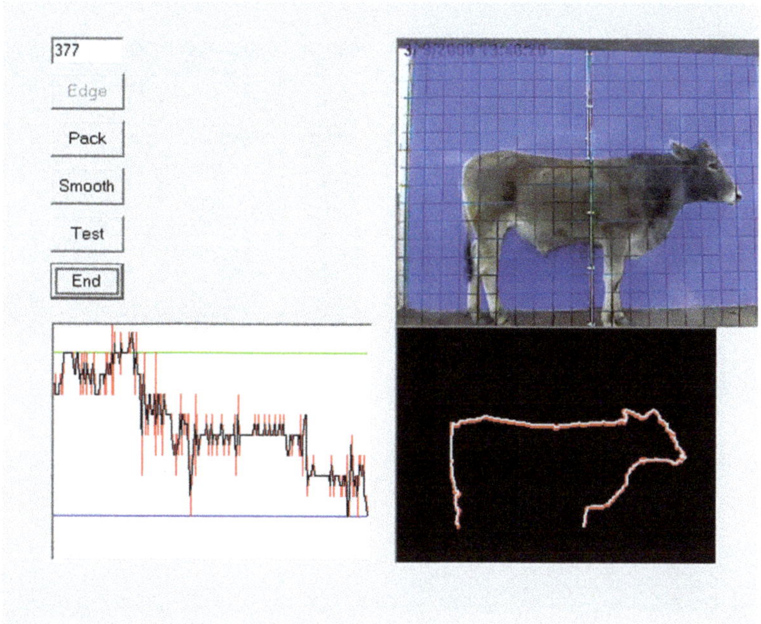

Figure 11 Tracing with a blue background.

Now the gradients are interpolated again to give the result shown in Fig. 12. The bottom right image has this time been reconstructed from the smoothed data.

After further smoothing, the values have been packed into an array of just 256 8-bit numbers. The result is shown in Fig. 13.

As can be seen, this gives a faithful representation of the outline from which identification or condition assessment can be performed. Just three more numbers are involved, to give the scale and starting coordinates.

This method has been used for identifying animals using a watering hole (Finch et al., 2006) and for assessing the condition of cattle (McCarthy et al., 2010a).

3.7 Lidar

A recent measurement method is becoming more widespread, the time-of-flight of a laser pulse. This first gained traction with a device marketed by SICK, but the method has escalated into a 3D scanner that is part of an Xbox game and into surveying devices with a kilometre range.

Since light is the basis of the measurement, it is perhaps reasonable to class it as vision.

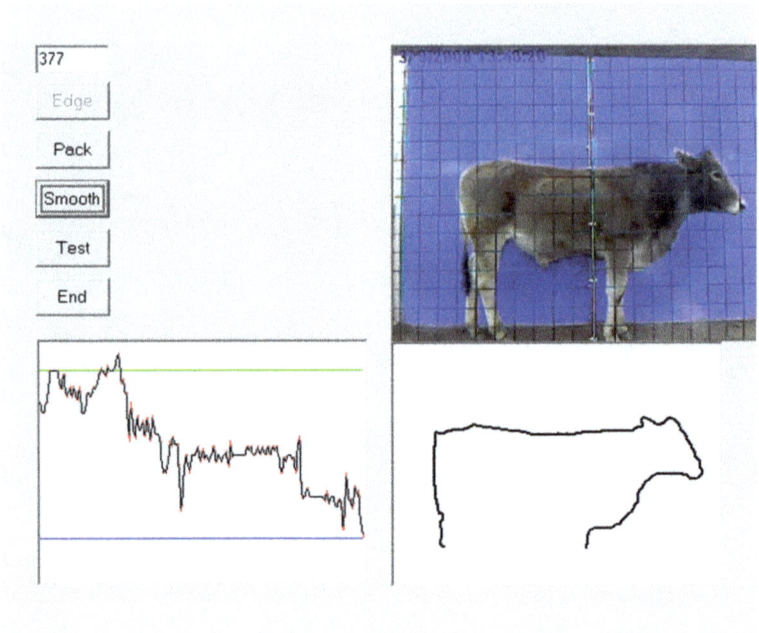

Figure 12 The gradients are interpolated to smooth them.

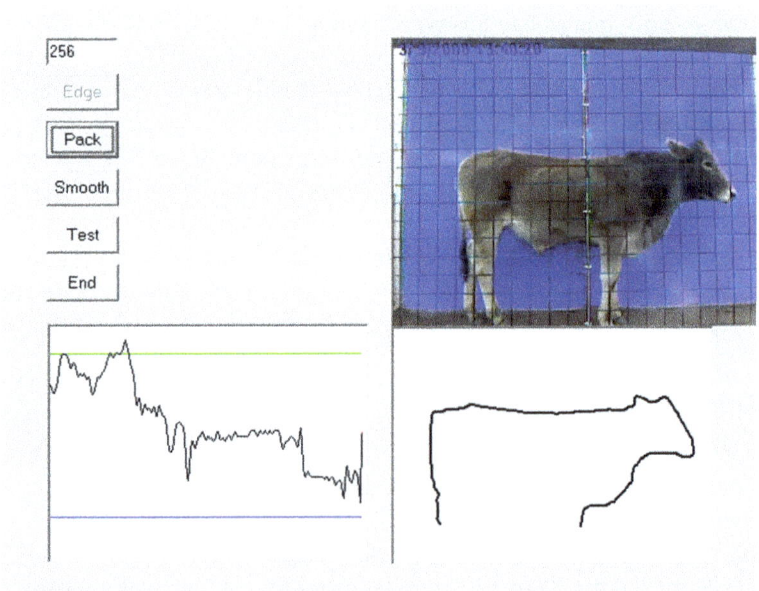

Figure 13 Data packed into 256 bytes.

In the SICK sensor (SICK, 2018), the laser beam is scanned by a rapidly rotating mirror. Originally designed as a detector to enforce a safety barrier, it became a favourite of researchers into autonomous vehicles. At a much lower price but with limited range is the Kinect 3D scanner (Microsoft, 2018) for the Xbox. It is no longer supported, but still has a role to play. Although out of the question for vehicle control, it is a strong possibility for a shape detector as part of a produce grading system.

3.8 Binocular vision

With two eyes it is possible to estimate distance – but over a very limited range. Essential for threading a needle, the separation of the human eyes will cause accuracy to be lost after a very few metres. When experimenters try to reproduce binocular range-finding, they come up with the problems of calibrating the match between the cameras and ensuring that their axes are exactly aligned. But there are easier ways.

A snake preparing to strike will wave its head from side to side, creating a baseline much wider than the separation of its eyes. If you move your own head from side to side you will see near objects moving relative to those in the distance. Thus a more promising strategy for distance estimation is to use the video stream from a single moving camera. Since displacements will be small from frame to frame, objects that have been identified can be tracked. Frames can be aligned to fix some distant point, so that the movement of nearer objects will signify their distance.

3.9 Overview of the tools

Cameras can gather image data either in isolation, in pairs or in multiples like the eyes of a spider. The image data can either be processed at once on the spot, or can be transmitted or stored – or both – as done in a drone. It can provide guidance for a human operator, or can yield control signals for an autonomous system.

To process the vision data, there are now many low-cost 'embeddable' computers on a single card. One of these, the 'BBC Micro:bit', has been provided free to all year seven schoolchildren in the United Kingdom, but still has sufficient processing power to perform many industrial control tasks. Another favourite of hobbyists, the Arduino, can support a camera, though its processing ability is somewhat limited. More substantial is the processing core of an Android tablet or smartphone.

There are few locations that lack an internet connection, though the data costs of satellite-based internet can be high. Nevertheless, remote reporting of progress or problems is easily performed by autonomous vehicles either by means of internet or cell phone.

4 The tasks

4.1 Guidance and navigation

The introduction of GPS guidance has seen the appearance of fields of ruler-straight crops. But by partnering GPS with computer vision, further great improvements can be made. There is no problem when using GPS for planting, but when it comes to the task of chipping, hoeing out the weeds between two plant rows, the tractor will be guided by the locations of where the seedlings were thought to have been planted, rather than their actual locations now. Any errors will be doubled between the two processes. The farmer will be keen to chip within a centimetre of the newly emerging crop, to remove weeds as thoroughly as possible from between the crop rows.

At the time of our first research, precision GPS carried a heavy price tag and vision guidance aimed at being much cheaper. Even today, with many new constellations of satellites, centimetre grade GPS is not cheap; but there is a method of using the cheapest 'bushwalker-grade' signal to enhance vision.

The vision strategy was extremely simple (Billingsley and Schoenfisch, 1995, 1997). Instead of processing the image for shapes or boundaries, the green emerging plants are simply regarded as 'blobs' (Fig. 14).

Their green pixels can be regarded as data points through which a 'line of best fit' can be drawn. But to do this, it is necessary to consider only the pixels that belong to a particular row. Accordingly, a window or 'viewport' is set up around each row to be fitted (Fig. 15).

And a line of best fit constructed from the pixels within it (Fig. 16).

Several rows are treated in this way to obtain a group of lines with a vanishing point (Fig. 17).

From the skewing of the lines, the lateral displacement of the tractor can be perceived (Fig. 18).

From the displacement of the vanishing point, the heading of the vehicle can be deduced, thus these two state variables can now be applied to steering. The viewports are moved instantaneously from frame to frame, while the vehicle responds in its own good time.

Figure 14 Emerging green shoots.

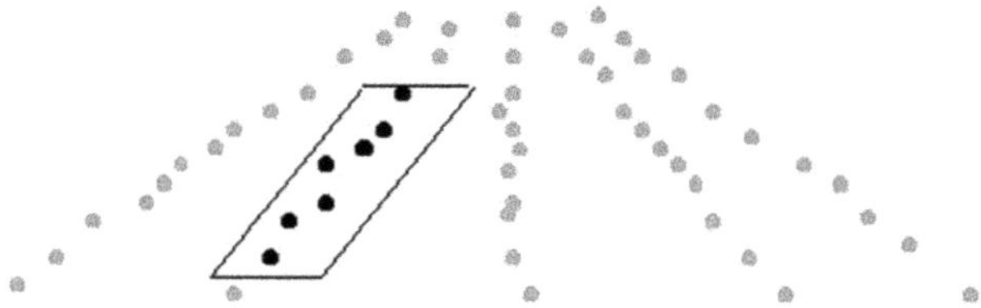

Figure 15 A viewport contains data from a single row.

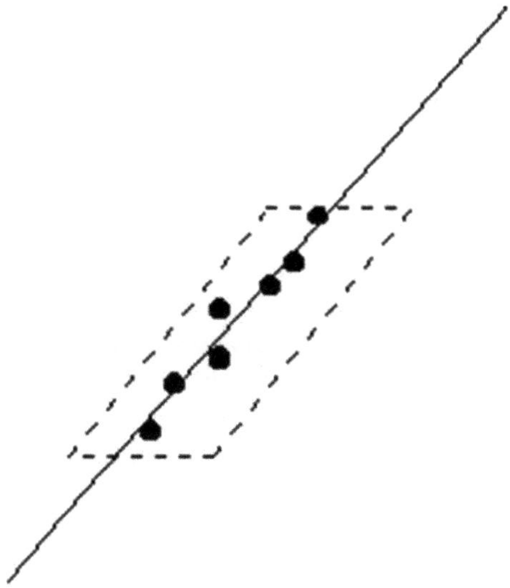

Figure 16 A line of best fit is drawn through these pixels.

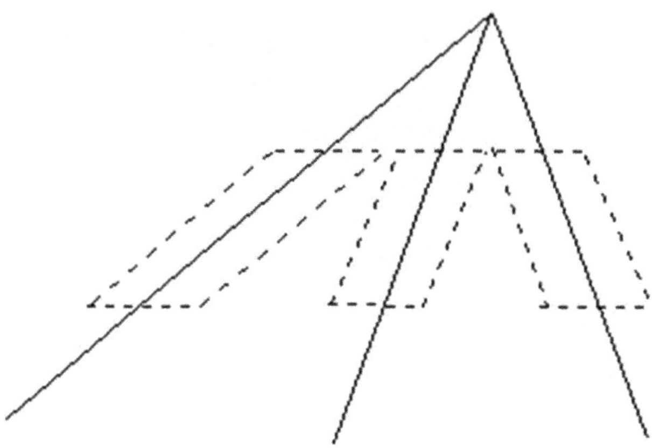

Figure 17 Two or more rows define a vanishing point.

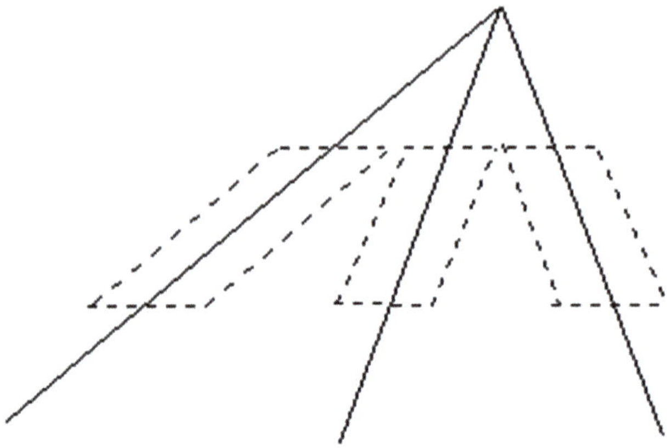

Figure 18 Heading and displacement can be deduced from the rows.

For precision tasks such as chipping, it is the position relative to the crop that is of importance, not the absolute global location. Even the cheapest of GPS receivers can measure movement by the change in phase of the 20 cm carrier wave, rather than the 300 m binary code from which 'coarse acquisition' position is computed. Thus if the GPS is one on which the binary value of the velocity can be accessed, its integral gives displacement with a great degree of accuracy. With the GPS displacement initially calibrated from the visual fit of the rows, this can take over guidance if there are barren patches where the rows can no longer be detected.

Figure 19 shows a frame taken from the video stream of a tractor being visually guided by rows of triticale. The viewports are shown in blue, with the detected green plants in red.

Figure 19 A video frame from a visually steered tractor.

4.2 Identification of pests and weeds

Once again, we have the option of identifying by shape or by colour. We can go further and use multispectral vision for the analysis. However the simplest detection systems simply look for something green (McCarthy et al., 2010b).

It is not enough to look at the green pixel value, since white also contains a strong value of green. The value of 'green minus red' will probably give an adequate discrimination, though this will also be influenced by brightness. When the harvesting is complete and the field is supposed to contain only stubble, anything that shows up as green will be sprayed.

But the more interesting task is to discriminate between weeds and a growing crop. Colour sensing involves examining the ratio of the reflectivities at different wavelengths. If a pair of wavelengths can be found where this ratio for a weed is very different from the ratio for a plant, then we have a workable sensor.

As mentioned earlier, the colour signals can be among the conventional red, green and blue, but it is more likely that the infrared or ultraviolet wavelengths will be involved. The spectral shaping can be performed in several ways. One is to procure a specially coloured filter to put in front of the lens of a broad-spectrum monochrome camera, such as one conventionally used for intruder detection. Signals from two such cameras with different filters can be compared, but it can be simpler to use a single camera with a pair of filters that are switched by rotating them. Simpler still is to cover half of the 'view' with each filter, so that as the vehicle moves forward each plant is seen first under one filter, then under the other.

Another technique is to weed at night. Freed from the tyranny of the Australian sun, the crop can be illuminated with an LED light source. This can flash between two wavelengths for alternate camera frames. With an autonomous vehicle, weeding can proceed while the farmer sleeps.

When the brightness levels are scaled so that sugar and panic give equal levels at 550 nm in the green, panic is 1.6 times as bright as sugar in the infrared at around 850 nm. A colour webcam has a blue signal that is also sensitive to infrared light between 800 and 900 nm, so by using under-run tungsten illumination, discrimination should be easy. As an example of this, Fig. 20 shows a frame from a video in which pixels for which the blue channel exceeds a threshold have been replaced with yellow.

Another candidate for multispectral analysis is the observation of crops from the air. Drones are being used increasingly for capturing images of farmland and from such data streams the progress of the crop and the presence of weeds can be seen (Swarmfarm, 2016).

Identification by shape can be a more complicated matter, unless it is the shape of the whole plant that is examined, rather than a leaf. When a plant

Figure 20 Panic with brighter infrared pixels highlighted.

stands tall, such as a thistle, it can be discerned by its simple contrast against the sky. When it is surrounded by other foliage, however, edge-tracing might be necessary for determining its shape.

But a simpler technique can give a more thorough solution for weeds in crops. Seedlings are already planted at precisely regular intervals in ruler-straight rows. Anything that is not precisely on this grid must be a weed. With a little more effort, those regular intervals might be synchronised to result in a rectangular grid. A sideways-looking camera or precision GPS can provide that synchronisation. Now it might be possible to perform chipping across the rows in addition to chipping up and down them.

4.3 Identification of ripe produce for harvesting

The detection of ripe fruit is an appealing one for researchers (Thendral and Suhasini, 2015). Conventional RGB vision is usually enough to discriminate a ripe fruit from a green one, but an efficient picking mechanism is harder to find. On the other hand, a vision system can readily estimate the amount of ripe fruit that a tree is bearing (Cheng et al., 2017).

As labour for picking becomes scarce, it seems probable that more will be done in future to train the plant to make the fruit accessible, as is done in viticulture, rather than the chaotic presentation of oranges, apples and similar fruit.

When location is less of a problem, such as for melons, the choice is between selective harvesting in several stages or picking everything and using post-harvest selection to pick out the produce that is ripe. For selective picking, a multispectral approach that includes infrared or ultraviolet examination seems likely to be the most promising.

4.4 Post-harvest grading and sorting

Machine vision has already been in use for several decades. A simple two-colour comparison without imaging has been used to detect fragments of shell remaining on pecan nuts. Similar sorting by colour is routinely used within lines that also grade by size or weight.

With the addition of imaging, blemishes can be detected (Fig. 21). Multispectral analysis has also been used for grading fodder.

Some years ago, an unpublished project for grading broccoli ran into the problem of singulating the heads. When held in position, the shape, colour and stem length were easy to determine, but extraction of the heads from floating in a cold saline bath was too unreliable for implementation. Today, the machine vision equipment is small and simple enough that it could be mounted on the vehicle that gathers the produce. Here, manual pickers placed heads on conveyor belts that carried the produce inwards to a vehicle that drove slowly down the field. They could just as easily have been placed in a chain of cups that carried the heads past a vision grading station on the vehicle.

Figure 21 A grading line.

4.5 Packing

As manipulators for packing become more common, vision can be a great asset for determining shapes to be packed. Simple fruit such as apples present no problem, but items such as broccoli heads must be arranged more carefully. The two styles of packaging are point-of-sale packs and transportation packs for produce being exported by air freight. Conservationist activities to reduce plastic waste are likely to require reconsideration of present packing technology.

4.6 Applications to livestock

With regard to applications to livestock, the imagination can run wild. The shape analysis methods already mentioned can be used for species identification for access to watering points (Finch et al., 2006) or for monitoring the fattening progress of cattle as they walk through a laneway that has RFID identification and the visual means to capture a silhouette. It can be a valuable supplement to walk-over weighing (Beer Central, 2017).

Mustering and shepherding might call for a 'robot sheepdog' (Vaughan et al., 1998) for which machine vision would be an essential. Vision also gives a means of monitoring the movement and clustering of poultry (Zaninelli et al., 2018).

In butchery, X-ray imaging can be used for detecting any metallic foreign bodies in the cuts of meat or for detecting bone-structure for optimising the division of a carcass.

5 Future trends and conclusion

The cost of electronics, cameras and computing devices was once a major consideration for automation. Now it has fallen below that of the simplest machinery. Autonomous farming machines are not just a possibility, they are surely inevitable. Large machines are indicated when a human driver is needed, but when machines are autonomous the economy of scale need not apply. If several low-cost small machines can perform the task of a large one, the resulting redundancy can give protection against machine failure, while further machines can be added incrementally to an expanding operation.

There are conflicting factors, however. In most developed countries it is the shortage of farm labour that is the dominant problem. Australia has to rely on 'back packers', visitors of whom many regard harvesting fruit as a way to subsidise an adventure holiday. In underdeveloped countries, however, labour is abundant but the cost of machinery is paramount. Once again, the paradigm of multiple small, cheap machines can be the answer to both types of problem.

The concept of 'swarm' machines is attracting some trending research effort (Albani et al., 2017), but small autonomous machines can demonstrate their advantages without being synchronised. For tasks such as chipping and selective weed spraying, the loads to be carried are small, but this is not the case with harvesting. However a precedent is set by the procession of haul-out bins that meet up with a cane harvester to unload it as it goes. The mutual cooperation of a variety of autonomous machines is a high probability.

Another likely future is the integration of farming functions with those of marketing. Supermarket chains are already integrated with many farming enterprises, while automatically capturing data on how much of each product is purchased by the consumer. Such data could lead to automated real-time decisions on how much to harvest, how much attention to lavish on the crop or whether to leave it in the ground or on the tree. At present, when a crop is plentiful the shelf price falls dramatically, tailoring demand to fit the supply. It might be found more lucrative to fit the supply to the demand.

But it is in the technology that there can be the greatest variability. Cameras have already fallen to penny prices, driven by their demand as entertainment components of a smartphone. Applications requiring vision are likely to use them in large numbers rather than singly, giving views in many directions. But then the question arises, should a processor be associated with each camera, or should the multiple video signals be sent to a single processor? My bet would be on multiple processors, since already the NVIDIA screen driver uses an array of processors to deliver images at the speeds that gamers need.

One thing seems sure. The greatest technological advances will come from a requirement to supply the needs of a mass market, rather than from a desire to improve agriculture. As self-driving cars become common, then the norm, then compulsory, so vision sensing will develop to supply their needs of safety. Research for the improvement of agriculture will lean heavily on the ability to borrow from that technology. Sensing and data capture are at the heart of all progress towards automation. Machine vision and image analysis will continue to have an essential part to play.

6 Where to look for further information

This chapter seeks to give a light overview of the techniques and possibilities for the application of machine vision to agriculture. Other chapters in this book cover the topics in much greater detail. You will also find more information in Bac et al. (2014), Billingsley (2011), Dunn et al. (2003), Gongal et al. (2015), Kamilaris and Prenafeta-Boldú (2018), Rachmawati et al. (2016), Reina et al. (2016) and Yandun Narvaez et al. (2017).

7 References

Albani, D., Nardi, D. and Trianni, V. 2017. Field coverage and weed mapping by UAV swarms. In: *2017 IEEE/RSJ International Conference on Intelligent Robots and Systems (IROS)*. IEEE, pp. 4319–25.

Bac, C. W., Van Henten, E. J., Hemming, J. and Edan, Y. 2014. Harvesting robots for high-value crops: state-of-the-art review and challenges ahead. *Journal of Field Robotics* 31(6), 888–911. doi:10.1002/rob.21525.

Beef Central. 2017. Adoption starts to grow for walk-over weighing systems, February 2017. Available at: https://www.beefcentral.com/production/adoption-starts-to-grow-for-walk-over-weighing-systems/(accessed June 2018).

Billingsley, J. 2011. Machine vision in agriculture. In: Gliński, J., Horabik, J. and Lipiec, J. (Eds), *Encyclopedia of Agrophysics*. Springer, Germany, pp. 434–6. ISBN: 978-90-481-3584-4.

Billingsley, J. and Schoenfisch, M. 1995. Vision-guidance of agricultural vehicles. *Autonomous Robots* 2(1), 65–76.

Billingsley, J. and Schoenfisch, M. 1997. The successful development of a vision guidance system for agriculture. *Computers and Electronics in Agriculture* 16(2), 147–63. doi:10.1016/S0168-1699(96)00034-8.

Cheng, H., Damerow, L., Sun, Y. and Blanke, M. 2017. Early yield prediction using image analysis of apple fruit and tree canopy features with neural networks. *Journal of Imaging* 3, 6. doi:10.3390/jimaging3010006.

Dunn, M., Billingsley, J. and Finch, N. 2003. Machine vision classification of animals. In: *10th Annual Conference on Mechatronics and Machine Vision in Practice*, 09–11 December 2003, Perth, Australia.

Finch, N. A., Murray, P. J., Dunn, M. T. and Billingsley, J. 2006. Using machine vision classification to control access of animals to water. *Australian Journal of Experimental Agriculture* 46(7), 837–9 doi:10.1071/EA05325.

Gongal, A., Amatya, S., Karkee, M., Zhang, Q. and Lewis, K. 2015. Sensors and systems for fruit detection and localization: a review. *Computers and Electronics in Agriculture* 116, 8–19. doi:10.1016/j.compag.2015.05.021.

Hunt, K. 2018. Concurrent chain code extraction, website of UW-La Crosse. Available at: http://charity.cs.uwlax.edu/projects/chaincode/chaincode.html (accessed June 2018).

Kamilaris, A. and Prenafeta-Boldú, F. X. 2018. Deep learning in agriculture: a survey. *Computers and Electronics in Agriculture* 147, 70–90. doi:10.1016/j.compag.2018.02.016.

McCarthy, C., Billingsley, J., Finch, N., Murray, P. and Gaughan, J. 2010a. Cattle liveweight estimation using machine vision assessment of objective body measurements: first results. In: *28th Biennial Australian Society of Animal Production Conference*, 11–15 July 2010, Armidale, Australia.

McCarthy, C., Rees, S. and Baillie, C. 2010b. Machine-vision based weed spot spraying: a review and where next for sugarcane? *Conference Australian Society of Sugar Cane Technologists 2010*. Available at: https://eprints.usq.edu.au/7965/2/McCarthy_Rees_Baillie_ASSCT_2010_AV.pdf (accessed January 2019).

Microsoft. 2018. Kinect for Windows. Available at: https://developer.microsoft.com/en-us/windows/kinect (accessed June 2018).

OpenCV reference. 2018. Available at: https://docs.opencv.org/2.4/modules/refman.html (accessed June 2018).

Rachmawati, E., Khodra, M. L. and Supriana, I. 2016. Shape based recognition using Freeman chain code and modified Needleman-Wunsch. In: *2016 8th International Conference on Information Technology and Electrical Engineering (ICITEE)*, 2016, Yogyakarta, pp. 1–6.

Reina, G., Milella, A., Rouveure, R., Nielsen, M., Worst, R. and Blas, M. R. 2016. Ambient awareness for agricultural robotic vehicles. *Biosystems Engineering* 146, 114–32. doi:10.1016/j.biosystemseng.2015.12.010.

SICK. 2018. 2D LiDAR sensors LMS5xx. Available at: https://www.sick.com/au/en/detection-and-ranging-solutions/2d-lidar-sensors/lms5xx/c/g179651 (accessed June 2018).

Swarmfarm. 2016. A review of technologies that can be enabled by robotics to improve weed control in Australian Cotton Farming Systems. Available at: http://www.insidecotton.com/xmlui/bitstream/handle/1/4380/CRDC1615%20Final%20Report.pdf?sequence=1&isAllowed=y (accessed June 2018).

Thendral, R. and Suhasini, A. 2015. Enhanced machine vision system for ripe fruit detection based on robotic harvesting. *Advance Journal of Food Science and Technology* 7(11), 841–9. doi:10.19026/ajfst.7.2520.

Vaughan, R., Sumpter, N., Henderson, J., Frost, A. and Cameron, S. 1998. Robot control of animal flocks. In: *Proceedings of the 1998 IEEE International Symposium on Intelligent Control (ISIC)* held jointly *with IEEE International Symposium on Computational Intelligence in Robotics and Automation (CIRA) Intell*, 1998, Gaithersburg, MD, pp. 277–82.

Yandun Narvaez, F., Reina, G., Torres-Torriti, M., Kantor, G. and Cheein, F. A. 2017. A survey of ranging and imaging techniques for precision agriculture phenotyping. *IEEE/ASME Transactions on Mechatronics* 22(6), 2428–39. doi:10.1109/TMECH.2017.2760866.

Zaninelli, M., Redaelli, V., Luzi, F., Mitchell, M., Bontempo, V., Cattaneo, D., Dell'Orto, V. and Savoini, G. 2018. Development of a machine vision method for the monitoring of laying hens and detection of multiple nest occupations. *Sensors* 18, 132.

Chapter 2

Advances in actuation and control in agricultural robots

Pål Johan From, Norwegian University of Life Sciences, Norway and University of Lincoln, UK; and Lars Grimstad, Norwegian University of Life Sciences, Norway

1 Introduction

The basic idea behind precision farming is to collect sensory information in the field and act on this after processing the available data. As sensors become more accessible, better and cheaper, the amount of data they can possibly gather from a field is increased. To be able to take full advantage of these data, precise and reliable actuation is required. This has led to the development of several new ways that agricultural machinery can interact with the environment to improve overall performance.

Over the last several years, we have seen a wide variety of actuation methods that can be used in combination with large amounts of sensory data and higher computer power to apply advanced control and decision-making in the field. As a result, agricultural robots now move away from being pure

http://dx.doi.org/10.19103/AS.2019.0056.06

sensor-carrying platforms for gathering data in the field into becoming action-delivering platforms providing physical interaction with the environment.

After data is collected from the field, machine intelligence is used to determine the optimal action to increase yield, suppress pests or to harvest the crop. In this chapter, we will discuss some of the actuation methods that are most commonly used on intelligent agricultural robots in order to transform this optimal action into implement behaviour, motion, physical interaction or manipulation. The actuation methods consist of both traditional actuators that have been transformed into precision farming tools and novel actuators enabled by robotics and autonomous systems.

Examples of traditional actuation methods are electrical, hydraulic and pneumatic actuation, spray nozzles and mechanical solutions. All these have been developed more or less independently of the advances of precision agriculture, but have been adopted by the precision farming community wherever they are found useful. On the other hand, the need for more precision, speed and flexibility has led to several novel actuation methods, such as variable-rate nozzles for spraying, precise mechanical harvesters and robot manipulators for packaging, pruning and harvesting to name a few.

There are plentiful challenges and design considerations to keep in mind when developing actuators for the agricultural domain. The spatial and temporal variations in agriculture are huge. No two farms are alike, and no two plants appear the same. This adds a complexity when physically interacting with the environment. Tools and implements developed for structured environments like factory floors or assembly lines, where one would normally find automation and robotics at a large scale, can therefore not be directly transferred to the unstructured environments found on farms. Implements found in agriculture thus needs to deal with this variation, for example through compliant behaviour or by specifically adapting to each and every plant or scenario.

In addition to the variety, the implements are to operate for hours on end in an extremely harsh environment involving dust, water, mud and direct sunlight. This imposes additional requirements on the mechanical design to obtain the necessary durability and reliability. As a result, we have seen a wide range of new actuator designs being developed to withstand this kind of harsh environment.

This chapter will give an overview of some of the actuators that are most commonly found in agricultural robotics. We present a short introduction to each actuator followed by a few example uses that we find in the agricultural domain.

2 Electric motors

Electric motors convert electrical energy into mechanical energy, and can be found in a wide range of devices in our daily lives. From tiny motors in wristwatches to giant ship motors working in the tens of megawatts, electric motors are available in many types, but can broadly be separated into two categories based on the current that powers them: direct current (DC) motors and alternating current (AC) motors. Motors can produce rotational or linear motion. However, linear actuators often use a motor of rotating type, and then convert the rotational motion into linear motion mechanically.

Electric motors are generally highly efficient, easy to control and have the added benefit that they can be used as generators. When an electric motor is used as a generator it converts mechanical energy back into electrical energy. An example is regenerative braking of an electric car. When the car brakes, the motor acts as a generator, converting kinetic energy into electrical energy, charging the car's batteries. Electric motors are also capable of delivering relatively high torque over a wide speed range, which means that they often are used with single-speed transmission, instead of the more complicated multi-speed transmission systems associated with internal combustion engines.

The motor most commonly found on farms and in industry is, by far, the induction motor, which is of the AC type. This motor, which also goes by the name asynchronous motor, is of a simple construction, which makes for low acquisition and maintenance costs. Drawbacks are somewhat difficult speed control and relatively low starting torque. The motor can be of single- or three-phase type, and is often found in centrifugal pumps, elevators, fans and similar devices.

As the name implies, DC motors run on direct current. This makes them ideal for use with batteries. The most common type of DC motor uses brushes to connect windings in the rotor (rotating part of the motor) to power, and have either electromagnets or permanent magnets in the stator (static part of the motor). These motors are inexpensive and easy to run, but the brushes are prone to wear and generate electrical noise.

Brushless DC (BLDC) motors have three sets of windings in the stator and permanent magnets in the rotor. Instead of brushes, BLDC motors may have built-in Hall effect sensors for electrical commutation. These sensors may also be used for course closed-loop control. In addition to not having the drawbacks affiliated with brushes, BLDC motors have very good power-to-weight ratios. They are, however, more costly and require a more sophisticated driver compared to their brushed counterpart.

A stepper motor is another type of DC motor. This motor moves in fixed steps. This type of motor is therefore often used for open-loop position control, and is found in devices like 3D printers.

For most practitioners, motor control has become relatively simple. Most motors have an advanced low-level controller that allow the user to give simple commands such as velocity or torque as input to the controller. The low-level controller, which can be integrated with the motor either internally or externally, will then make sure that the motor follows the velocity or torque command to an accuracy that is sufficient for most applications in agriculture.

3 Hydraulic actuators

A hydraulic actuator converts hydraulic energy into mechanical energy, and both linear and rotary motion is easily achieved. Hydraulic actuators are often used in heavy-load applications, and are commonly found on machines such as cranes and excavators, or in the wheels of heavy vehicles.

Linear hydraulic actuators are generally referred to as hydraulic cylinders and provide linear motion by driving a piston rod back and forth. Rotary actuators, generally referred to as hydraulic motors, can be uni- or bidirectional with fixed or variable displacement. There are also semi-rotary motors capable of high torque. These are normally limited to less than one full revolution.

Hydraulic actuators allow for precise motion control, and have excellent power-to-weight ratios. It is however important to note that although the actuators are compact compared to their power output, components such as pumps, reservoirs and hoses add to the total size and complexity of the system. To avoid leaks and contamination, hydraulic systems require regular maintenance. The efficiency of hydraulic systems is in general poor, as much energy tends to be lost to heat.

Hydraulic systems are a common sight on farms. Tractors use hydraulics to raise and lower implements, and often also have auxiliary hydraulic valves, called remotes, for connecting implements to the tractor's hydraulic system. Tools such as manual car jacks and pallet jacks also work by means of hydraulics, and allow heavy objects to be lifted with only human energy as input.

4 Pneumatic actuators

Pneumatic systems make use of gas, often pressurized air, to power machines and tools. Both linear and rotary motion is easily achieved. Benefits of pneumatic systems include high-speed, reduced fire hazard compared to hydraulics and low initial cost. Air is also free and available pretty much anywhere. In addition, exhaust air can be vented to the environment.

On farms, pneumatic systems can be found in some precision planters and fertilizer spreaders, as well as in workshops where pressurized air is commonly used for powering tools such as drills or impact wrenches.

In both automated and manually operated systems in the food industry, pneumatic actuators are commonly used, as they are well suited for clean environments. Robot arms used in pick-and-place applications, for example in the food industry, are often fitted with vacuum grippers for moving objects. Compared to many other robotic grippers, vacuum grippers are simple and inexpensive. There is, however, limitation to the range of surfaces any one gripper can suck onto, which makes grippers less versatile. The gripper will also leave marks on some types of surfaces. Some experimental fruit picking robots use vacuum grippers to grip vegetables like capsicum (Lehnert et al., 2017).

As air is a compressible fluid, accurate control of pneumatic actuators is more difficult to achieve than with both hydraulic and electric actuators. Pneumatic systems are also sensitive to moisture, which means that air must be dried and excess moisture drained throughout the system.

5 Nozzles and metering orifice

Nozzles and metering are some of the most commonly used actuators in conventional agriculture. Traditionally, these are used to broadly apply pesticides such as herbicides, fungicides and insecticides to plants and the field, as well as for liquid fertilizers. Variable-rate application is also attaining more attention, and precise spraying according to accurate prescription maps has huge potential in terms of efficiency, resistance, the environment and so on. This technology is the backbone of many applications in precision agriculture.

Spray patterns needed for herbicides, insecticides and other pesticides are different from the ones used, for example, for fertilizers. While liquid fertilizer normally requires a metering orifice, pesticides are normally applied using a flat-fan nozzle.

Spraying of liquid pesticides is normally done by dissolving the product in a fluid, and then by increasing the pressure of the liquid using a pump, it is forced through a nozzle. There are numerous types of nozzles available; each gives a characteristic shape, droplet size, tip flow and uniformity. The flow is normally controlled either by the choice of nozzle, or by the pressure applied. However, for fixed orifice nozzles and meters, the range in which the flow can be controlled is very limited. One can typically double the output flow by increasing the liquid pressure four times, but this is not a sufficient range for many applications, and as a result the driver will often have to slow down to apply a sufficient amount. This slows down the operation.

While fixed orifice nozzles have a very limited variation in the application rate and spraying speeds, variable-orifice nozzles can change the flow by changing the size of the orifice, and in some cases change from fully open to closed almost instantly. There are many different nozzles that address this inherent limitation in flow rate adjustment capabilities that can be obtained using hydraulic pressure only.

One way to increase the range of the flow is to use metering assemblies using elastomer (EPDM) metering, which adjusts the flow by changing the diameter of the orifice as it expands with pressure. This will result in a wider range of flows because the increasing pump pressure will not only increase the liquid pressure, but also the orifice opening. This is a relatively simple system with no moving mechanical parts or control needed. A similar conceptual idea is used in spring-loaded mechanisms that in many of the same ways increase the size of the orifice with increased pressure.

Pulse width modulation is also used to change the rate of flow. In this case, the liquid flow rate is controlled by using an intermittent, that is by opening and closing the nozzle very rapidly, typically around 10 Hz, the amount of fluid is controlled. The duty cycle (DC) then gives the flow; a DC of 100% means that the nozzle is completely open, and 50% means it is open half the time. These nozzles are rather expensive, but can obtain a range of five times with relatively constant droplet size and characteristics.

To obtain a wider range of the flow, direct injection of undiluted pesticide can be used. Air pressure then injects the undiluted pesticide into the liquid, and by varying the fluid and air pressure, the rate can be varied with a factor of several 10s.

5.1 Nozzles and metering orifice in precision farming

In precision farming systems, we have accurate information about the location of weeds and pests, and we may also have spatial and even temporal information about plant growth. This can be used to apply pesticides and fertilizers only where required. This has resulted in a need for more precise actuators capable of adjusting the rate on the go.

Crop sensing systems are used to apply variable-rate nitrogen (N) based on an estimate of fertilizer need. This has been shown to enhance performance substantially over traditional methods, as less fertilizer is used, and the fertilizer is applied where it is most needed. Tractor-mounted systems can for example use nitrogen sensors to change the rate of fertilizer application. For liquid fertilizer application, variable-orifice nozzles are commonly used since they provide a much larger nozzle flow range compared to traditional fixed orifice nozzles (Fulton, Taylor and Sharda, 2016) (Fig. 1).

In fertilizer application, variable-rate application will speed up the operation. In cases where the speed of the tractor needs to be reduced when using conventional systems to apply a sufficient amount of input to the field, variable-rate application allows the driver to maintain constant speed throughout the operation and hence increasing the efficiency of the operation.

Very recently, we have also seen tractor-mounted systems for precise pest control. These are typically advanced vision systems that give the position of a weed, on which nozzles rapidly turn on and off to apply the product only to the weed, thus avoiding applying pesticides on the plants and resulting in a substantial reduction in the use of chemicals.

Figure 1 The precision weeder from Blue River Technologies uses advanced machinery to selectively spray weeds in the field.

Several robotic systems are also being developed. Most of these use the same principle, with a camera system to identify the weed, and then applying pesticides only on the weed. Robotic systems are currently rather slow, but have the advantage that they can operate without interruption (Fig. 2).

Figure 2 The AgBot II from Queensland University of Technology has both a precision spot sprayer and a mechanical weeder, both taking input from a vision system.

6 Thermal methods

Heat is used in many different forms in agriculture, especially for weed control. Flaming is widely used in many areas of food production as a means to kill weeds and insects. Propane is the most common fuel used in flaming, and can generate temperatures up to 1900°C. The flaming process ruptures the cell walls and will kill the weeds within days. Young seeds are more susceptible to damage, and the ability to regrow depends on the plants' energy reserves, environmental conditions such as soil moisture and competition from neighbouring plants (Upadhyaya and Blackshaw, 2018).

Hot water and steam have also been used in weed control, avoiding the danger of fires caused by flaming or burning. When the weeds are exposed to the heat, they will degrade in a matter of days, but several applications may be necessary, as the roots are often not directly affected by the treatment (Upadhyaya and Blackshaw, 2018).

There are also other ways to transfer energy that has been applied to weeding. Microwave radiation is electromagnetic radiation in the 300 MHz-300 GHz range and can transfer large amounts of energy to plants. Energy is transferred to the plants by microwave application, causing water molecules to oscillate resulting in the conversion of electromagnetic energy into heat energy and thus can be used to kill weeds (Barker and Craker, 1991).

Another efficient method for transferring energy is by using lasers (Xiong et al., 2017). Lasers can be used to cut off the stem of the weed, or simply to burn the weed entirely. How much time is needed to kill a weed depends on the laser intensity, but reports show that a weed can be killed in less than 1 s with a strong class 3 or class 4 laser. This method requires an exact localization of the weed, typically using advanced machine vision, and is promising for precision applications. These lasers may cause harm to humans even if a person only looks at the target point and not the light source itself. These systems must therefore be completely enclosed if humans are present in the field. For the same reason, laser is a promising technology to be used along with advanced unmanned weeding robots.

7 Optical-based management

Optical methods involve using natural or artificial light to obtain certain beneficial properties in plant production. This light can be in the visible part of the spectrum, like red, blue and green, or in the non-visible part, like UV. Although these methods have been used to a certain extent in agriculture, they have not reached their full potential, as there has not been a good way to get the technology into the field. We expect the use of this kind of technology to grow as more automation and robotics techniques are moving into the agricultural domain.

7.1 Fungal and disease management

Optical methods are well known for their efficient fungal and disease management. In particular, UV light has been shown to eliminate many forms of fungus and diseases, and is extensively used in sterilizing water, medical equipment and so on. Many cultivars suffer from high levels of fungus pressure, especially when grown under plastic (Gadoury et al., 2017). Strawberry is one such example, and UV-C has therefore been tested extensively on strawberries using both conventional tractors (Gadoury et al., 2017) and autonomous robots (Grimstad and From, 2017). Very good results were obtained when UV-C light was applied at night on open strawberry fields. This has been shown to drastically reduce the outbreak of mildew, which is a big problem in strawberry production, especially in dry areas and in polytunnels (Onofre et al., 2017; Suthaparan et al., 2017) (Fig. 3).

Robots are ideal for this task for several different reasons (Grimstad and From, 2017). Firstly, humans should not be exposed to UV light as it can damage the skin and eyes. Secondly, it has to be applied at night to have full effect. Applying the light at night means that we can reduce the intensity of the light to a level that does not affect the plants and still have a very good effect on for example mildew. Finally, this is a rather time-consuming operation, and it has been shown to be difficult to find workers that are willing to do this work at

Figure 3 The Thorvald robot applying UV for mildew protection in strawberries. The robot is completely autonomous and applies the treatment at night for improved efficiency.

night. For these reasons, we believe that robots will be an important enabler to get this technology into the fields.

7.2 Antioxidant enrichment

In addition to having a positive effect on fungus and diseases, research has shown that UV light treatment of plants also has a positive effect on the nutrients in fruit and berries. As an example, strawberries that are exposed to UV-C have shown to have increased levels of antioxidants content (Oviedo-Solís et al., 2017). The effects seem to be larger when applied post-harvest, but also pre-harvest application of UV-C has shown positive results on certain cultivars of strawberries (Xie et al., 2014).

8 Robotic manipulators

Robotic manipulators have traditionally only been used in structured environments like factory floors. As robotic systems are becoming more agile and flexible – and advanced vision and intelligent decision-making to a larger extent can be considered off-the-shelf technology – robotic arms are now moving into the agricultural domain. Robotic manipulators can be used for a wide variety of applications, such as weeding, pruning, harvesting, sampling and so on.

8.1 Serial manipulators

Serial manipulators are open chain arms, that is the end effector/tool is connected to the base through a single series of links. Serial manipulators have a large workspace and a good reach, which make them useful in many agricultural applications. Even though faster and more precise robots exist, serial robots are typically precise enough and fast enough for most applications in agriculture. Serial manipulators can have revolute joints, like most industrial manipulators, or linear prismatic joints. Serial manipulators with six or more revolute joints can generate a full 6 degrees of freedom (DoF) motion, which means that the end effector can reach any position and any orientation within its workspace. This is useful for performing complex manipulation tasks. Several tasks in agriculture require dexterity and advanced manipulation due to the complex environment. Serial robots with only linear actuators can only generate linear motions and no rotations, so these robots can take the end effector to a certain position in space, but not change its orientation. While some tasks require full 6-DoF motion with orientation to perform an operation, other tasks benefit from having the orientation fixed.

8.2 Parallel manipulators

Parallel robotic arms are characterized by the end effector being connected to the base through more than one kinematic chain. The Delta robot is the most commonly used parallel arm in many areas of industrial production. Parallel manipulators are generally faster and more accurate than their serial counterparts, which comes at the expense of a smaller workspace and reach (Fig. 4).

8.3 Mobile manipulator systems

In agriculture, there is normally a need to move around the farm, which requires the robotic arms to be mounted on some sort of moving base. Many examples of mobile agricultural robots that have robotic arms for manipulation are available. These systems require coordinated motion between the arm and the base (From, Gravdahl and Pettersen, 2014). As the robots operate in the field, there are two main ways to coordinate the motion of the base and the arm. One simple way is to drive the robot to a certain location in the field, stop the robot and let the robotic arm perform manipulation as if mounted on a fixed base. This is by far the simplest approach, as it resembles the traditional robots that operate in structured environments from a fixed base. Alternatively, one might choose to use coordinated motion between the arm and the base (From, Gravdahl and Pettersen, 2014). In this case, we can perform manipulation as the platform moves in the field, resulting in a more efficient operation. This will, however, require advanced and accurate control of the robotic arm and the base, and intelligent coordinated motion between the two.

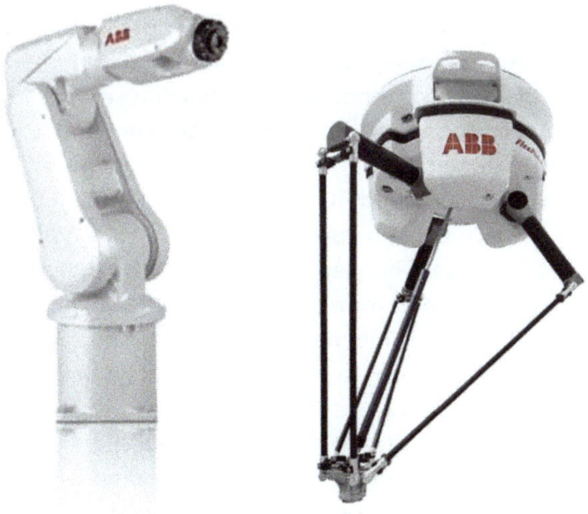

Figure 4 Serial and parallel manipulator arms from ABB.

The combination of the mobility offered by the mobile platform and the manipulability offered by the robotic arm creates huge possibilities in agriculture. These systems can perform many of the tasks that are performed in agriculture today with far higher accuracy and precision, resulting in a dramatic reduction in the inputs needed for food production. In addition, these systems have a huge potential in replacing human labour performing repetitive, tedious and heavy tasks, and operations in the field, such as planting, harvesting and plant care.

8.4 Vision-guided manipulation and visual servoing

Visual servoing, or vision-based robot control, uses vision sensors to generate feedback signals that are used to control a robotic arm or mobile robot to help the robot obtain its final goal. Typically, a camera is mounted either in a fixed location or on the robot arm or base, and is used to give the relative location of the tool with respect to the goal. The location of the robot with respect to the target is then used to generate motion commands for the robot.

There are two major developments that just over the last few years have made visual servoing an extremely important tool in agricultural robotics.

- The first is the development of more advanced cameras. RGB (red-green-blue) cameras have been available for a long time and are extremely cheap. Recently, the development of RGB-D cameras has resulted in a completely new area in robotics. RGB-D is a camera that in addition to RGB also gives the distance from the camera to the object for each pixel in the picture. This depth information has been shown to be very useful for example in accurately determining the position of fruit for harvesting. There are many ways to obtain the depth data. Triangulation is one commonly used method where one can use two or more cameras and identify corresponding points in the different images, and then use triangulation to obtain the depth. The main advantage of this approach is that it only requires standard RGB cameras. Time-of-flight cameras, on the other hand, use a dedicated image sensor to capture both the RGB image and the depth image. The depth is obtained by measuring the time it takes for the light to travel from the camera to the object and back. Finally, structured light sensors illuminate the objects with a specially designed light pattern, and the depth is obtained from how this structured light is projected onto the object.
- The second advancement that has made visual servoing so useful is the great improvements in image analysis and computer vision. Machine learning, and in particular deep learning, can be used to obtain extremely fast and accurate solutions from images. This development, together with increased computational power and the introduction of the graphics

processing unit (GPU) to run these algorithms, allows us to swiftly identify objects from an image and to obtain very accurate location of the object.

The combination of cheap and good cameras and advanced image analysis lay the ground for precise manipulation in the fields. This has been applied to fruit harvesting and manipulation. Once the fruit has been identified, the location of the fruit is determined in the three-dimensional space. The difference in position between the fruit and the manipulator tool is then used to generate a motion to the robot arm so that the tool moves towards the fruit until it finally picks it. Visual servoing is also applied to pruning, post-harvest manipulation and packing of the produce, robotic pollination and several other tasks.

9 Control for precision agriculture

A control system takes sensory data as input, processes this data using one of many available control schemes and finally sends a control signal to an actuator that gives the desired motion or actuation. In this section, we will briefly discuss how sensory data is processed and communicated to the actuator to obtain the desired responsive motion.

9.1 Control schemes

The control system is fundamental for fast and responsive actuation in agriculture (see below). Responsive behaviour means that the system senses the world using sensors such as cameras, laser scanners, pressure or humidity sensors, to mention a few, and uses this input to generate control signals that make actuators act on the real world based on the current state. This is normally called feedback control or closed-loop control.

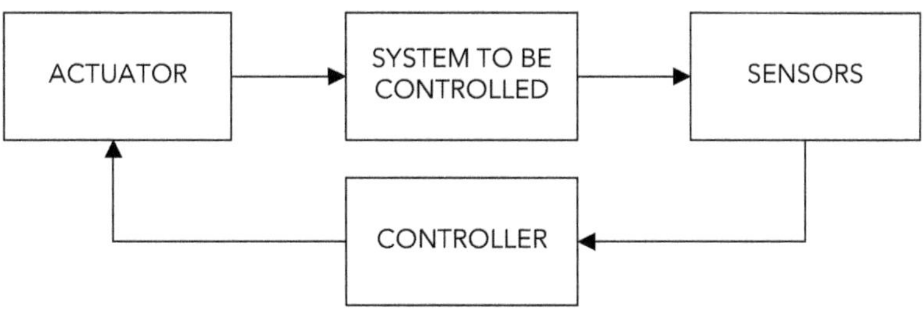

The most basic controllers take the difference between the desired state, or reference, and the measured state to calculate the current error of the system. The error is thus obtained by comparing the actual output of the system with the desired output, and this error is then used to calculate the actuation that will take this error to zero.

If the relationship between the error and the calculated control signal is proportional, we get what is known as a proportional controller. The controller u then takes the form $u = K_p e$, where K_p is a constant proportional gain and e is the error, that is the difference between the measured and desired outputs. We can also get the control signal by the integral or derivative of the error. The combination of proportional (P), integral (I) and derivative (D) action is known as a PID controller. The PID controller is implemented by adding the proportional, integral and derivative of the error with different gains. This is one of the most widely used controllers due to its robustness and simple implementation. Different variations of the PID controller are used in motor control, liquid flow, water level control and in many other areas.

More advanced control methods have also been developed and there is a wide variety of control strategies available. Control methods based on Lyapunov theory use a Lyapunov function to test whether a dynamical system is stable, that is the control brings the energy to zero. Adaptive controllers are another important group of controllers, which identify the process parameters and system characteristics online. This leads to a series of controllers that obtain good robustness by modifying the controller gains. Model-predictive controllers (MPCs) use a mathematical model of the system to predict the future states, and calculate the control signal based on this prediction. This can give enhanced performance in many different cases, especially when good and accurate mathematical models are available.

Optimal control is a large group of methods that identify the control signal based on a cost function and the optimization of this. Two of the most fundamental controllers in this field are the linear-quadratic regulator and the linear-quadratic-Gaussian controller, which solve the control problem by finding the optimal solution subject to a quadratic cost function. These controllers are linear feedback controllers to non-linear systems, and of much use in a wide variety of applications, including non-linear systems.

Intelligent controllers is a class of systems that uses more advanced artificial intelligence approaches to solve the control problem. These controllers are becoming more popular as the field of artificial intelligence is developing.

9.2 Microcontrollers

Microcontrollers are small computers that are found 'embedded' inside a wide range of devices, from digital wristwatches and microwave ovens, to tractors and robots. A typical microcontroller is inexpensive, has low power consumption and is programmed to perform one specific task for its entire lifespan. One such task may be to continuously read data from a pressure sensor and displaying the data on an LCD screen. Another task may be to read data from a temperature sensor, and based on the reading automatically turn on or off a heating element to reach or maintain a desired temperature.

Microcontrollers are typically embedded in larger machinery and systems to perform a specific task. The use of microcontrollers to control specific processes will increase. They can for example be used to control implements that perform a very specific operation such as opening and closing a valve for pesticide applications, turning on and off a motor or other similar tasks.

9.3 Communication

Sending information from sensors to the controller and from the controller on to the actuators can be done in many ways and using different standards. What standard you choose mainly depends on the needs, especially for speed, robustness and flexibility.

The CANbus was developed for the car industry as the data communication standard in the 1980s and was designed to take data from sensors and pass it on to actuators. The main advantage of the CANbus is the robustness, as competing signals are dealt with by introducing priority in what unit gets to send its package first. The bit rate of CAN is 1–2 Mb/s for relatively short network distances. The ISOBUS is the de facto standard in communication between the tractor and the implements today. ISO 11783 – Tractors and machinery for agriculture and forestry – Serial control and communications data network – is a standardized communication protocol that allows farmers to use equipment from different manufacturers with different tractors. CANopen is for example used for automation of industrial plants. The CANbus is thus very useful when the amount of data is relatively low, and the need for robustness is high. Examples found in agriculture are motor control based on encoder information, breaking and simple implement control.

For many operations in agriculture, far more data is needed. This is the case for sensors such as 3D laser scanners and high-definition cameras. In this case, a high-speed network such as Ethernet can be used. For intelligent control algorithms, one might need data such as images to be processed in real time. This requires far higher bandwidth. For robotics, for example, intelligent control using a wide variety of high-definition sensors will increase the need for bandwidth. Ethernet can for example provide several Mb/s, and in some cases several Gb/s. These fast communication protocols will therefore be used more frequently in the area of agricultural robotics in the future.

10 Case study: automated strawberry production

In this section, we will look at how different actuators are used to automate strawberry production. Some examples are taken from systems that have already been implemented and are in operation today, while others are examples that we envision for future use. We will discuss table-top production of strawberries in tunnels.

Once the strawberry plants have been planted, the farmer needs to start an intensive programme of irrigation, fertilization and plant care. Fertilizers are normally applied through the irrigation system, which is controlled using electrically controlled pumps and valves. Several farms have automated irrigation systems that are also used for fertilization. Advanced systems that take into account weather, temperature and measurements from the field are used to control the irrigation on the most advanced farms. Even though these systems do not require extremely fast or accurate pumps and valves, they need to be extremely reliable as irrigation is one of the key parameters for optimal growth.

Throughout the season, there is a need for an intensive plant care programme. Strawberries are vulnerable to several different fungal diseases, such as mildew and grey mould. The main tool used against these diseases is applying fungicides. Fungicides are normally applied using tractor-mounted sprayers, but we envision that this will be done by robots in the near future, which is likely to improve both precision and timely application. The fungicides are applied based on the temperature and humidity, and also to a large extent on observations in the field. Some farmers, especially when polytunnels are used, also use water by showering the plants with very fine droplets of water, which helps to keep fungal disease pressure manageable. Electric pumps and specially designed nozzles are then used to automate this treatment, which normally happens at pre-defined intervals. One final tool against fungal diseases is optical methods. UV treatment is not used to fight mildew and other fungi today, but extensive testing has shown this to work very effectively against fungus. As the use of robots become more widespread, we expect this treatment to become more common. As resistance against fungicides has become one of the main challenges in strawberry production, it is very important to have several different methods available.

As the strawberries ripen, the main challenge is to automate the harvesting. There are several different research projects that use advanced machine vision and robotic grippers to solve this problem, but still no commercial solution. Most systems try to estimate the position of the stem, and cut this using a scissor-like device without actually touching the berry (Hayashi et al., 2010). The main challenge with this approach is the extremely hard vision problem of identifying the location and orientation of the stem. Other approaches grasp the fruit and tear or snap it off (Dimeas et al., 2015). The problem of locating the strawberry is a far easier machine vision task than localizing the stem, while the drawback is that touching the strawberry may bruise the fragile berries. One way to deal with this is to enclose the strawberry using a hollow gripper so that the position is exactly known, and then cut the stem so that the strawberry falls into a container, like in Xiong, From and Isler (2018) (Fig. 5).

Figure 5 An example of a vehicle-manipulator system where a linear serial manipulator is mounted on a moving base. The system is used for strawberry harvesting.

11 Summary

Advanced actuation is the backbone of precision agriculture. With the development of advanced sensing and intelligent decision-making and control, these actuators will enable a revolution in agriculture that will give us higher yields and lower environmental impact. Actuators come in many different shapes and forms, but common for virtually all actuators used in agriculture is that more precision and intelligence is constantly being introduced throughout the entire value chain.

We have seen examples of how existing implements are augmented with advanced vision and intelligence to obtain more precision in the applications. Examples of this are variable-rate spraying, which uses sensors to adjust the amount of pesticides or fertilizer that is needed in the farm. Instead of spraying the entire field, we can change the application rate depending on the need and obtain substantial reductions in input use.

We also see the development of entirely new types of implements that we have not seen in the field to date. Examples are advanced robotic arms for harvesting and extreme-precision spot sprayers. These systems have been enabled by novel sensors and actuation, in combination with machine learning techniques and increased computational power. This group of sensors, which are either being entirely developed for the agricultural domain, or developed for other areas but taken into use in the agricultural domain, have the potential to completely change the way we produce food today with a transition towards extreme-precision agriculture. Robots and self-driving systems will be one of the main enablers in getting this technology into the fields.

In both cases above, we see that the introduction of advanced actuators that are more precise and that can be integrated with and controlled by intelligent computers and sensing systems are important enablers to get these precision systems into the field. We envision these implements to be the main components of precision agriculture and will play a vital role in increasing the food production to completely new levels, decrease the use of inputs and hence runoffs, and dramatically improve environmental aspects related to modern food production.

12 Future trends in research

We see the overall trend in agricultural machinery taking two rather distinct paths. Firstly, we see the urge to develop larger, heavier and faster systems. These are generally tractor-mounted implements that target the larger markets in the United States, Brazil and Australia, as well as large farms elsewhere. As tractors become bigger and faster, this affects the way we design the implements. Implements and actuators need to be developed to be extremely quick, for example to do precision weeding on a tractor driving at full speed. This requires sufficiently high computational power and efficient algorithms to be able to make decisions in matters of milliseconds as well as fast and accurate actuators that can turn on and off or change the application rate with extreme speeds. As an example, some very impressive dot-spraying weeding systems have appeared only in the last few years, although none of these are in widespread use in agriculture today. A wide range of other precision tools that do not require the same speeds are starting to gain widespread use in agriculture, but also these systems will probably improve performance as they become faster and more accurate.

The other path that we see future research is taking is the development of smaller and more precise tools that are generally mounted on mobile robots. These systems take advantage of the fact that robots will allow for continuous day and night operations, reducing the need for extreme speeds. Robots that operate 24 h, and in many cases will work in swarms, reduce the need for these high speeds, but the focus is rather to develop low cost systems that allow for swarm implementation and precision tools that give a high return on investment for the farmer. These systems are now reaching a level where they create value to farmers that one day will supersede the investment costs, and we therefore believe we will see a rapid introduction of these robots into agriculture over the next few years.

Agriculture is extremely diverse in virtually every sense. We therefore believe that there is room for implements mounted on both tractors and robots in the future. In both cases, the drive towards faster and more precise actuation in the field will push food production towards more sustainable operation with less use of inputs and far higher yields than we see in the fields today.

13 Where to look for further information

There is an extensive literature on actuators available. For a generic introduction, see for example Gomis-Bellmunt and Campanile (2009), while for more specific treatments of for example electric motors see Hughes and Drury (2013), pumps see Nesbitt (2006) and valves see Nesbitt (2011). Asaka and Okuzaki (2014) give an overview of some soft actuators that may be of use in many agricultural applications, as well as some important literature on soft robotics, such as Verl et al. (2015). For a special treatment on weeding see Upadhyaya and Blackshaw (2018). An extensive introduction to robotics can be found in Siciliano and Khatib (2016) and a more mathematical approach in Murray, Li and Sastry (1994). For a more in-depth treatment of probabilistic robotics in the areas of perception and control, see Thrun, Burgard and Fox (2005).

For control theory and feedback systems we refer to Aström and Murray (2010) and for a general introduction Ogata (2011). For a more extensive treatment, including non-linear control methods we refer to Isidori (2013) and Sastry (2013). MPC can be studied in Camacho and Alba (2013) and Allgöwer and Zheng (2012), while an introduction to machine learning techniques can be found in Abu-Mostafa, Magdon-Ismail and Lin (2012) and Bishop (2006). A special treatment of deep learning is found in Goodfellow, Bengio and Courville (2016).

14 References

Abu-Mostafa, Y., Magdon-Ismail, M. and Lin, H.-T. (2012). *Learning from Data: A Short Course*. Place of publication unknown: AMLbooks.

Allgöwer, F. and Zheng, A. (2012). *Nonlinear Model Predictive Control*. Basel: Springer.

Asaka, K. and Okuzaki, H. (2014). *Soft Actuators: Materials, Modeling, Applications, and Future Perspectives*. London: Springer.

Aström, K. and Murray, R. (2010). *Feedback Systems: An Introduction for Scientists and Engineers*. Princeton: Princeton University Press.

Barker, A. and Craker, L. (1991). Inhibition of weed seed germination by microwaves. *Agronomy Journal*, 83(2), 302-5.

Bishop, C. (2006). *Pattern Recognition and Machine Learning (Information Science and Statistics)*. New York: Springer.

Camacho, E. and Alba, C. (2013). *Model Predictive Control*. London: Springer.

Dimeas, F., Sako, D., Moulianitis, V. and Aspragathos, N. (2015). Design and fuzzy control of a robotic gripper for efficient strawberry harvesting. *Robotica*, 33, 1085–98.

From, P., Gravdahl, J. and Pettersen, K. (2014). *Vehicle-Manipulator Systems - Modeling for Simulation, Analysis, and Control*. London: Springer.

Fulton, J. P., Taylor, R. K. and Sharda, A. (2016). Performance of variable-orifice nozzles for liquid fertilizer applications. *Applied Engineering in Agriculture*, 32(3), 347–52. doi:10.13031/aea.32.11428.

Gadoury, D., Bierman, A., Rea, M., Stensvand, A., Arupillai, S. and Onofre, R. (2017). Design and operational considerations for use of UV and red light for suppression of plant diseases under field conditions. *Phytopathology*, 107(12), 5–6.

Gomis-Bellmunt, O. and Campanile, L. (2009). *Design Rules for Actuators in Active Mechanical Systems*. London: Springer.

Goodfellow, I., Bengio, Y. and Courville, A. (2016). *Deep Learning*. Massachusetts: The MIT Press.

Grimstad, L. and From, P. (2017). The Thorvald II agricultural robotic system. *Robotics*, 6(4), 24.

Hayashi, S., Shigematsu, K., Yamamoto, S., Kobayashi, K., Kohno, Y., Kamata, J. and Kurita, M. (2010). Evaluation of a strawberry-harvesting robot in a field test. *Biosystems Engineering*, 105(2), 160–71.

Hughes, A. and Drury, W. (2013). *Electric Motors and Drives: Fundamentals, Types and Applications*. Oxford: Elsevier.

Isidori, A. (2013). *Nonlinear Control Systems*. London: Springer.

Lehnert, C., English, A., McCool, C., Tow, A. and Perez, T. (2017). Autonomous sweet pepper harvesting for protected cropping systems. *IEEE Robotics and Automation Letters*, 2(2), 872–1.

Murray, R., Li, Z. and Sastry, S. (1994). *A Mathematical Introduction to Robotic Manipulation*. Boca Raton: CRC Press.

Nesbitt, B. (2006). *Handbook of Pumps and Pumping*. Oxford: Elsevier.

Nesbitt, B. (2011). *Handbook of Valves and Actuators*. Oxford: Elsevier.

Ogata, K. (2011). *Modern Control Engineering*. London: Pearson Education.

Onofre, R., Leandro Cordova, L., Tadesse, B., Stensvand, A., Gadoury, D. and Peres, N. (2017). Resistance-proof approaches to suppression of strawberry powdery mildew (*Podosphaera aphanis*). *Phytopathology*, 107, 2–3.

Oviedo-Solís, C., Sandoval-Salazar, C., Lozoya-Gloria, E., Maldonado-Aguilera, G., Aguilar-Zavala, H., Beltrán-Campos, V., Beltrán-Campos, V. and Ramírez-Emiliano, J. (2017). Ultraviolet light-C increases antioxidant capacity of the strawberry (*Fragaria × ananassa*) in vitro and in high-fat diet-induced obese rats. *Food Science and Nutrition*, 5, 1004–14.

Sastry, S. (2013). *Nonlinear Systems: Analysis, Stability, and Control*. New York: Springer.

Siciliano, B. and Khatib, O. (2016). *Springer Handbook of Robotics*. Berlin: Springer.

Suthaparan, A., Solhaug, K., Stensvand, A. and Gislerød, H. (2017). Daily light integral and day light quality: Potentials and pitfalls of nighttime UV treatments on cucumber powdery mildew. *Journal of Photochemistry and Photobiology*, 175, 141–8.

Thrun, S., Burgard, W. and Fox, D. (2005). *Probabilistic Robotics*. Cambridge: The MIT Press.

Upadhyaya, M. and Blackshaw, R. (2018). *Non-chemical Weed Management: Principles, Concepts and Technology*. Academic Press.
Verl, A., Albu-Schäffer, A., Brock, O. and Raatz, A. (2015). *Soft Robotics: Transferring Theory to Application*. London: Springer.
Xie, Z., Charles, M., Fan, J., Charlebois, D., Khanizadeh, S., Rolland, D., Roussel, D., Deschênes, M. and Dubé, C. (2014). Effects of preharvest ultraviolet-C irradiation on fruit phytochemical profiles and antioxidant capacity in three strawberry (*Fragaria × ananassa* Duch.) cultivars. *Journal of the Science of Food and Agriculture*, 95, 2996–3002.
Xiong, Y., Ge, Y., Liang, Y. and Blackmore, S. (2017). Development of a prototype robot and fast path-planning algorithm for static laser weeding. *Computers and Electronics in Agriculture*, 142, 494–503.
Xiong, Y., From, P. and Isler, V. (2018). Design and evaluation of a novel cable-driven gripper with perception capabilities for strawberry picking robots. *IEEE International Conference on Robotics and Automation (ICRA)*.

Chapter 3

Advances in communication systems in agricultural robotics

Christopher Wiegman, Santosh Pitla and Scott Shearer, The Ohio State University, USA

1 Introduction

The idea of utilizing unmanned agricultural ground vehicles (UAGVs) is not a new concept with "driverless tractors" prototypes using leader cable guidance systems dating back to the 1950s (Li et al., 2009). Since then, substantial progress has been made in numerous areas toward automating modern field production practices including navigation, obstacle detection/avoidance, and path and task planning (Fig. 1). Advances in sensing, communication, and control technologies coupled with Global Navigation Satellite Systems (GNSS) and Geographical Information Systems (GIS) are driving the transition from simple, off-road mechanical machines to intelligent machines. Controller area networks (CAN) and GNSS are contributing to the accelerated transition of tractors to highly automated machines (Stone et al., 1999; Shearer and Pitla, 2013).

Robust and fault tolerant control architectures are essential to ensure safe and productive operation of UAGVs. Behavior-based (BB) control approaches are helping to provide UAGVs with the intelligence to break complex world problems down into a series of simple behaviors (Arkin, 1990; Brooks, 1986). BB architectures are software frameworks commonly deployed in cases where it is difficult to model or characterize the real world. Unpredictable settings are characteristics of field production environments making BB architectures ideal for UAGVs. Behaviors can be either reactive or deliberative. Reactive behaviors

http://dx.doi.org/10.19103/AS.2019.0056.07

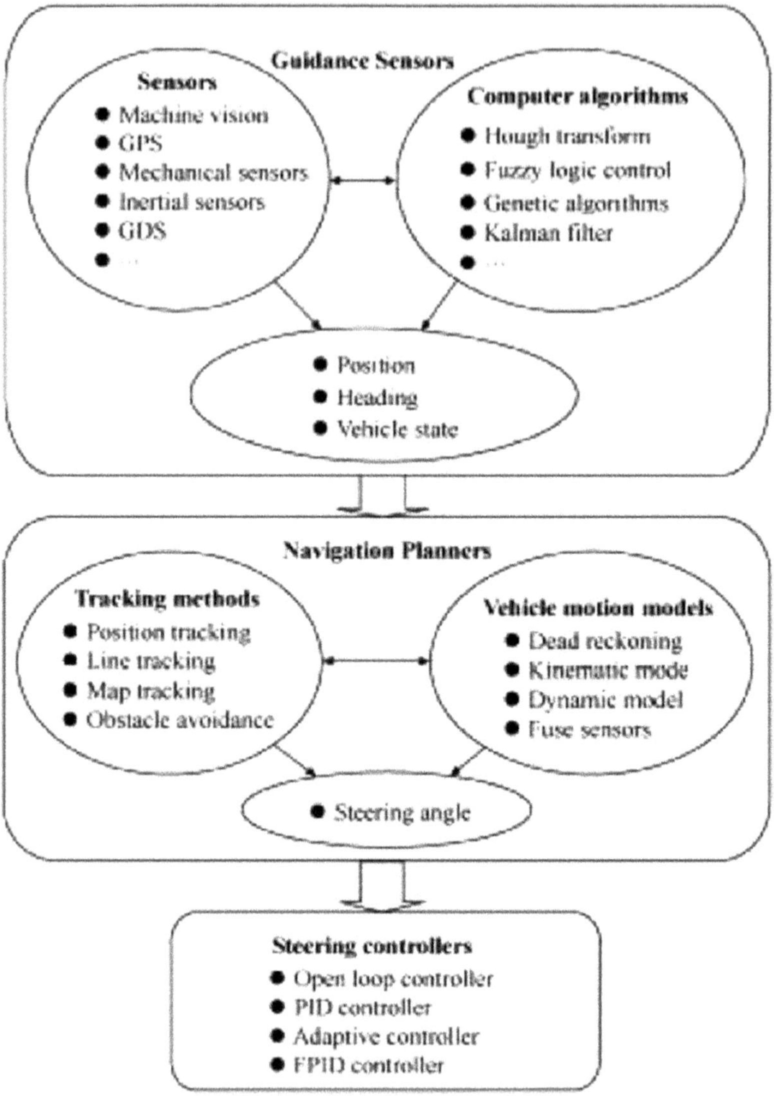

Figure 1 Framework of past progress in autonomous vehicle navigation. Note the lack of research into specific communication architectures. Source: adapted from Li et al. (2009).

respond to sensor output which trigger control actions. Deliberative behaviors are predefined steps executed to achieve certain outcomes. The combination of reactive and deliberative behaviors is critical to achieving robust and fault tolerant intelligence in UAGVs (Konolige et al., 1997; Blackmore et al., 2004).

Yavuz and Bradshaw (2002) have proposed a control structure that includes fuzzy logic and modular controls with integrated reactive and deliberative controls.

Today, robotic control architectures include hierarchical or deliberative, reactive and hybrid components. Using BB principles, Blackmore et al. (2001, 2002) have proposed an architecture for behavioral control in agricultural robots. Mott et al. (2009) proposed a control architecture with a bridge between high- and low-level components. Zeitzew (2006) reported on the successful deployment of mid-level control architecture for an UAGV stadium mower and peat moss harvester. Mid-level control was developed in light of high-level control communication failures, and to ensure safe transition to vehicle shutdown.

2 The need for communication systems

However, despite this push for autonomy, there is still a need for communication systems to link individual robots with external entities such as other robots or human supervisors. The need for reliable and robust wireless communications with and between the mobile land-based agricultural vehicles is a critical and often overlooked element in the push for automation in agriculture. Current academic research utilizes existing communication protocols and technologies, such as XBee, that functions well enough to support many current robotics applications (Zhang et al., 2010). Although this approach has proven effective for short-term gains, further advances and commercialization of autonomous agricultural field machinery will require more reliable, robust, resilient, and secure communication architectures.

Despite several decades of research, little attention has been paid to optimal communication architectures in agricultural robotics. Wireless networking is a complex task involving numerous steps with specific functions to successfully connect devices with a diversity of options at each step. Each option has unique implications that will affect its efficacy upon deployment in an agricultural setting. Achieving more robust communication architectures has a range of benefits, such as increasing productivity through enhanced coordination on multiple autonomous field machines, near real-time performance monitoring, repair and maintenance scheduling, and cloud connectivity to access advanced computing resources (Zhang and Noguchi, 2017; Steward et al., 2015).

Particular communication needs in agricultural robotics include:

- the need to facilitate human-machine coordination in the current generation of agricultural robots;
- the continuing need for overall supervision and control especially as a safety measure;

- the need to allow individual robots to coordinate operations with other robots in coordinated multi-robot systems (MRS) and robot fleets; and
- to facilitate data exchange.

The rest of the section explores some of these themes in more detail.

The first generation of UAGVs will typically be operated without an operator inside the tractor cabin. However, cyber-physical systems (CPS) will be essential to connect the operator/supervisor with the machine. Examples of these UAGVs include autonomous grain cart systems developed by Kinze Manufacturing, Inc. (Williamsburg, IA) in partnership with Jaybridge Robotics (Cambridge, MA), or the AutoCart System developed and marketed by SMARTAG (Ames, IA). Either system can be summoned by the operator of the combine harvester during the grain unloading phase (Held, 2014). The combine operator can also initiate four autonomous grain cart functions remotely (follow, unload, park, and idle). Similar systems have been developed by other agricultural equipment manufacturers which transform traditional manned machinery to autonomous equipment. Deere and Company (Moline, IL) have, for example, developed peat moss harvesters based on three-point hitch attachments to a traditional manned tractor (Johnson et al., 2009). Semi-autonomous tractors have also been developed for orchard work (i.e., spraying and mowing) (Zeitzew, 2006; Moorehead et al., 2009). Sahi (2016) has predicted a significant upswing in the purchase of supervised UAGVs or tractors, projecting a worldwide market topping $30 billion by 2024 (Fig. 2).

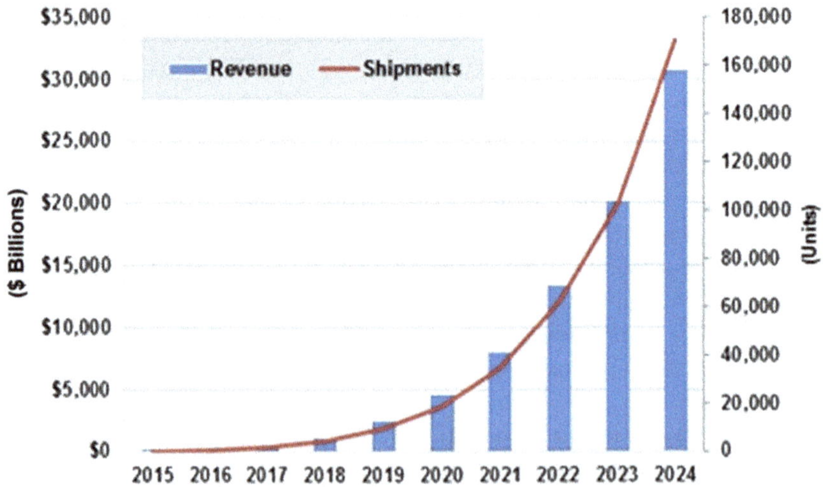

Figure 2 Driverless tractor revenues and shipments worldwide 2015–24. source: www.tractica.com.

While communication technologies are important in facilitating these technologies, the primary driving factor for wireless communication is safety. Modern systems require a "human in the loop" to monitor UAGVs, either riding in a cab or nearby to monitor robot operations (Zhang and Noguchi, 2017; Steward et al., 2015). Given a robust communication system, humans can be removed from the operator's cab and placed in a remote monitoring role, only intervening if there is a problem and the machine cannot solve itself. As one might suspect, the autonomous vehicle legal landscape will drive the development of wireless communications in agriculture. For example, current Nevada state law in the United States requires the ability of a human to intervene in the case of an emergency (Kohanbash, et al., 2012). This will most likely serve as precedence for all US state regulations governing autonomous vehicles in the future. The main implication is that not only is it unsafe but could become illegal for completely offline, unmonitored operations. Therefore, supervised autonomy with a "human in the loop" will prevail as the dominant control architecture for land-based, unmanned agricultural autonomous agricultural field machinery. Stentz et al. (2002) have developed a system where humans supervise autonomous tractors. They concluded that "the real challenge for agricultural operations is safeguarding the people, the environment, and the machines. Until this problem is solved, we cannot field unmanned vehicles." Managing safety concerns when utilizing this kind of equipment is challenging. Even with redundant safety features to mitigate liability, the ability to stop a faulted UAGV (large or small) for unanticipated events remains a significant engineering challenge.

Managing safety risks thus represents a major impediment to marketing fully autonomous field machinery. Even more important is which entities assume or share liability. In the near term, tractors will have human operators relegated to a *babysitting* role as equipment size and power levels continue to increase. CPS will enable removal of the operator from the field and relocation to a central control facility. Until machinery manufacturers, producers, and insurance companies reach consensus on how to apportion and control risk, it seems likely therefore that the most significant development will be for humans in the loop to transition from controlling a single machine to managing many from a central location.

Another key role for communication technologies is the coordination of multiple UAGVs for achieving desired field operations. Lowenberg-DeBoer (2002) has concluded: "Autonomous farm equipment may be in our future, but there are important reasons for thinking that it may not be just replacing the human driver with a computer. It may mean a rethinking of how crop production is done. In particular, once the driver is not needed, bigger is no longer better. Crop production may be done better and cheaper with a swarm of small machines than with a few large ones." Inter-UAGV communications are essential for successful deployment of such MRS. Balch and Arkin (1994) found that inter-agent communication significantly improves performance. Fung et al. (1994), for example, has utilized wireless communications to broadcast georeferenced location data amongst different machines. Rude et al. (1997) developed a wireless inter-agent communication network termed IRoN that included implicit as well as explicit communications. More recently, Wilke and Braunl (2001) reported on the development of a flexible wireless communication network that supported message transmission between individual agents in unison with a central workstation. Modest cooperation between UAGVs requires implicit communication while dynamic UAGV cooperation requires explicit communication.

Emmi et al. (2014) have described a system architecture for robots working in fleets to improve reliability as well as minimize complexity and costs. They reviewed a range of solutions, from fully distributed to integrated systems run from a central computer. Conesa et al. (2015) have reported on a multilevel supervision systems for a fleet of autonomous tractors. Target field activities included selective weed control via spray application and flaming. A recent example of cooperating UAGVs is a Combine (grain harvester) working with Grain Carts I and II (see Posselius et al. 2016) (Fig. 3). Grain Carts I and II receive command-based messages from the Combine instructing them to navigate along specific paths for off-loading grain. Continuous machine-to-machine (M2M) communications between the Combine and Carts is required during this coordinated activity. Grain Carts I and II maintain their trajectories relative to the Combine (Pitla, 2012). As Cart I is fully loaded, it departs for the edge of the field while Cart II takes its place and the Combine continues unloading. The states of all UAGVs are communicated via M2M and to a central monitoring station.

Another important area for communications is data handling. Currently, we have two technological advancements in agriculture, machine automation and digital agriculture, developing alongside each other (Kaloxylos et al., 2014; Rehman et al., 2014). Precision agriculture technologies include variable rate distribution of inputs, phenotyping, and crop health monitoring. This has resulted in a deluge of data being collected by various sensors resulting in what is known as digital agriculture. CPS are reshaping many industries and markets including aviation, communications, energy, infrastructure, health care, manufacturing, robotics, and transportation to name a few (Khaitan and McCalley, 2015). CPS are becoming mainstream for the generation and management of agronomic and machine data in agriculture (Weltzien, 2016; Jacobs et al., 2017). Wolfert et al. (2017) recognizes the potential when they state "... it can be expected that farm management and operations will drastically change by access to real-time

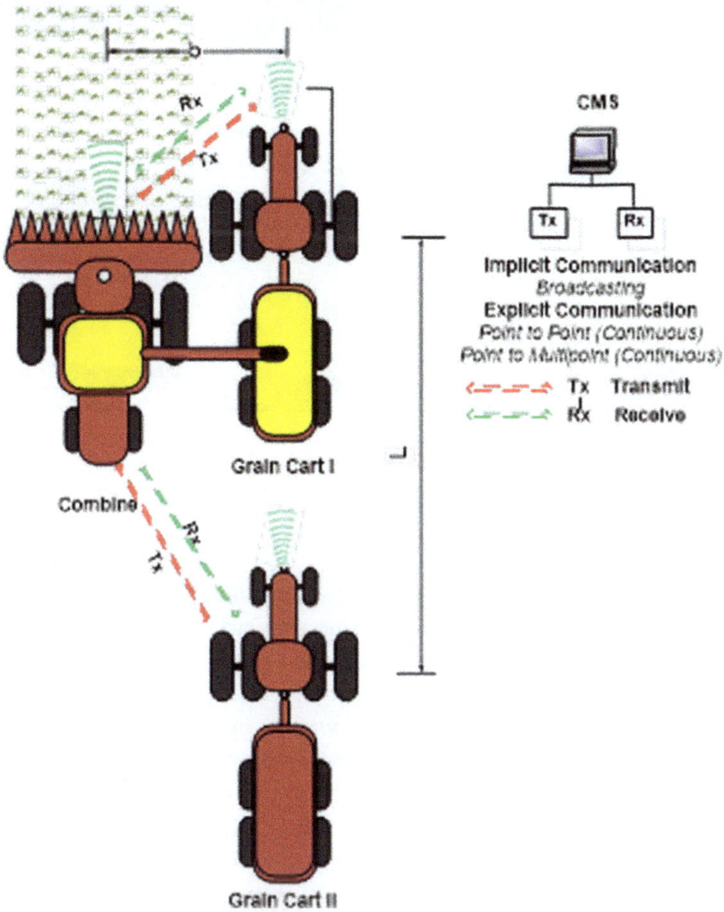

Figure 3 Coordinated vehicle operation for accomplishing biomass harvest, accumulation, and transfer operations. Source: adapted from Pitla (2012).

data, real-time forecasting and tracking of physical items and in combination with IoT developments in further automation and autonomous operation of the farm."

The deployment of UAGVs via CPS will help hasten the use of large, complex data sets to realize the benefits of precision agriculture. Through such channels digital agriculture will generate massive volumes of both structured and unstructured data from both UAGVs and small Unmanned Aerial Systems (sUAS). Typical sensor data includes, but is not limited to, machine, agronomic, weather, soil (electrical conductivity, compaction, chemical composition, etc.), biotic stress, and phenotype data (Fig. 4).

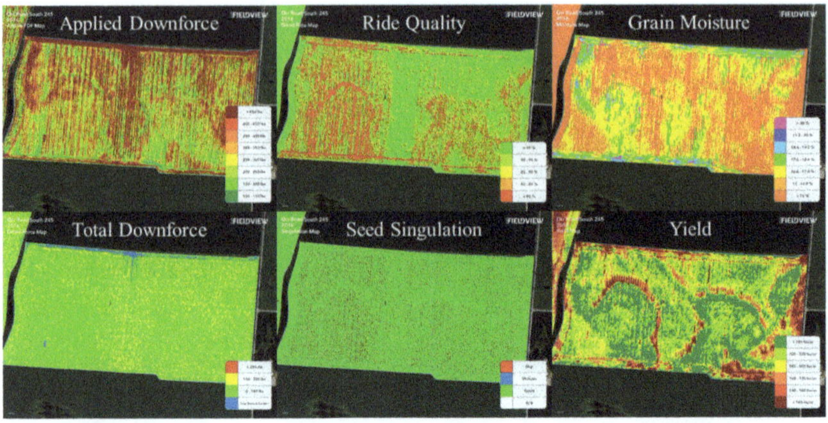

Figure 4 Map resulting from cloud-connected field machinery from sensors including pressure transducer–applied downforce, load cell–total downforce, accelerometer–ride quality, NIR reflectance–seed singulation, dielectric measurement–grain moisture, and force impetus–yield.

Increasingly, the private sector and academic researchers are finding value in coupling in-field automation and sensing with cloud computing. The central theme is to transform data into information via a suite of algorithms in the cloud. These algorithms would then be able to provide near real-time interventions to UAGVs to improve on-farm efficiency as well as decision support systems for farmers to achieve their economic and environmental goals. Digital agriculture will also have a significant positive impact on the environment by promoting targeted application of nutrients and pesticides.

Cloud computing comes with the ability to enable innovation by improving human-machine interfaces. Typically, cloud computing includes a software-defined infrastructure that can robustly ingest and store data for retrieval and analysis by a connected suite of algorithms (across different languages and platforms). This high level of abstraction enables end users to focus on the details of using UAGVs for precision agriculture management. The command center will consist of a high-end workstation with a windows-based operating system. Temporal geospatial data collected at the command center will be communicated to the cloud via application programming interfaces (APIs). Primary geospatial analysis of these data will occur both at the command center as well as in the cloud using GIS. This network will require robust communication technologies.

3 Introduction to wireless communication

This section provides an introduction to key concepts in wireless communication which allow comparison between different systems.

3.1 The Open Systems Interconnect model: communication layers

Wireless communication systems utilize standards to format data and control its flow. The most common model is the Open Systems Interconnect (OSI) which breaks communication into five fundamental layers (Olsson, 2014) (Fig. 5):

- physical layer;
- data link layer;
- network layer;
- transport layer; and
- application layer.

The physical layer is the radio link which converts bits of data into radio signals that are transmitted over the air and subsequently received. For example, in Bluetooth, ZigBee, and IPv6 over Low Power Wireless Personal Area Networks (6LoWPAN) technologies, IEEE 802.15.4 standard governs this process (e.g., radio frequency, bandwidth etc.) (Lethaby, 2017).

The data link layer provides linkages with connected devices by detecting and correcting any errors that may occur in the physical layer during transmission and reception and handles data framing. The network layer handles addressing and routing of data through the network. A common example is the Internet Protocol (IP) that provides IP addresses to each device to direct data packets to one and other.

The transport layer creates the specific communication sessions between applications running on networked devices. Transmission Control Protocol (TCP) is the dominant protocol in use today; however, others such as User Datagram Protocol (UDP) are still common. Finally, the application layer handles data formatting and ensures it is transported in application optimal schemes. One common example is Hypertext Transfer Protocol (HTTP). A diagram of this

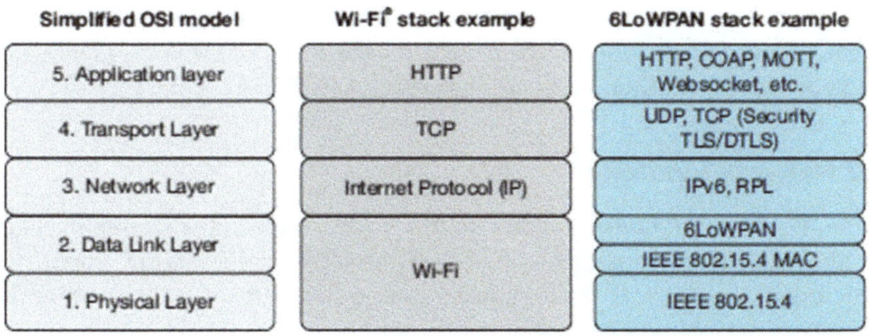

Figure 5 The OSI communication system model with Wi-Fi and 6LoWPAN implementation examples. Source: adapted from Olsson (2014).

simplified OSI model is shown in Fig. 5 with example implementations in a Wi-Fi and 6LoWPAN network (Olsson, 2014).

The primary area of focus for the remainder of the chapter will be on various technologies that address the physical and data link layers. However, it is important to be aware of the additional requirements of a functioning network. The physical and data link layers physically transmit data from one device to another which is the ultimate goal. However, the other layers influence how that data is formatted, what network architectures can be used, and how local networks connect to the internet. All of this information combined will influence how effective any communication architecture is at transmitting data and enabling autonomous agricultural equipment.

An example of these upper-level effects can be seen in the differences between the two common transport layer protocols, UDP and TCP. TCP is the most widely used because it is extremely reliable and has very low rates of data loss. However, this reliability comes with the cost of being top heavy, meaning it is computationally more complex and demanding. UDP on the other hand is simpler and is optimal for low latency networks that can tolerate high rates of data loss (Olsson, 2014). This has implications with respect to their use in an autonomous vehicle setting: first, UDP would be optimal for the efficient control as the low latency would lead to a responsive system but the increased data loss rates could lead to an unreliable and ultimately unsafe operation. One noted hole in the autonomous research area are studies addressing these areas to optimize a network for this very specific application.

The final comment in respect to the specific communication models pertains to the various options in the upper-layer selection and the effect on the physical and data link layers. An example would be in contrasting Wi-Fi with 6LoWPAN. Wi-Fi is inherently built to utilize HTTP running on TCP, making connecting to the internet seamless. 6LoWPAN, however, utilizes Internet Protocol version 6 (IPv6) which requires an additional conversion step to IPv4 before the devices connected to the network can be connected to the internet. Although this is not insurmountable, as there are numerous commercially available resources, depending on the use case and deployment situation, the additional complexity could be undesirable (Olsson, 2014; Lethaby, 2017).

3.2 Network topology

Another common consideration in the construction of a communication architecture is the network topology. Topology refers to the connected structure of the network controller or access point (AP) and connected nodes or devices. Topology is a product of the data link and physical layers. Common topologies in use today include star, mesh, and cluster tree (Rahman, 2015; Olsson, 2014). Different physical layers can have different topologies (Fig. 6).

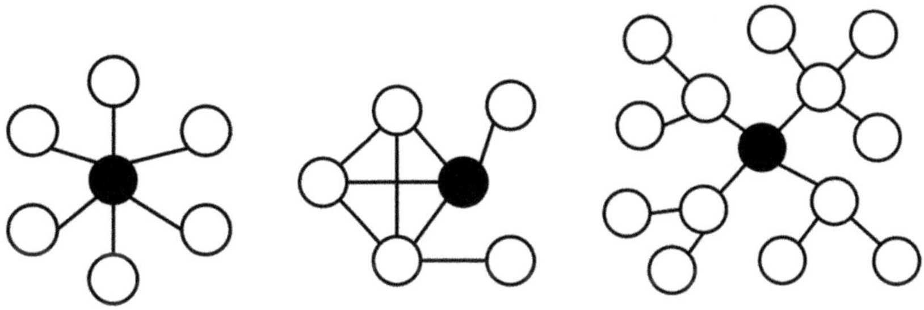

Figure 6 Star, mesh, and cluster tree network topologies. The filled in notes represent the network controllers or access points while the empty circles represent connected devices.

For example, Wi-Fi utilizes a star topology whereas ZigBee can have either mesh or cluster tree arrangements (Rahman, 2015). This becomes important as the network topology exhibits various strengths and weaknesses that influence its effectiveness. For example, a mesh topology is preferred for IoT applications because of its flexibility. Data can be routed through a variety of paths depending on the linkage of the network nodes. If one node goes down, the data stream can be routed along, or hop through, other network nodes thereby avoiding a complete network failure. This flexibility does not apply to a star topology but comes at a cost as transmission time increases as the number of nodes along the path increases.

3.3 Communication technologies

The following sections describe common communication technologies such as cellular, Multiple-Input and Multiple-Output (MIMO) broadband wireless networks, Wi-Fi (trademark name for wireless local area networking (WLAN) of devices based on the IEEE 802.11 standards), Bluetooth (trademark name for wireless technology standard for data exchange over short distances using short-wavelength ultra-high frequency (UHF) radio waves from 2.400 to 2.485 GHz from fixed and mobile devices), ZigBee (trademarked name for wireless communications based on IEEE 802.15.4 specification for a suite of communication protocols for low-power digital radios), Z-Wave (trademark name for wireless, mesh network communications protocol design for home automation), and XBee DigiMesh (trademarked name for peer-to-peer mesh network operating at 2.4 GHz).

These are just a small sample of common IoT technologies that are most commonly used today; however, it is by no means an exhaustive list. Other technologies such as Thread (trademarked name for IPv6-based, low-power mesh network for IoT), WirelessHART (trademarked name for wireless network based on the Highway Addressable Remote Transducer Protocol (HART)), Sigfox (trademarked name for low-power wireless network to connect devices that continuously transmit small quantities of data), and other proprietary technologies could hold potential. However, these technologies are not as widespread and therefore harder to evaluate for inclusion in this specific application. All current technologies have been developed and deployed in areas such as the automotive industry, building automation and consumer electronics, and therefore possess certain advantages and disadvantages that can be evaluated for use in agriculture.

4 Person-to-machine communication

The primary purpose of person-to-machine (P2M) communication for UAGVs is to send task-specific instructions to initialize autonomous operations from remote locations. Secondary purposes include safety and the ability to remotely monitor UAGV performance. As stated in the introduction, current legal precedent for autonomous operation mandates the ability of user to intervene in the event of a system threat to person or property; a dedicated P2M link serves this purpose. Additional benefits include the ability to constantly measure progress and efficiency. A robust communication link with the machine enables monitoring of machine conditions such as oil temperature, wheel slip, fuel level, and so forth and utilizes this information for task adjustment or maintenance records (Blackmore et al., 2004; Steward et al., 2015).

Offline operations do not seamlessly enable this level of real-time monitoring which decreases opportunities for improvement compared to online operations. This monitoring will require streaming of the vehicle information as well as a video feed, meaning the link will need to have high bandwidth requirements with reasonably low latency. Depending on the operation, that is passive monitoring or active intervention and control, priorities will change. For passive monitoring, the bandwidth is priority, whereas for active control, low latency is most important.

A final advantage of a dedicated P2M communication link is for enabling Real-Time Kinematic Global Positioning System (RTK-GPS) for autonomous navigation. Although various autonomous navigation techniques exist utilizing technologies such as dead-reckoning and machine vision, RTK-GPS is one of the most accurate and precise, assuming a reliable correction signal is present (Li et al., 2009). The accuracy of modern RTK-GPS unit is a function of station-dependent errors such as satellite and station clock bias, ionosphere and troposphere interference as well as others. Various techniques such as using a Virtual Reference Station (VRS) and network area corrections (FKP) have been developed to model the influence of these confounding factors and adjust the calculated position accordingly (Landau et al., 2003). The difficulty comes when getting these corrections to the GPS receiver. Methods exist to employ either VRS or FPK over various communication architectures such as cellular networks or Local Area Networks (LAN) (Millner et al., 2006; Hwang et al., 2012).

The remainder of Section 4 details two methods for P2M communication:

- Cellular networks
- WLAN

WLAN uses MIMO broadband radio antennas. Both methods have various advantages and disadvantages in their application to autonomy in agriculture. The largest hurdle for these architectures is range as the goal is to maintain connection to multiple devices from a remote central location. Field operations can cover substantial areas and therefore pose a significant challenge in connecting them all together under a single network architecture.

4.1 Cellular networks

Cellular networks fall into a wireless category known as Wireless Wide Area Network. These networks function by utilizing radio base stations that transmit and enable connection to devices inside of its "cell" or specific coverage area. These coverage areas vary based on the technology and signal specification used on that base station (Pauli et al., 2010; Pabst et al., 2004). Common examples include Universal Mobile Telecommunications Network/High Speed Packet Access (UMTS/HSPA) third-generation (3G) and fourth-generation (4G) long-term evolution (LTE) standards. These technology standards govern the radio frequencies used as well as the packet formatting and routing. Although these standards and their utilization inside of the base station network system is exceedingly complex, there are a number of basic issues which affect utilization with autonomous vehicles such as agricultural robots.

The first is network speed, notably uplink and downlink speeds. For 3G, speeds range from 384 kbps to 2.0 Mbps (UMTS) and 600 kbps to 10.0 Mbps (HSPA), while 4G is 3.0 to 10.0 Mbps (RS Components, 2015). Although the upper end of 3G data rate would suffice, 4G is preferred to handle the passive monitoring, mainly due to the high bandwidth requirements of live streaming video.

Next is operational frequency, ranging from 900 MHz to 2.1 GHz. Latency information for these networks is difficult to come by as the Federal Communications Commission (FCC) in the United States has continually refused to accept a standard latency metric. Fifth-generation (5G) is the most recent incarnation of wireless networks which likely alleviates many of the limitations of 3G and 4G (Gammel et al., 2017). The greatest potential of 5G is for expansion into the spectrum allocations well beyond that of previous generations. Not only does this substantially increase data speeds, but also resolves propagation loss and coverage density problems. Although these capabilities stand to make cellular networks a serious option for use in P2M communication for UAGVs, full deployment of 5G networks is in its infancy.

3G, LTE networks have been in use in agriculture for many years in the United States as evidenced by commercial offerings like Field View by Climate Corporation, MyJohnDeere.com, and FarmMobile Puck. These products stream various "mission critical" data to the cloud and provide data visualization capabilities via web- and tablet-based applications. These are powerful tools enabling near real-time monitoring of field operations, and they also provide insight to the use of cellular networks in agriculture. However, the current networks are limited in coverage. The 2018 FCC Broadband Deployment Report details the access of 10 mbps downlink and 1.0 Mbps uplink LTE networks to be limited to approximately 70% in rural areas with provider density (Fig. 7). Therefore, a significant proportion of US producers do not have access to the necessary cellular networks. Further, the FCC report details the lagging connectivity of rural areas not just in mobile networks, but in terrestrial (fixed) broadband deployment as well, as shown in Fig. 8. Although utilizing existing communication networks seems like an ideal solution, limited availability in some areas of the country remains a problem.

Another consideration is the level of expense associated with these networks. Given the monthly service charges on a device basis, operational expenses will continually mount year after year and in proportion to fleet size. One of the largest benefits of UAGVs is the productivity increase via

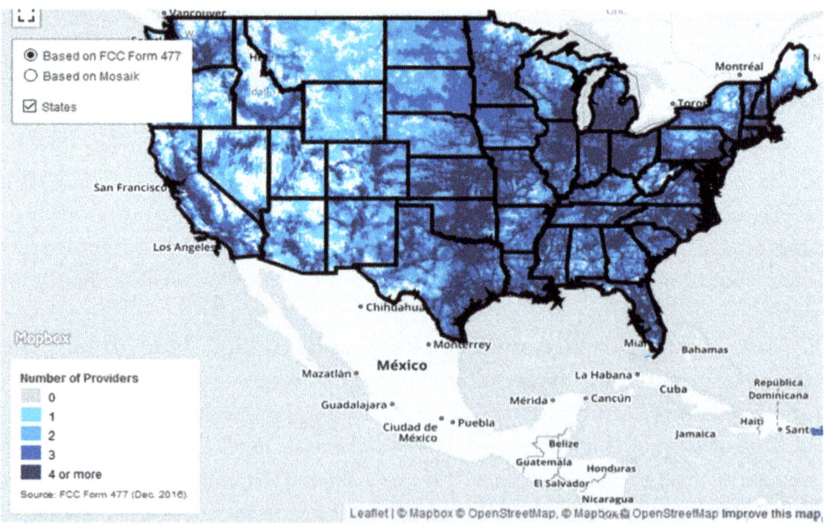

Figure 7 LTE deployment provider density according to FCC 2016 report.

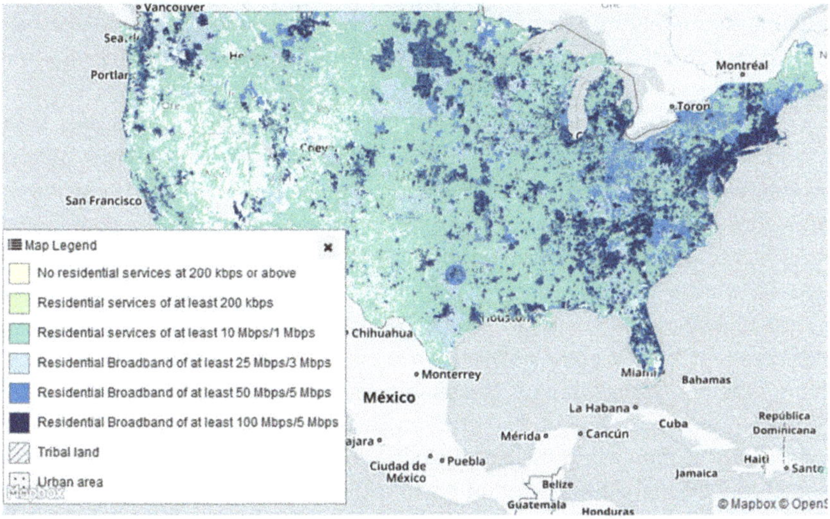

Figure 8 Fixed terrestrial broadband internet access deployment according to FCC 2016 report.

low-power/GVW vehicles which also reduces soil compaction, the ability to amortize technology costs over more acres given its 24 × 7 operation, and potential labor savings. These advantages would potentially be offset by high operational costs.

4.2 MIMO broadband radio antenna network

Stemming from an expansion Wi-Fi WLAN capabilities, the use of MIMO broadband radio antennas for point to multi-point communication architectures provides substantial potential as a reliable, robust, and cost-effective option for P2M communication. MIMO is a relatively new technology where an antenna array is used to enable duplexed communication on multiple channels over a segment discrete bandwidth (Catreux et al., 2002). Typical channel width is around 40 MHz in the 2.4 and 5 GHz bands depending on the antenna model (Ubiquiti Networks, 2014, 2017).

The use of the 5 GHz band is significantly more vulnerable to non-line of sight conditions than 2.4 GHz (Pabst et al., 2004). However, the broadband capabilities and increased channel availability make them much more resilient to environmental noise as channels with below-optimal signal to noise plus interference ratio can be avoided, therefore ensuring reliable communication. Bandwidth can reach as high of 350 Mbps depending on signal quality. The ability to hop channels will continue to prove invaluable in the current crowded spectrum (Ubiquiti Networks, 2017). The range of these systems is listed as being 10.0+ km Visual Line of Sight, which proves sufficient for many agricultural operations. Further, these antennas are lightweight weighing under 10.0 kg, making them more acceptable for larger, dispersed (non-contiguous) operations.

Current unpublished work at the Ohio State University utilizing these antennas has shown substantial promise. A communication network utilizing Ubiquiti brand antennas has been incorporated into a mobile field lab for various data-intensive projects. The antennas were deployed on both UAGVs and sUAS for control and data acquisition tasks, respectively, with substantial success. Figure 9 represents a potential set up for this communication architecture. Variable sector antennas are connected to the command center and represent the APs of the network while omnidirectional antennas are placed on the mobile unit themselves. The variable sector antennas yield a higher gain than their omnidirectional counterparts resulting from beamforming of the signal, thereby increasing the potential range of the network. Figure 10 shows an omnidirectional antenna mounted to a small UAGV and Fig. 11 shows the variable sector antenna serving as the AP in the bed of a truck. The advantage of omnidirectional antennas is that, despite their lower signal gains, they ease network deployment, with 360° amplification rather than a directed signal, making it ideal for mobile units.

This network architecture is relatively easy to set up and operate and comes with relatively low overhead cost with each antenna radio unit costing around $250.00 depending on manufacturer and product grade. As previously stated, it operates in a similar fashion to a Wi-Fi network. In the Ubiquiti network, at least one node (in this case the variable sector antenna(s)) is specified as the APs, and the end nodes (the omnidirectional antennas) connect to the network APs. The network can be expanded to cover larger areas via inclusion of additional APs causing the end nodes to automatically hop to the AP with the strongest signal. Specific devices such as the task

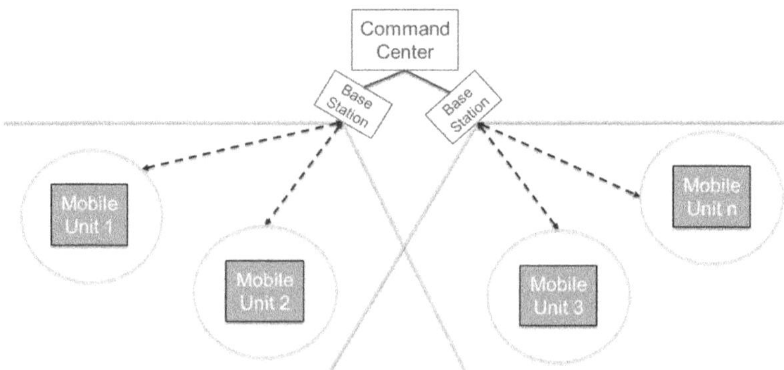

Figure 9 Potential architecture utilizing variable sector and omnidirectional MIMO Broadband Antennas for P2M communication.

Figure 10 UAGV with MIMO broadband omnidirectional antenna for wireless P2M communication.

Figure 11 MIMO broadband with variable sector antenna access point for wireless P2M communication.

controller on the UAGV are connected via ethernet to the omnidirectional antenna mounted to the unit. The Ubiquiti system supports both UDP and TCP transport layer protocols enabling a level of flexibility from the layer side and through the use of a gateway and HTTP network protocol for internet access (Ubiquiti Networks, 2017). This means that every device on the farm, even those miles away, can be connected to the internet at a single remote location.

5 M2M communication

The goal of separating M2M and P2M is that they each have different requirements depending on the autonomous architecture being deployed. If each unit is operating completely independently, then a M2M communication network is not useful. However, production efficiency potentially increases with the number of units working collaboratively. In these situations, various swarm or leader-follower systems are likely to be employed.

P2M architectures could be adapted to serve this M2M purpose. However, they are subject to higher latency because of increased communication distance. Switkes et al. (2017) describes a system for vehicle platooning (multiple trucks closely following each other) utilizing a P2M architecture (cellular) to initiate the autonomous behavior and then a subsequent direct M2M communication link (e.g. Wi-Fi or Bluetooth) to execute behavior. As stated previously, various autonomous architectures will have different requirements. P2M communication requires more bandwidth for task assignment or video streaming for monitoring but can tolerate increased latency.

Conversely, M2M communication will be needed for multiple machines operating in close proximity while executing tasks, thereby requiring reduced latency to avoid machine collisions and ensure safety (Zhang and Noguchi, 2017). Further, there is not as high of a bandwidth requirement for M2M communication as only machine- and operation-specific data will need to be shared, not video or similar media. In addressing latency concerns specifically, the star communication topology for some P2M technologies includes end nodes linking back to a central AP. This means that any communication must first go through that AP before returning to the end destination address. The M2M technologies covered in this section employ local star or mesh topologies enabling a direct communication link between devices, eliminating at least one 'leg' of the trip and therefore reducing latency.

The second benefit of a dedicated M2M network is that it serves as a redundant communication system which increases system reliability and safety. Catastrophic failures are not acceptable for these operational conditions (Blackmore et al., 2004). Any system needs to be able to withstand total failure and exhibit a graceful degradation of performance during a partial system failure. Redundant systems help to achieve this as they do not duplicate other systems such as the P2M, but offer an alternative method of condition assessment. If the primary P2M communication link fails, the functionality of that device is not totally lost as the user can still route data requests and attain information about the failure through the M2M network through a nearby machine. Therefore, complete functionality is not lost and the system only experiences a reduction in functionality (Blackmore et al., 2004).

Wi-Fi, Bluetooth, ZigBee, and Z-Wave are a few of the current technologies being deployed for M2M communication in other IoT applications. Their topologies and physical characteristics will be discussed along with the advantages and disadvantages associated with each.

5.1 Wi-Fi

Wi-Fi was originally created as the replacement to the popular wired ethernet standard and has been integrated into numerous devices such as laptops, TVs, and smartphones. One reason for the large success of Wi-Fi is the interoperability of the programs and the ubiquity of the system. Any device can easily be connected to the internet without the use of custom gateways and in a cost-effective manner (Lethaby, 2017).

Wi-Fi physical and data link layers are governed by the IEEE 802.11 standard. 802.11n is the most common today, specifying the use of MIMO antennas and the 2.4 and 5 GHz bands (Rahman, 2015). As will be common in the listing of these technologies, the ability to utilize the 5 GHz band represents a large advantage. Many devices already utilize the crowded 2.4 GHz region of the spectrum, and 5 GHz is much more resilient to noise influences due to increased bandwidth availability (Cisco, 2018). This represents a large advantage for Wi-Fi over the other options.

Bandwidth under the 802.11n specification are usually around 150-200 Mbps with the maximum of 600 Mbps. The latest iteration of the specification, 802.11ac, greatly increases this bandwidth to a range of 500 Mbps to 1 Gbps (RS Components, 2015). This increase in speed is accomplished by increasing channel width from 40 to 80 MHz, or even 160 MHz, as well as increased number of MIMO spatial streams from four with 802.11n to eight with 802.11ac. An additional benefit of the new 802.11ac standard is inclusion of "sounding." All devices with antennas utilize beamforming to direct the energy to the receiver. However, sounding enables the receiver work with the transmitter to form a beamforming matrix that enables the transmitter to precisely steer the energy toward the receiver (Cisco, 2018). All of this helps to increase bandwidth at greater ranges than standard 802.11n specification.

Wi-Fi networks typically employ a star topology with networked devices connecting to a single AP. These APs can then be networked to increase the size and coverage of the WLAN. In UAGVs, this network technology is most useful for a leader-follower application where the leader has the AP with followers connecting to enabling direct communication between the two. However, this is a local star topology, so followers cannot communicate directly with each other; they still have to communicate through the AP. As mentioned by RS Components, 2015, the range of the signal is dependent upon AP quality but can reach out to 50 m which should prove suitable for most leader-follower operational schemes. This technology is also scalable, with some manufactures claiming support of as many as 250 devices simultaneously while others utilize consumer grade APs handling no more than 50 (Lethaby, 2017).

5.2 Bluetooth

Bluetooth is another extremely popular technology for device communication. Since its introduction in 1994, it has been incorporated into a wide range of wearable devices associated with smart phones as well as in the automotive industry and computers. The physical layer was originally covered under the IEEE 802.15.1 layer which is no longer maintained. Bluetooth has two common versions, classic and LE (low-energy) both operating in the 2.4 GHz band. Classic uses 79 1 MHz wide channels while LE uses 40 2 MHz channels (Lethaby, 2017).

Originally, this Personal Area Network technology was limited to a star topology with a range around 3–10 m and data transfer rates around 1–2 Mbps for up to eight devices (Rahman, 2015). However, more recent standard released in 2016, Bluetooth 5, enables a significant increase in range (around 50+ m) (RS Components, 2015). The standard also specifies a choice of four data rates; 2.0 Mbps, 1.0 Mbps, 500 kbps, and 125 kbps with lower rates offering longer range. In 2017, the Bluetooth Special Interest Group released mesh topology specifications that are compatible with Bluetooth 5 standards enabling Bluetooth use for many to many communications (Lethaby, 2017).

Bluetooth offers substantial potential in autonomous networking. Considering its rapid development and huge variety in products, it is finding its way into numerous applications. Range is one of its most limiting factors as well as difficulty with scale-up potential. However, Bluetooth is one of the fastest evolving technologies as the extremely diverse deployment of its products pushes its continued development. As new versions of the technology evolve, these deficiencies will continue to be addressed.

One of the most significant benefits of Bluetooth that has led to large wide-range adoption is the extremely low-power requirements. However, for UAGV applications, power consumption is not an issue, largely because of the prevalence of Internal Combustion engines as the power source. The bandwidth is suitable for machine data communication, and new specifications enabling mesh topology serve as a large advantage for reliable and redundant connectivity. Perhaps a more significant advantage is the short connection and reconnection times between devices. Depending on the network configuration, device connection can take as little as 3.0 ms in LE applications with the average join time being under 3.0 s (Rahman, 2015; Tomar, 2011). Considering the flexibility of UAGV operations, this could prove valuable for rapid start-up and task execution as multiple devices can be added to the network quickly. Bluetooth also supports UDP and TCP, and can be connected to the internet via a gateway for IP support.

5.3 ZigBee

ZigBee is an open standard designed by the ZigBee Alliance, a group of more than 400 companies from various industries. The ZigBee Alliance is responsible for the network and application layer, while IEEE 802.15.4 defines the physical and data link layers, meaning it can operate at either 2.4 GHz or in the sub 1 GHz range (Lethaby, 2017). ZigBee was designed to operate with a mesh topology, providing the ability to scale up the network geographically through multi-hop operations as well as increase reliability and fault tolerance through back-up paths in the mesh. Data rates are lower than Bluetooth at 250 kbps and a range of around 100 m. ZigBee also defines up to three types of nodes: coordinator, router, and end nodes. First, a coordinator node is responsible for forming the network and routing traffic. Router nodes act as an intermediate node, relaying data from other devices and finally end points. Every network needs one coordinator node and can have multiple routers which can also act as end nodes (Daintree Networks, 2008).

In assessing the benefits of utilizing ZigBee in agriculture, one of the largest advantages is that it provides a complete solution that enables device interoperability between different manufacturers. The ZigBee protocol suite includes standard commissioning, security, network, and device-management procedures. This guarantees end-to-end device interoperability from the start of operation and through integration with various data collectors or hubs (Lethaby, 2017). Compared with other technologies, ZigBee appears to be one of the easiest to incorporate into custom operations and possesses the greatest device flexibility. It is substantially simpler than other alternatives, with a protocol stack estimated to be a quarter of Bluetooth or Wi-Fi. It is this simplicity that helps lend to its low cost, interoperability, and lower maintenance. ZigBee also has very fast network join times of around 30 ms on average, meaning device it can handle rapid changes in network membership (Tomar, 2011). The primary disadvantages of ZigBee stem from its low bandwidth and limited range. Although 250 Mbps is enough for mission critical data such as control and navigational data, it is unable to handle every message on both tractor and implement CAN bus which would be desirable in monitoring and situations. The range of the system is usually acquired via hopping and in real applications, "dummy" nodes are often incorporated for range extension. Although the mesh topology enables direct peer-to-peer communication, the need for a dedicated coordinator does limit application flexibility.

5.4 6LoWPAN

6LoWPAN is an IP-based networking protocol built on the IEEE 802.15.4 standard. Unlike an application protocol technology such as ZigBee or Bluetooth, 6LoWPAN is a network protocol that defines a layer between the 802.15.4 data link layer and the standard TCP/IP stack, as shown in Fig. 2. It enables small, low-powered devices to directly communicate with any other IP-based server or device on the internet. Since it is built on IEEE 802.15.4, 6LoWPAN implementations exist for both 2.4 GHz and sub 1 GHz (915 MHz) bands and uses a mesh network topology (Olsson, 2014).

In operating with the 802.15.4 specification, 6LoWPAN inherits all of the benefits of a mesh network topology as discussed in the preceding section. However, what separates 6LoWPAN from technologies such as ZigBee, Bluetooth, or proprietary networks such as DigiMesh is that it does not require a gateway to connect to an IP-based system like the internet. These sometimes complex application gateways must understand any application profiles that may be used in the network and any changes in application protocols on wireless nodes must be accompanied by changes in the gateway (Olsson, 2014). If the M2M vehicle network is to be connected to the internet, 6LowWPAN overcomes a potentially difficult hurdle to overcome. Network application protocols are liable to change with the season or operation as control architectures change to accompany different equipment, that is tractors with planters in the spring and combines with tractors and grain carts in the fall.

5.5 Radio Frequency Identification

Radio Frequency Identification (RFID) originated in 1945 in the Soviet Union as an espionage tool and similarly in the United Kingdom as IFF (Identify Friend or Foe) to identify allied aircraft in the Second World War (Jia et al., 2012). An RFID system is composed of two components, tags and readers. The tag is simply an antenna attached to a microchip and communicates with the reader via radio waves to identify that object. There are two types of tags, passive and active. Active devices are connected to a power source, usually a battery, which powers the chip and broadcasts the signal back to the reader. Passive tags are powered by the signal transmitted from the reader by the tags antenna and used to power the chip (Zhu et al., 2012; Want, 2006). In general, these tags operate in three general frequencies, UHF, high frequency (HF), and low frequency (LF). Typically, near-field connections utilize the HF and LF ranges (<100 MHz) where the UHF is used for far-field applications and operates in the 2.4 GHz range. The RFID chips operating in the UHF range generally have an increased range to around 20–30′ but use more power (Want, 2006; Zhu et al., 2012).

RFID chips and readers are generally point to point; a means for a central reader to gather information about the tagged object. These devices are categorized into different classes to describe the capabilities of the unit with functionality increasing with class number as described by EPCglobal (Want, 2006; Jia et al., 2012). Class 0/1 describes the most basic functionality of a passive radio capability with a 96-bit tag whereas Class 4 are active tags enabling peer-to-peer communications and additional sensing capabilities. Currently, RFID technology is used in a large variety of industries from automotive industry for anti-theft devices, logistics companies, retailers such as Wal-Mart to Old Navy, and even in the medical and defense fields. It fills numerous applications in monitoring of material and persons and giving an ordered sequence of location data or in the monitoring the conditions of a processes or condition (Jia et al., 2012).

In applications to autonomous vehicle communication, the greatest potential lies in identity preservation of the various inputs and outputs of the overall process. RFID tags are most often used in conjunction with other peer-to-peer communication topologies in that the readers communicate over a network such as the internet (Sundaram et al., 2010). Using tags to label common inputs such as fertilizer, pesticides, seed, and even fuel enables automatic tracking and verification of what is currently in use as well as a record of what has been applied. RFID provides a simple means for both M2M and P2M verification of what is on each machine as it moves through a field or when it stops for re-supply. At harvest, all grain handling equipment can be tagged with RFID chips and used to track grain as it exits the field moved to a grain storage unit such as an elevator and even up to the point of sale, giving the producer and purchaser a detailed electronic record of the grain.

6 Security issues

Network security is and will continue to be a significant issue when all wireless networks are in use. Although the threat of machine hacking is less likely to be as large of a concern in the rural deployment of these UAGVs as compared to self-driving cars in urban areas, it is still a concern that needs to be addressed. All of the wireless methods presented here have well-defined security protocols as described by their specifications. It is important to remember that there are multiple areas for weakness, whether it is in the joining of the network, or in the transmission of messages.

Given the wireless transmission methodology, any one in range of the signal can eavesdrop, thus highlights the need for rigorous encryption protocols such as Advanced Encryption Standard or AES. ZigBee, 6LoWPAN, Wi-Fi, and Bluetooth all utilize 128 bit AES. ZigBee, having the most unique architecture of the three, utilizes three types of keys to manage network security, Master, Network, and Link to manage what devices can access the network and encrypt transmitted data (Olsson, 2014; Lethaby, 2017; Daintree Networks, 2008). 6LoWPAN can employ an additional transport layer security running on top of TCP for an additional layer of security. However, it is important to note that this additional key requires increased computational resources such as a hardware encryption engine with advanced cipher suites (Olsson, 2014). Wi-Fi utilizes AES for message encryption and Wi-Fi Protected Access version 2 WPA2 is the most common network protection protocol. Bluetooth was historically notorious for low security until Bluetooth 4.2 specification introduced in 2014 added multiple industrial strength protocols. These additional security measures include elliptic curve cryptography, AES, and CCM or cipher block chaining message authentication code cryptography. The final comment with security is like everything, it comes at a cost (Lethaby, 2017). The better the encryption, the greater the computational cost and time.

7 Summary

UAGVs represent substantial potential with their ability to optimize crop yields and increase sustainability, making them a high priority with many academic and corporate institutions. These systems need to be safe and robust while operating in suboptimal conditions compared to other autonomous systems. The variability encountered in the field, such as ground conditions, weather, and the sheer size of the operations, highlight just a few of the challenges they face. Reliable communication networks will help to achieve safe and reliable operations in these conditions. P2M technologies such as cellular networks or MIMO broadband WLAN antennas will enable constant contact with the equipment and the ability to intervene should the need arise. Their high bandwidth connection over an extended range enables remote monitoring of operations via both video and machine data streaming to ensure optimal operational efficiency. The incorporation of M2M networks with technologies, such as Bluetooth, ZigBee, Wi-Fi, and 6LoWPAN as well as other proprietary brands such as DigiMesh XBee, enable direct communication between active UAGVs in cooperative operations. This system redundancy increases operational safety as it helps to mitigate the probability of a catastrophic communication failure. Although utilization of a single technology is usually preferable, existing systems do not possess all of the needed traits to ensure success such as high bandwidth, long range, and low latency. Therefore, a combination of technologies is required along with the added safety inherent with redundancy.

Wireless networking is a rapidly changing, complex field. New technologies are constantly evolving to accommodate the growing level of device connectivity associated with IoT. This represents an abundance of opportunity for automation in food production. However, it is important to keep in mind that none of these technologies were developed specifically for use in agriculture and will involve some level of adaptation for implementation. Future technologies, such as 5G cellular networks, represent substantial potential, but epitomize this challenge of implementation in a rural setting. Regardless of these challenges, wireless communication remains an often overlooked but crucial key to realizing the goal of truly autonomous vehicles in agriculture.

8 References

Arkin, R. C. 1989. Motor schema-based mobile robot navigation. *The International Journal of Robotics Research* 8(4), 92–112. doi:10.1177/027836498900800406.

Arkin, R. C. 1990. Integrating behavioral, perpetual, and world knowledge in reactive navigation. *Robotics and Autonomous Systems* 6(1), 105–22.

Balch, T. and Arkin, R. C. 1994. Communication in reactive multiagent robotic systems. *Autonomous Robots* 1(1), 27–52. doi:10.1007/BF00735341.

Blackmore, B. S., Have, H. and Fountas, S. 2001. A specification of behavioral requirements for an autonomous tractor (keynote address). In: *6th International Symposium on Fruit, Nutrition and Vegetable Production Engineering Conference*. Institut fur Agrartechnik Bornim, Potsdam-Bornim, Germany, pp. 25–36.

Blackmore, B. S., Have, H. and Fountas, S. 2002. A proposed system architecture to enable behavioral control of an autonomous tractor (keynote address). In: Zhang, Q. (Ed.), *Proceedings of the Automation Technology for Off-Road Equipment*. ASABE, St. Joseph, MI, pp. 13–23.

Blackmore, B. S., Fountas, S., Tang, L. and Have, H. 2004. Systems requirements for a small autonomous tractor. *Agricultural Engineering International: the CIGR Journal of Scientific Research and Development*. Manuscript PM 04 001.

Brooks, R. A. 1986. A robust layered control-system for a mobile robot. *IEEE Journal on Robotics and Automation* 2(1), 14–23. doi:10.1109/JRA.1986.1087032.

Catreux, S., Erceg, V., Gesbert, D. and Heath Jr., R. W. 2002. Adaptive modulation and MIMO coding for broadband wireless data networks. *IEEE Communications Magazine* 40(6), 108–15. doi:10.1109/MCOM.2002.1007416.

Cisco. 2018. 802.11ac: The Fifth Generation of Wi-Fi. Cisco Technical White Paper.

Conesa, J. M., Gonzalez-de-Soto, M., Gonzalez-de-Santos, P. and Ribeiro, A. 2015. Distributed multi-level supervision to effectively monitor the operations of a fleet of autonomous vehicles in agricultural tasks. *Sensors* 15(3), 5402–28. doi:10.3390/s150305402.

Daintree Networks. 2008. Getting Started with ZigBee and IEEE 802.15.4.

Emmi, L., Gonzalez-de-Soto, M., Pajeres, G. and Gonzales-de-Santo, P. 2014. New trends in robotics for agriculture: integration and assessment of a real fleet of robots. *The Scientific World Journal* 2014, 1–21. doi:10.1155/2014/404059.

FCC. 2018. *2018 Broadband Deployment Report*. Federal Communications Commission, Washington DC.

Fung, C., Eren, H. and Nakazato, Y. 1994. Position sensing of mobile robots for team operations. In: *Proceedings of the 10th Annual IEEE Instrumentation and Measurement Technology Conference*. Hamamatsu, Japan, pp. 185–8.

Gammel, P., Pehlke, D. R., Brunel, D., Kovacic, S. J. and Walsh, K. 2017. *5G in Perspective: A Pragmatic Guide to What's Next*. Skyworks Solutions Inc.

Held, J. 2014. Kinze autonomous harvest system. Available at: https://www.kinze.com/kinze-adds-new-features-to-its-autonomous-harvest-system.

Hwang, J., Yun, H., Suh, Y., Cho, J. and Lee, D. 2012. Development of an RTK-GPS positioning application with an improved position error model for smartphones. *Sensors* 12(10), 12988–3001. doi:10.3390/s121012988.

Jacobs, G., Schlüter, F., Schröter, J., Feldermann, A. and Strassburger, F. 2017. Cyber-physical systems for agricultural and construction machinery-current applications and future potential. In: Jeschke, S., Brecher, C., Song, H. and Rawat, D. B. (Eds), *Industrial Internet of Things*. Springer Series in Wireless Technology. Springer International Publishing, Switzerland, pp. 617–45.

Jia, X., Feng, Q., Fan, T. and Lei, Q. 2012. RFID technology and its applications in internet of things (IOT). In: *2012 2nd International Conference on Consumer Electronics, Communications and Networks (CECNet)*. IEEE. doi:10.1109/cecnet.2012.6201508.

Johnson, D. A., Naffin, D. J., Puhalla, J. S., Sanchez, J. and Wellington, C. K. 2009. Development and implementation of a team of robotic tractors for autonomous peat moss harvesting. *Journal of Field Robotics* 26(6–7), 549–71. doi:10.1002/rob.20297.

Kaloxylos, A., Groumas, A., Sarris, V., Katsikas, L., Magdalinos, P., Antoniou, E., Politopoulou, Z., Wolfert, S., Brewster, C., Eigenmann, R., et al. 2014. A cloud-based farm management system: architecture and implementation. *Computers and Electronics in Agriculture* 100, 168–79. doi:10.1016/j.compag.2013.11.014.

Khaitan, S. K. and McCalley, J. D. 2015. Design techniques and applications of cyberphysical systems: a survey. *IEEE Systems Journal* 9(2), 350–65. doi:10.1109/JSYST.2014.2322503.

Kohanbash, D., Bergerman, M., Lewis, K. M. and Moorehead, S. J. 2012. A safety architecture for autonomous agricultural vehicles. In: *ASABE Annual International Meeting*, Dallas, Texas. Paper No.: 121337110.

Konolige, K., Myers, K., Ruspini, E. and Saffiotti, A. 1997. The Saphira architecture: a design for autonomy. *Journal of Experimental and Theoretical Artificial Intelligence* 9(2–3), 215–35. doi:10.1080/095281397147095.

Landau, H., Vollath, U. and Chen, X. 2003. Virtual reference stations versus broadcast solutions in network RTK – advantages and limitations. Submitted to GNSS. April. Graz, Austria.

Lethaby, N. 2017. *Wireless Connectivity for the Internet of Things: One Size Does Not Fit All*. Texas Instruments.

Li, M., Imou, K., Wakabayashi, K. and Yokoyama, S. 2009. Review of research on agricultural vehicle autonomous guidance. *International Journal of Agricultural and Biological Engineering* 2(3), 1–25.

Lowenberg-DeBoer, J. 2002. Liability problems for autonomous equipment? Site-specific management center. *Newsletter*. February.

Millner, J. C., Asmussen, H. J. and Andreola, R. 2006. Delivery of networked GPS corrections for machinery guidance. In: *Controlled Traffic Farming Conference*, pp. 86–98.

Moorehead, S., Ackerman, C., Smith, D., Hoffman, J. and Wellington, C. 2009. Supervisory control of multiple tractors in an orchard environment. In: *Proceedings of the 4th IFAC International Workshop on Bio-Robotics, Information Technology, and Intelligent Control for Bio-Production Systems*, 10–11 September, Champaign, IL.

Mott, C., Ashley, G. and Moorehead, S. 2009. Connecting high level and low level controllers on robotic vehicles using a supporting architecture. In: *Proceedings of the 4th IFAC International Workshop on Bio-Robotics, Information Technology, and Intelligent Control for Bio-Production Systems*, 10–11 September, Champaign, IL.

Olsson, J. 2014. *6LoWPAN Demystified*. Texas Instruments.

Pabst, R., Walke, B. H., Schultz, D. C., Herhold, P., Yanikomeroglu, H., Mukherjee, S., Viswanathan, H., Lott, M., Zirwas, W., Dohler, M., et al. 2004. Relay-based deployment concepts for wireless and mobile broadband radio. *IEEE Communications Magazine* 42(9), 80–9. doi:10.1109/MCOM.2004.1336724.

Pauli, V., Naranjo, J. D. and Seidel, E. 2010. *Heterogeneous LTE Networks and Inter-Cell Interference Coordination*. Nomor Research GmbH, Munich, Germany.

Pitla, S. K. 2012. Development of control architectures for multi-robot agricultural field production systems. Ph.D. Dissertation. University of Kentucky, Lexington.

Posselius, J. H., Foster, C. H., Pitla, S. K., Shearer, S. A., Luck, J. D., Sama, M. P. and Zandonadi, R. S. 2016. Control architecture for multi-robot system. US Patent Number: 9527211.

Rahman, A. B. 2015. Comparison of internet of things (IoT) data link protocols. Available at: https://www.cse.wustl.edu/~jain/cse570-15/ftp/iot_dlc/index.html.

Rehman, A., Abbasi, A. Z., Islam, N. and Shaikh, Z. A. 2014. A review of wireless sensors and networks' applications in agriculture. *Computer Standards and Interfaces* 36(2), 263–70. doi:10.1016/j.csi.2011.03.004.

RS Components. 2015. 11 internet of things (IoT) protocols you need to know about. Available at: https://www.rs-online.com/designspark/eleven-internet-of-things-iot-protocols-you-need-to-know-about.

Rude, M., Rupp, T., Matsumoto, K., Sutedjo, S. and Yuta, S. 1997. IRoN: an inter robot network and three examples on multiple robots motion coordination. In: *Proceedings of the 1997 IEEE/RSJ International Conference on Intelligent Robot and Systems. Innovative Robotics for Real-World Applications. IROS '97*, pp. 1437–44.

Sahi, M. 2016. Are driverless tractors on the right track? Available at: https://www.tractica.com/robotics/are-driverless-tractors-on-the-right-track/ (accessed on 5 March 2017).

Shearer, S. A. and Pitla, S. K. 2013. Field production automation. In: Zhang, Q. and Pierce, F. J. (Eds), *Agricultural Automation: Fundamentals and Practices*. CRC Press, pp. 97–124.

Stentz, A., Dima, C., Wellington, C., Herman, H. and Stager, D. 2002. A system for semi-autonomous tractor operations. *Autonomous Robots* 13(1), 87–104. doi:10.1023/A:1015634322857.

Steward, B. L., Tang, L. and Han, S. 2015. A design framework for off-road equipment automation. In: *Proceedings of the 2015 Conference on Autonomous and Robotic Construction of Infrastructure*. Ames, IA, pp.180–96.

Stone, M. L., McKee, K. D., Formwalt, C. W. and Benneweis, R. K. 1999. ISO 11783: electronic communications protocol for agricultural vehicles. ASAE Distinguished Lecture Series No. 23. ASABE, St. Joseph, MI, pp. 1-17.

Sundaram, D., Zhou, W., Piramuthu, S. and Pienaar, S. 2010. Knowledge-based RFID enabled web service architecture for supply chain management. *Expert Systems with Applications* 37(12), 7937–46. doi:10.1016/j.eswa.2010.04.041.

Switkes, J. P., Gerdes, J. C., Lyons, D. F., Boyd, S. N. and Berdichevsky, E. 2017. Vehicle platooning systems and methods. Patent No: US 9645579.

Tomar, A. 2011. Introduction to Zibgbee technology. *Element14*. Available at: https://www.cs.odu.edu/~cs752/papers/zigbee-001.pdf.

Ubiquiti Networks. 2014. airMax ac Sector 2x2 MIMO BseStation Sector Antenna. Ubiquiti Networks Datasheet.

Ubiquiti Networks. 2017. Rocket 2 AC Prism airMAX ac BaseStation with airPrism Technology. Ubiquiti Networks Datasheet.

Want, R. 2006. An introduction to RFID technology. *IEEE Pervasive Computing* 5(1), 25–33. doi:10.1109/MPRV.2006.2.

Weltzien, I. C. 2016. Digital agriculture - or why agriculture 4.0 still offers only modest returns. *Landtechnik* 71(2), 66–8.

Wilke, P. and Braunl, T. 2001. Flexible wireless communication network for mobile robot agents. *Industrial Robot: an International Journal* 28(3), 220–32. doi:10.1108/01439910110389380.

Wolfert, S., Ge, L., Verdouw, C. and Bogaardt, M. J. 2017. Big data in smart farming - a review. *Agricultural Systems* 153, 69–80. doi:10.1016/j.agsy.2017.01.023.

Yavuz, H. and Bradshaw, A. 2002. A new conceptual approach to the design of hybrid control architecture for autonomous mobile robots. *Journal of Intelligent and Robotic Systems* 34(1), 1-26. doi:10.1023/A:1015522622034.

Zeitzew, M. 2006. Autonomous utility mower. In: *Proceedings of the Autonomous Technology for Off-Road Equipment*. Bonn, Germany, pp. 247–56.

Zhang, C. and Noguchi, N. 2017. Development of a multi-robot tractor system for agriculture field work. *Computers and Electronics in Agriculture* 142, 79-90. doi:10.1016/j.compag.2017.08.017.

Zhang, X., Geimer, M., Noack, P. O. and Grandl, L. 2010. Development of an intelligent Master-slave system Between agricultural vehicles. In: *2010 IEEE Intelligent Vehicles Symposium University of California*, San Diego, CA, 21–24 June, pp. 250–5.

Zhu, X., Mukhopadhyay, S. K. and Kurata, H. 2012. A review of RFID technology and its managerial applications in different industries. *Journal of Engineering and Technology Management* 29(1), 152–67. doi:10.1016/j.jengtecman.2011.09.011.

Chapter 4

Human-robot collaboration in agricultural robots

Yael Edan, Ben-Gurion University of the Negev, Israel

1 Introduction

Agricultural robots are being developed for many agricultural tasks such as field cultivation, planting, harvesting, pruning and spraying (Edan et al., 2009; van Henten et al., 2013; Bechar and Vigneault, 2017; Baxter et al., 2018). However, in practice, current working agricultural robotic systems are limited despite intensive R&D and many feasible technical solutions that have proven successful (Blackmore, 2016; Bontsema et al., 2014; Duckett et al., 2018). Fully robotized farms are not yet available, mostly due to the unstructured, dynamically changing and undefined agricultural environment that demands a high level of sophistication and complicates the development process (Bac et al., 2014). Furthermore, each crop and task requires tailored designed developments (Bac et al., 2014). Introducing semi-autonomous operations in which a human operator cooperates with the robot can enable improved performance while overcoming the complexity that current autonomous robots face (Adamides et al., 2017a). This chapter will present initial work in the domain of human-robot collaboration (HRC) in agriculture. Additionally, we provide directions on how HRC should be further adopted to enable successful robotization of agricultural operations.

http://dx.doi.org/10.19103/AS.2019.0056.22

Collaboration between a human and a robot ensures a simplified, flexible and robust system that can react and cope with the dynamic and complex agricultural conditions. A human-robot collaborative system uses the advantages of both human and robot and overcomes their limitations. Humans have superior perception capabilities and can easily adapt to changing environmental and object conditions (Fong et al., 2001; Rodriguez and Weisbin, 2003). Their acute recognition capabilities enable humans to deal with flexible, vague, changing and a wide scope of definitions. Robots can achieve high accuracy, consistency, reliability and yield (strength and speed) while providing continuous 24/7 operations (Holland and Nof, 1999). Since the complex tasks can be performed by the human the robotic system can be simplified. Another specific added value of such human-robot collaboration is to remove the human operator from the tedious and harsh agricultural environment. Moreover, by collaboration, the workload of the human operator is reduced.

Vast research has been conducted in the area of HRC focusing on interaction (HRI – roles and modalities, Goodrich et al., 2008; Jaimes and Sebe, 2007; Sim and Loo, 2015) and levels/models of collaboration/automation/autonomy (Sheridan and Verplank, 1978; Cherubini et al., 2015).

Interaction is the process of communicating when working together to accomplish a certain goal (Goodrich et al., 2008). Human-robot interaction (HRI) is 'the field of study dedicated to understanding, designing, and evaluating robotic systems for use by or with humans' (Clarkson and Arkin, 2007) or 'the study of the humans, robots and the ways they influence each other' (Fong et al., 2003). It is a multidisciplinary field in which researchers from areas of robotics, human factors, cognitive science, natural language, psychology and human-computer interaction work together to understand and shape the interactions between humans and robots. The interaction depends on the *interaction role* and the *autonomy level* (also denoted as level of collaboration/automation or intervention level). The role defines the relation between the human and the robot and the tasks of each. There are different roles that a human may have when interacting with a robot: supervisor, operator, teammate, mechanic/programmer and bystander (Scholtz, 2003).

Autonomy Level/Amount of Intervention (sometimes noted as *levels of collaboration/automation*, LOA): the different levels at which a human operator can control an automatic process. Several taxonomies have been proposed of LOAs (Endsley, 1988; Kaber et al., 2000; Sheridan and Verplank, 1978) with the most common taxonomy defining ten levels from fully manual to fully autonomous with intermediate levels (Sheridan and Verplank, 1978; Sheridan, 1995). The autonomy level measures the percentage of time that the robot is carrying out its task on its own; the amount of intervention required measures the percentage of time that a human operator must be controlling the robot.

These two measures sum up to 100%. There is a continuum for robot control ranging from teleoperation to full autonomy; the level of HRI measured by the amount of intervention required varies along this spectrum. Constant interaction is required at the teleoperation level, where a person is remotely controlling a robot. Less interaction is required as the robot has greater autonomy.

When discussing HRI, the task to be accomplished sets the tone for the system's design and use, so the task must be identified as part of the system's classification (Yanco and Drury, 2004). A critical component of successful HRC is the *interface design* (Eliav et al., 2011) which deals with two main aspects. The first aspect deals with perception (Doisy, 2014), what information should be provided, and when and how should this information be presented? The second aspect deals with the control of the robot related to the decision regarding the robot's current location, perception regarding the location it should have been in and moving the robot to the designated destination (Fong et al., 2003). Several modalities are commonly applied to different robot control aspects and must be assessed for their efficiency and usability (Jevtic et al., 2015).

2 Interaction roles

There are different roles that a human may have when interacting with a robot: *supervisor*, *operator*, *teammate*, *mechanic/programmer* and *bystander* (Scholtz, 2003). A supervisory role is taken by a human when it needs to monitor the behaviour of a robot, but does not need to directly control it. For example, a supervisor of an unmanned tractor may tell the tractor where it should move, then the tractor plans and carries out its task. An *operator* needs to have more interaction with a robot, stepping in to teleoperate the robot or needing to change the robot's behaviour. For example, guiding a robot along the row (e.g. Murakami et al., 2008; Adamides et al., 2017b; Bechar et al., 2000) and noting it when and where to spray (e.g. Adamides et al., 2017a). A *teammate* works with a robot to accomplish a task. An example of this would be a human harvester placing picked fruits on an in-field transportation robot that conveys the picked fruit to a storage room (e.g. Baxter et al., 2018; Cheein et al., 2015) if working in sequential mode of collaboration. Sometimes teammates work together on the same task but in parallel (e.g. robots and humans both harvesting sweet peppers, Elkoby et al., 2014). A *mechanic or program*mer needs to physically change the robot's hardware or software; this is the common practice in all autonomous agricultural robotics to date. A *bystander* does not control a robot but needs to have some understanding of what the robot is doing in order to be in the same space. For example, a farmer who walks into the greenhouse while a robot is performing its tasks (e.g. harvesting, spraying, weeding) must be able to avoid the robot safely.

The time-space Computer-Supported Cooperative Work taxonomy (Yanco and Drury, 2004) defines four categories based on whether collaborators are using robotic systems at the same time (synchronous) or different times (asynchronous), while in the same place (collocated) or in different places (non-collocated). Fong et al. (2001) distinguishes between direct HRI and teleoperation. In direct HRI the robot and the human interact directly (e.g. proximal/physical interaction). If the robot(s) and the human(s) working together to accomplish some task/goal are not collocated (i.e. in time and/or space), then the interaction is called teleoperation (Sheridan, 1992). These distinctions are important for the interface design and are also referred to in the levels of automation noted below with specific agricultural tasks detailed.

3 Levels of collaboration

The ultimate goal for HRI is to develop and use efficiently robots such that human skills and abilities become more productive and effective, while freeing humans from routine or dangerous tasks (Rahimi and Karwoski, 1992). Interaction, the process of humans working collaboratively with robots to accomplish a goal, emerges from the confluence of autonomy, information exchange, teams and task shaping. For a fully autonomous robot the interactions may consist of high-level supervision and direction of the robot with the human providing goals and with the robot maintaining knowledge about the world, the task and its constraints. Sheridan and Verplank (1978) divide automation into ten levels, from fully autonomous, with no human intervention to fully manual. Parasuraman et al. (2000) developed a model for types and levels of automation that supplies an objective basis for determining the degree of automation for each system.

Bechar and Edan (2003) defined four HRC levels for target recognition tasks in unstructured environments: (1) HO - the human operator unaided, detects and marks the desired target - compatible with level 1 on Sheridan's scale; (2) HO-Rr - the human operator marks targets, aided by recommendations from an automatic detection algorithm, that is the targets are automatically marked by a robot detection algorithm, the human acknowledges the robot's correct detections, ignores false detections and marks targets missed by the robot - compatible with 3-4 on the Sheridan scale; (3) HO-R - targets are identified automatically by the robot detection algorithm; the human operators, assignment is to cancel false detections and to mark the targets missed by an automatic robot detection algorithm - compatible with 5-7 on the Sheridan scale; and (4) R - the targets are marked automatically by the system (robot) - compatible with Sheridan level 10. Analytical (Bechar et al., 2009) and simulation (Oren et al., 2012) analyses demonstrated that collaboration of human operator and robot increases detection by 4% when compared to an HO alone and by 14% when compared to a fully autonomous system. In addition, when compared to the human alone, detection times of integrated systems are reduced by 20% and false alarms are decreased.

In order to quantify performance and determine the best collaboration level for a given set of parameter values, an objective function that includes operational and time costs was developed (Bechar et al., 2009). The findings indicate that the best system performance, the optimal values of performance measures and the best collaboration level depend on the task, the environment, human and robot parameters, and the system characteristics. This work has been expanded for dynamic real-time switching between collaborative levels and for evaluating the influence of humans' reaction times on performance of integrated human-robot target recognition (Tkach et al., 2009). The switching algorithm developments enable smooth real-time adaptation of the combined human-robot system to many possible changes of the environment, HO and robot performance. Analyses revealed that increasing the switching execution frequency greatly improves the score achieved by the switching algorithm and increases system performance (Tkach et al., 2009). Analysis of the model developed to evaluate the effect of reaction time on the HRC level reveals an extreme threshold selection in two cases: when human sensitivity decreases and when the cost of time increases (Yashpe et al., 2009). The above research was conducted on field data; however, studies were limited to HRC in laboratory studies and simulations.

Similar collaboration levels were also implemented for guidance tasks (Bechar and Edan, 2000). Results indicated that incorporating the human did not always improve performance, it depended on the guidance strategy (see details in Section 5).

The only known field experiments of HRC LOA were performed on an operational robotic sprayer (Berenstein, 2016). Three levels of HRC for detection and marking of the targets were developed and implemented (see Fig. 1) based on Sheridan's 10 levels of HRC (Sheridan, 1992) and the above work in agricultural target detection (Bechar and Edan, 2003). Experimental results of the HRC levels indicate that the highest hit rate 92.66% is achieved with the collaboration level in which the robot marks the targets and the human supervises (can add, remove and correct computer marking, defined as collaboration level 3), these results are statistically significant. The lowest FA was achieved with the collaboration level in which the computer suggests and the human decides (defined as collaboration level 2), and 15 s switching time. Results revealed also that it takes time for the user to learn the system and training is necessary. In this system, learning occurred after presenting 30 images to the user.

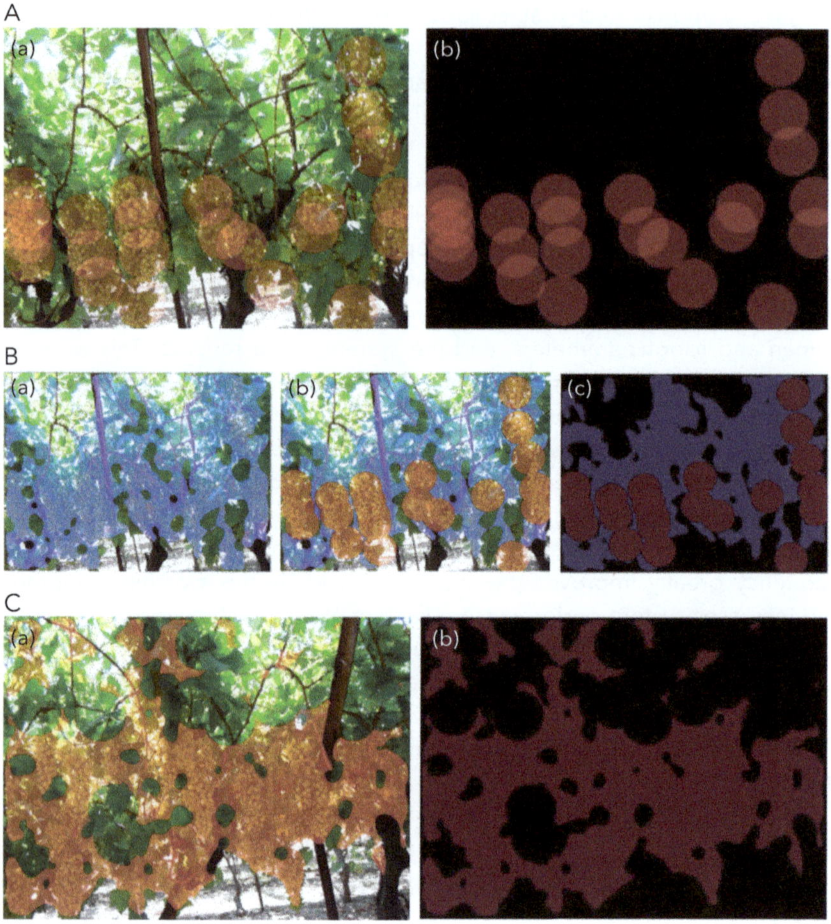

Figure 1 (A) Collaboration level 1: (a) user marked and (b) binary image for analysis; (B) collaboration level 2: (a) robot suggested mark, (b) human mark area and (c) binary image to be sprayed; (C) collaboration level 3: (a) robot marked image and (b) binary image to be sprayed. Source: Adapted from Berenstein (2016).

4 Interface design

A user interface is needed to enable the user to interact with the robot - to communicate sensor data transmitted (or available to be transmitted) from the robot(s) to the human(s) and to provide means of control from the human(s) to the robot(s) (Yanco and Drury, 2004). Research questions associated with this interaction include (Adamides, 2016):

1 What information should be provided, and when and how should this information be presented? Providing too much information and

too many interaction options will confuse the user. Providing too little information and too few interaction options won't allow the user to understand the robot behaviour or to deal with complex situations. When users have too little information and too few interaction options they will not be able to understand the robot behaviour or to deal with complex situations, such as changes in the environmental conditions. Even with the correct information and the proper interaction options the problem is not solved, since the way the information and the interaction options are presented can influence greatly how they are perceived by the user.

2 How should the farmer guide the robot's operation (e.g. moving along a pathway, identifying targets to be harvested/sprayed, performing site-specific operations e.g. spraying, fertilizing), what modalities should be used for which operations and when and how should they be used?

3 How should the interface design be evaluated (Olsen and Goodrich, 2003)? Reducing interaction effort without diminishing task effectiveness is the goal of HRI design. The first metrics are those that measure task effectiveness. Task effectiveness is some measure of how well a task is actually performed. At the end of the day we care mostly about getting some task done. In driving or navigation scenarios we might measure effectiveness as the time required to get from point A to point B. In target detection tasks we could measure the time to find all targets or the number of targets found in a given amount of time. However, task effectiveness measures do not shed any insight on how to improve the human-robot interface or how that interface might be modified to increase the effectiveness. The interface should be designed so as to reduce interaction effort and improve performance. Interaction effort is more than just the time required to manipulate input devices. In most scenarios, interaction effort is dominated by cognitive rather than physical effort, both of which must be measured.

Adamides et al. (2015) presented usability guidelines for an agricultural teleoperated robot based on a focused literature review, teleoperation interface design guidelines, user-centred methods and limited field experiments. Extended work was thereby focused on HRI for semi-autonomous spraying (Adamides, 2016). The goal was to evaluate the selection input device (mouse vs. Wii Remote vs. digital pen) for marking grape clusters. Results indicated usability preference for the mouse and the digital pen. Later, in a field experiment, the usability of different interaction modes for agricultural robot teleoperation was also investigated. Specifically, two different types of peripheral vision support mechanism, two different types of control input devices, two different types of output devices and the overall influence of the user interface on observed and perceived usability of a teleoperated agricultural sprayer were examined. Participants were more effective (i.e. had less collisions and sprayed more grape clusters), both in spraying and in robot path guidance, when they had multiple views, as compared to when they had a single view. Participants required significantly less time to complete the task when they had the single view as compared to when they had multiple views. The PC keyboard was found to be significantly superior to a PS3 gamepad controller in terms of time to complete the task, perceived usability, perceived workload index and perceived sense of presence. Results indicated usability preference for a mouse and the digital pen. Placing the camera on top of the end effector spray provided improved performance (see Fig. 2 for the proposed system). Similarly, placing a camera at the rear-top of the robot provides peripheral vision and enables the operator to locate obstacles around the robot wheels, thus improving location and surroundings awareness. A monitor was preferred over a head-mounted display.

Interface design for an HRC system operating with different LOA was designed by Berenstein (2016). The focus was on interface design (Berenstein et al., 2012) and on evaluating remote target marking methods (Berenstein and Edan, 2012b). To maximize the target detection HIT rate, the best marking method is applying a constant diameter method. The preferable marking method to minimize false alarms is marking the clusters by free hand. An example of the interface is presented in Fig. 3.

Figure 2 Proposed solution for camera placement for spraying robot. Source: Adapted from Adamides (2016).

Murakami et al. (2008) developed a teleoperated system for a crawler robotic vehicle that aimed to navigate the robot along a pre-defined path. Two-degree cooperation between the robot and the remote operator was set, including direct and supervisory control. The study shows that direct control was difficult for the operator due to the high delay time between the operator command and the robotic response. The supervisory mode showed the ability to travel straight with a maximum lateral error of 0.3 m.

A co-located operator remotely controlled transportation robots to the pickers' locations (Baxter et al. 2018). A small pilot study revealed positive user experience with the robots, no details of the design were provided probably since it was noted as very limited initial study.

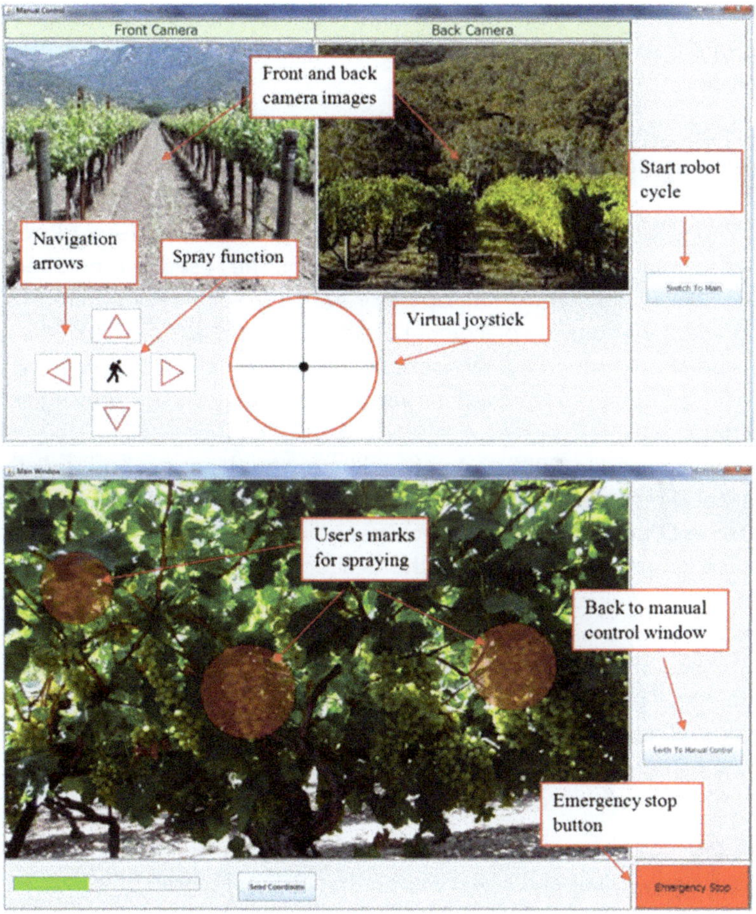

Figure 3 Example of interface for HRC sprayer control: manual control window (top) and main window (bottom). Source: Adapted from Berenstein (2016).

5 Tasks

The following HRC tasks have been implemented in agricultural settings to date.

Detection: Bechar and Edan (2003) provide proof of the advantage of HRC in target recognition tasks. According to their research, collaboration of human and robot increases detection by 4% when compared to an HO alone and by 14% when compared to a fully autonomous system. In addition, when compared to the human alone, detection times of integrated systems are reduced by 20%. The above research was conducted on simulation analyses and field data of melon locations in the field; however, studies were limited to HRC in laboratory studies and simulations. Field experiments with a robotic sprayer between a remote robot and the human showed that the hit and the false alarm rate was improved (hit rate increased by 13.4% and false alarm rate decreased by 19.5%) compared to a fully autonomous operation (Berenstein, 2016).

Navigation: Stentz et al. (2002) developed semi-autonomous operations for tractor guidance. Murakami et al. (2008) developed a teleoperated system for a crawler robotic vehicle that aimed to navigate the robot along a pre-defined path. Two degree cooperation between the robot and the remote operator was set, including direct and supervisory control. They used an omnidirectional camera and a field map for the operator to observe the teleoperated vehicle during teleoperation. The placement of a camera on the top-back of the robot enhanced the surroundings awareness. The placement of a camera on the end effector sprayer improved target identification, thus improving activity awareness.

Bechar and Edan (2000) tested two different guidance methods for off-road guidance: directional guidance and waypoint guidance. Directional guidance denoted as D, pointing out the approximate direction of advance; and waypoint guidance denoted as W, drawing the desired course of advancing along the path. For each guidance method a HO strategy and a HO-robot collaboration (HO-Rr) strategy was tested. Experiments were conducted to compare the performance of each strategy on field acquired images. Results indicated that for directional guidance, the HO-Rr strategy increases the time required to fulfil the task, and requires more time per operation as compared to the HO strategy. For waypoint guidance, HO-Rr strategy significantly reduces the required time, the number of null and net operations issued by the HO.

Harvesting: Recent work introduced side-by-side working of humans and robots in berry, olives and sweet pepper harvesting focused on safety aspects (Baxter et al., 2018), and optimal HRCs (Cheein et al., 2015; sweet peppers - Elkoby et al., 2014).

The use of a robot for in-field transportation of picked fruit was investigated for greenhouses (Baxter et al, 2018) and olive groves (Cheein et al., 2015). The robots fetched and carried crates of picked fruit and delivered them to refrigerated storage units, freeing the pickers to focus on picking. The robot's worked on a complementary task to the human in a sequential teammate mode.

Elkoby (2016) presented a simulation model for analysing optimal HRC for sweet pepper harvesting in greenhouses. The robots and humans work as parallel teammates performing the identical harvesting task. Results revealed the advent of collaboration and provided the optimal number of robots and humans for specific robotic performance values.

Spraying: Adamides (2016) provided important design aspects (Adamides et al., 2012) and usability experimentation of user interfaces for robot teleoperation (Adamides et al., 2013, 2014a,b) and tele-manipulation (Fig. 4) for the specific tasks of navigation, target identification and spraying. Berenstein (2016) developed a human-robot collaborative framework (Fig. 5) for targeted spraying of grape clusters focusing on sprayer design, collaboration levels (Berenstein and Edan, 2012a), interface design (Berenstein et al., 2012) and their combined evaluation (Berenstein and Edan, 2017). Both researches which were evaluated intensively in field conditions with an operational robotic system provided valuable insights into the need to advance HRC R&D and its significance in improving performance.

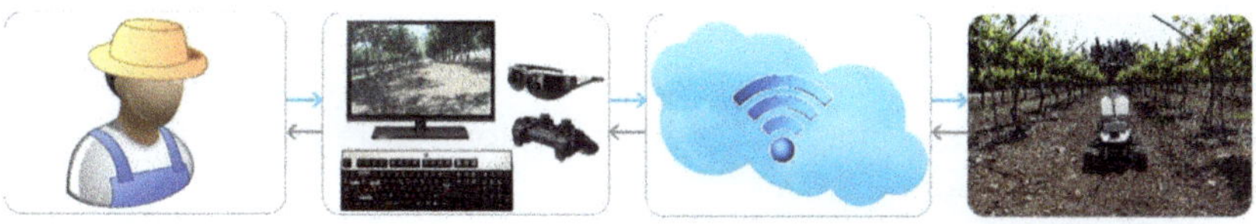

Figure 4 Robot teleoperation scheme. Source: Adapted from Adamides (2016).

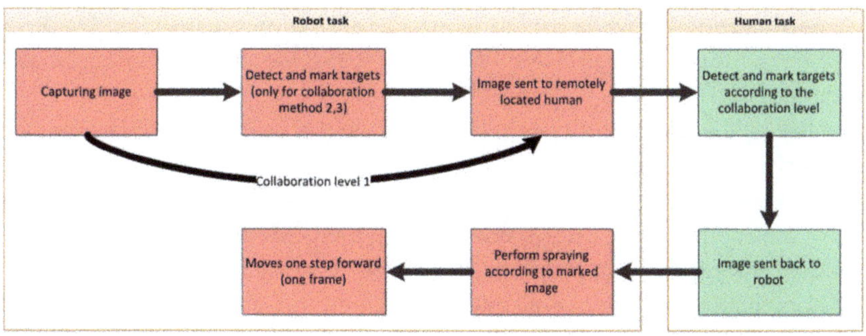

Figure 5 Human-robot collaborative working diagram. Source: Adapted from Bernstein (2016).

Miscellaneous: Additional recent applications include a collaborative system for greenhouse stress management (Guo et al., 2018) and a co-robotic device for improving irrigation by robustly adjusting indoor drip irrigation emitters (Berenstein et al., 2018).

6 Summary and insights for HRI in agriculture

HRC is becoming common practice in industry due to the many benefits it provides. Bringing in HRI to the agricultural domain is a key factor to ensure robots penetrate into market. Due to the complicated agricultural environment which is characterized by unstructured, highly variable and dynamic objects and environment most robots to date are limited in their performance. Introducing an HO to assist the robot or work alongside it can provide several advantages: increased performance and robustness, reduced complexity of the agricultural robotic system and therefore leading to an economic alternative. Furthermore, it will allow for humans to team up with robotic systems to perform tasks that humans cannot realistically accomplish alone. Even for autonomous and semi-autonomous systems, teleoperation in the agricultural domain will be an important default mode.

The HO can be linked with any of the three functional components of the robot – sensing, planning and acting. Most HRC operations to date in agriculture are in the sensing and acting phases. It's interesting to see that no HRC has been included to date in the decision-making phase despite the advent of the HO and its complexity in agriculture.

This chapter presented an overview of the limited work performed in HRC in agricultural settings. The main findings which can provide a basis for future developments are summarized below in the form of requirements, interface design and evaluation.

Example of HRI requirements for an agricultural robot is noted below.

- User interface (UI): ergonomic UI is required in order to create intuitive user experience while using the interface. The interface should be based on the most common WIMP (Windows, Icons, Menus and Pointers)-based interface.
- User is 'the boss': the system should allow the user to override algorithm decisions.
- Feedback: interface should include real-time feedback regarding the overall system state (fuel, battery power, pesticide amount left).
- User awareness: the interface should be designed in such a way for the human to have high overall awareness of the robot state at all times.
- Reliability: the system reliability should be extremely high so as to ensure continuous operation.

A taxonomy of HRI usability heuristics for agricultural robotics design (Adamides et al., 2015):

- Platform architecture and scalability: 'Provide the flexibility to iterate robotic and computing technological developments in the user interface of the HRI system'. The UI of an HRI system should be flexible to follow and benefit from developments in computing and robotic technologies.
- Error prevention and recovery: 'Provide information and alerts to avoid and recover from user errors'. The information provided by the UI should prevent user errors, and if a user makes a mistake, the UI should allow for its rectification. In contrast with undoing a 'Cut' operation in a word processor, a 'Cut' command to prune a tree through a teleoperated AgriRobot cannot be undone.
- Visual design: 'Provide an aesthetic, clear, and simple design of the user interface with the relevant information necessary'. Since the UI is the communication medium between the operator and the remote robot, it should provide the operator only relevant information (from video and other robot sensors) in a simple, consistent, effective and minimalist way. Specific examples include minimizing use of multiple windows, avoiding window occlusion, providing large video windows and displaying the robot's body in the interface.
- Information presentation: 'Provide the necessary information, in the right context, moment, and modality'. Controlling a remotely located robot is demanding on operators who need to integrate various sources of information coming from the robot cameras and sensors. Therefore, information presentation is of high importance to enhance situation awareness of the operators, and to bridge the gaps of execution and evaluation.
- Robot state awareness: 'The knowledge that the robot has about its own systems' situation and the information it gives to the operator about its health status and mode of operation'. The robot should be able to self-inspect its systems and take autonomous action or ask for user intervention. The HO should have a clear understanding about the robot's status and its activities. For instance, to support understanding of the camera(s) and their position(s), the overall mission and the current progress, and when multiple robots are available, use one to view another.
- Interaction effectiveness and efficiency: 'Provide efficient and effective interactions between human and robot'. In HRI, efficiency is measured in terms of the time required to complete a task; effectiveness is measured in terms of how well a task is completed.

- Robot environment/surroundings awareness: 'Provide spatial information about the robot's surroundings and the environment where it is operating'. Environment awareness is essential, because in field robot teleoperation it is important to have knowledge of the robot's whereabouts and the area covered, such as orientation, obstacles or why a robot is not moving. This can be accomplished through maps, orientation information (such as compass) and sensors that will provide the necessary information about the robot's surroundings.
- Cognitive factors: 'Use mental models and metaphors to lower the cognitive load'. Cognitive factors are characteristics that affect performance and learning. The UI of a teleoperated HRI system should be designed in such a way that it directs the user's attention to the task the robot is operating, improves learnability and provides fused information from the various sensors and cameras from the robot, in order to lower the cognitive load on the user.

6.1 HRI usability evaluation

The usability of different interface designs should be evaluated by measuring users' interaction effectiveness, interaction efficiency and overall satisfaction in field experiments. This should be conducted separately for each task.

For the robot navigation task, effectiveness was operationalized by the total number of collisions: fewer collisions, is more effective. Steinfeld et al. (2006) suggested to use the number of obstacles avoided as one of the effectiveness metrics in the navigation task. However, the number of actual collisions was used because in an agricultural field one might avoid obstacles along the path but still have collisions, that is with tree stems or support poles on the side. For the spraying task, effectiveness was measured by the number of grape clusters sprayed and the number of clusters missed. Efficiency was operationalized by time on task, which is the overall time required to complete the whole teleoperation task (navigation and spraying).

Subjective assessment of usability (i.e. perceived usability) was measured by the post-task ten-item System Usability Scale (SUS, Brooke, 1996). SUS is a post-study questionnaire that assesses the perceived usability of a system. It comprises ten statements to which participants rate their level of agreement on a five-point scale. Half of the statements are positively worded (e.g. 'I would imagine that most people would learn to use this system very quickly') and half are negatively worded (e.g. 'I found the system very cumbersome to use'). Based on a formula, a total SUS score is obtained from each user ranging from 0 (negative) to 100 (positive). An overall SUS score for the evaluated system can be obtained by averaging the users' SUS scores.

7 Future trends in research

The role of human workers in agriculture will not be eliminated by the introduction of robotic systems. Even if autonomous systems become feasible in more tasks, humans will still be needed for supervision and collaboration. By introducing a human into the operational loop, system design can be simplified leading to decreased system costs and improved performance and reliability. Furthermore, the advent of the human-in-the-loop is maintaining the 'green thumb' insight that expert farmers will never give up. Hence, HRC in agricultural robots is the future as noted by several recent reviews on agrobotics (Bac et al., 2014; Blackmore, 2016; Duckett et al., 2018).

Systems with partial autonomy need intuitive UIs. Usability is a key factor for success. Intuitive and simple to use systems and interfaces must be designed in such a way so that farmers can simply put the robot to perform agricultural operations and collect the relevant data when and where needed (Blackmore, 2016). The interface design should be transparent to the non-expert farmer. Engineers must co-design the interfaces with the end users using common human factor methods which must be applied as standard methods in agriculture to ensure design success.

Most human–robot collaboration to date has been in the perception and action tasks. Collaboration should be advanced also into the decision-making process. The robot can collect timely data with high spatial and temporal frequency, thereby providing valuable information to the farmer who can use it for decision-making for optimal farm performance by directing the robot to the right location at the right time. Furthermore, the human–robot collaboration can reduce the amount of low-level procedural decision-making by the farmer relieving him to attend other tasks and enhancing higher-level, strategic decision-making (Devitt, 2018).

The more the system is autonomous, the less an HO needs to know to operate it and the less the cognitive load. However, this also implies that farmers reduce over time their know-how of the farm processes (Devitt, 2018). This can affect decision-making in the long run and must be considered when advancing HRC. As noted by Devitt, '*Farmer knowledge should be continuously updated by active engagement of the farmer in the loop*'.

Safety and reliability is another important factor for agricultural robotics implementation. Safety standards should be adopted – the interface design should be safe to crops, to the operator's operation and to others. While substantial efforts have been invested in making robotics reliable, experience demonstrates that robots operating in unstructured environments are prone to frequent failures. Automatic diagnosis tools along with intuitive simple-to-use interfaces to enable untrained users to identify and solve robotic failures must be developed since the farmer will never risk not being able to enter the field.

Developing training materials for the farmers to use the technology should be advanced. This should include hands-on experimentation with the system in actual operating conditions including learning how to deal with unexpected events. Systems should be supported with online product manuals or personal assistance along with diagnosis tools. Too much information load and/or too much time consumption might cause the farmer to disregard the system overall. However, as farmers deal with larger farms and become more technology-oriented the actual employment of robotic technologies can attract additional personnel to the tedious farm work.

Future improvement of interface design could be explored and evaluated for different tasks (e.g. disease detection) and applications (e.g. UAV control). Some examples include an embedded representation of the robot's body in the UI displaying sensor information and robot direction in relation to the active camera views; embedded help explaining functionality and controls (heuristic 8), for example simplify and explain algorithmic settings for automated target identification; embedded tooltips and/or labels on the buttons related to user-defined targets; and personalization to individual farmer needs and preferences can be another development direction (Devitt, 2018).

In terms of UI technologies, with the emergence of new sensor technologies and 3D camera improvements, it would be worthwhile to develop UIs with augmented reality capabilities to investigate their effect on situational awareness of operators when using tele-robotics in complicated agricultural settings.

8 Where to look for further information

Additional references with relevant HRI research are presented in Section 10.

R&D in HRI is published in all leading robotics journals and conferences, for example *IEEE Transactions on Robotics*, *International Journal of Robotics Research*, *Autonomous Robots*, *Robotica*, *IEEE Robotics and Automation Letters-RAL*, *Journal of Field Robotics*, *Intelligent Robots and Systems* and more. See below for a list of relevant conferences.

http://www.ieee-ras.org/conferences-workshops
https://www.robotics.org/Upcoming-Events
https://scholar.google.com/citations?view_op=top_venues&hl=en&vq=eng_robotics

Specific human-robot interaction conferences and journals:

Conferences: IEEE IROS, IEEE ROMAN, ACM/IEEE HRI
Journals: *Journal of Human-Robot Interaction*, *Journal of Social Robotics*, *Paladyn Journal of Behaviour Robotics*, *Service Robotics*.

Human factors journals and conferences: *Applied Ergonomics*, *IEEE Transactions on Human-Machine Systems*, *IEEE Transactions on Systems, Man, and Cybernetics*, *Human-Computer Interaction*.

9 References

Adamides, G. 2016. User interfaces for human-robot interaction: Application on a semi-autonomous agricultural robot sprayer. PhD thesis. School of Pure and Applied Sciences, Open University of Cyprus, Cyprus.

Adamides, G., Berenstein, R., Ben-Halevi, I., Hadzilacos, T. and Edan, Y. 2012. User interface design principles for robotics in agriculture: The case of telerobotic navigation and target selection for spraying. *Proceedings of the 8th Asian Conference for Information Technology in Agriculture* 36, Taipei, Taiwan, September 2012, 8p.

Adamides, G., Katsanos, C., Christou, G., Xenos, M., Kostaras, N. and Hadzilacos, T. 2013. Human-robot interaction in agriculture: Usability evaluation of three input devices for spraying grape clusters. *Proceedings of the EFITA/WCCA-CIGR Conference Sustainable Agriculture through ICT Innovation*, June 2013, 8p.

Adamides, G., Katsanos, C., Christou, G., Xenos, M., Papadavid, G. and Hadzilacos, T. 2014a. User interface considerations for telerobotics: The case of an agricultural robot sprayer. *Proceedings of the SPIE 9229, Second International Conference on Remote Sensing and Geoinformation of the Environment (RSCy2014)*, 92291W, August 2014, 8p.

Adamides, G., Christou, G., Katsanos, C., Kostaras N., Xenos, M., Hadzilacos, T. and Edan, Y. 2014b. A reality-based interaction interface for an agricultural teleoperated robot sprayer. *Proceedings of the Second International Conference on Robotics and Associated High-Technologies and Equipment for Agriculture and Forestry (RHEA 2014) New trends in Mobile Robotics, Perception and Actuation for Agriculture and Forestry*, May 2014, pp. 367–76.

Adamides, G., Christou, G., Katsanos, C., Xenos, M. and Hadzilacos, T. 2015. Usability guidelines for the design of robot teleoperation: A taxonomy. *IEEE Transactions on Human-Machine Systems* 45(2): 256–62.

Adamides, G., Katsanos, C., Parmet, Y., Christou, G., Xenos, M., Hadzilacos, T. and Edan, Y. 2017a. HRI usability evaluation of interaction modes for a teleoperated agricultural robotic sprayer. *Applied Ergonomics* 62: 237–46.

Adamides, G., Katsanos, C., Constantinou, I., Christou, G., Xenos, M., Hadzilacos, T. and Edan, Y. 2017b. Design and development of a semi-autonomous agricultural vineyard sprayer: Human-robot interaction aspects. *Journal of Field Robotics* 34(8): 1407–26. doi:10.1002/rob.21721

Bac, C. W., van Henten, E. J., Hemming, J. and Edan, Y. 2014. Harvesting robots for high-value crops: State-of-the-art review and challenges ahead. *Journal of Field Robotics* 31(6): 888–911.

Baxter, P., Cielniak, G., Hanheide, M. and From, P. 2018. Safe human-robot interaction in agriculture. In companion of the *2018 ACM/IEEE International Conference on Human-Robot Interaction*, pp. 59–60.

Bechar, A. and Edan, Y. 2000. Human-robot collaboration for agricultural robot guidance. ASAE Paper No. 00-3135. ASAE, St. Joseph, MI.

Bechar, A. and Edan, Y. 2003. Human-robot collaboration for improved target recognition of agricultural robots. *Industrial Robot* 30(5): 432–6.

Bechar, A. and Vigneault, C. 2017. Agricultural robots for field operations. Part 2: Operations and systems. *Biosystems Engineering* 153: 110–28.

Bechar, A., Edan, Y., Meyer, J., Rotman, M. and Friedman, L. 2000. Human-robot collaboration for melons detection. ASAE Paper No. 00-3143. ASAE, St. Joseph, MI.

Bechar, A., Meyer, J. and Edan, Y. 2009. An objective function to evaluate performance of human-robot systems for target recognition tasks. *IEEE Transactions on Systems, Man and Cybernetics, Part C* 39(6): 611–30.

Berenstein, R. 2016. A human-robot cooperative vineyard sprayer. PhD thesis. Department of Industrial Engineering and Management, Ben-Gurion University of the Negev, Beer Sheva, Israel.

Berenstein, R. and Edan, Y. 2012a. Human-robot cooperative precision spraying: Collaboration levels and optimization function. Paper No. 84, SYROCO 2012, 10th IFAC Symposium on Robot Control.

Berenstein, R. and Edan, Y. 2012b. Evaluation of marking techniques for a human-robot selective vineyard sprayer. *International Conference of Agricultural Engineering (CIGR-AgEng)*, Valencia, Spain.

Berenstein, R. and Edan, Y. 2017. Human-robot collaborative site-specific sprayer. *Journal of Field Robotics* 34(8): 1519–30. doi:10.1002/rob.21730.

Berenstein, R., Ben Halevi, I. and Edan, Y. 2012. A remote interface for a human-robot cooperative vineyard sprayer. Paper No. 1167. *11th International Conference on Precision Agriculture*, 15–18 July, Indianapolis, USA.

Berenstein, R., Fox, R., McKinley, S., Carpin, S. and Goldberg, K. 2018. Robustly adjusting indoor drip irrigation emitters with the Toyota HSR Robot. IEEE ICRA 2018.

Blackmore, S. 2016. Towards robotic agriculture. *Proceedings of the SPIE 9866, Autonomous Air and Ground Sensing Systems for Agricultural Optimization and Phenotyping*, 986603, 8 June 2016.

Bontsema, J., Hemming, J., Pekkeriet, E., Saeys, W., Edan, Y., E., Shapiro, A., Hacover M., Hellström T. and Ringdahl, O. 2014. CROPS: High tech agricultural robots. *International Conference of Agricultural Engineering AgEng*.

Brooke, J. 1996. SUS-A quick and dirty usability scale. In: P. W. Jordan, B. Thomas, B. A. Weerdmeester and A. L. McClelland (Eds), *Usability Evaluation in Industry*. Taylor and Francis, London.

Cheein, F. A., Herrera, D., Gimenez, J., Carelli, R., Torres-Torriti, M., Rosell-Polo, J. R. and Arnó, J. 2015. Human-robot interaction in precision agriculture: Sharing the workspace with service units. *2015 IEEE International Conference on Industrial Technology (ICIT)*, pp. 289–95.

Cherubini, A., Passama, R., Fraisse, P. and Crosnier, A. 2015. A unified multimodal control framework for human-robot interaction. *Robotics and Autonomous Systems* 70: 106–15.

Clarkson, E. and Arkin, R. C., 2007. Applying heuristic evaluation to human-robot interaction systems. *FLAIRS Conference*, pp. 44–9.

Devitt, S. K. 2018. Cognitive factors that affect the adoption of autonomous agriculture. *Farm Policy Journal* 15(2): 49–60.

Duckett, T., Pearson, S., Blackmore, S., Grieve, B. and Smith, M. 2018. White paper – Agricultural robotics: The future of robotic agriculture. Other. EPRSC, 2018 International Robotics Showcase. Available from: http://eprints.uwe.ac.uk/36839

Edan, Y., Kondo, N. and Shufeng, H. 2009. Automation in agriculture. In: S. Nof (Ed.), *Springer Handbook of Automation*. Springer Verlag, Berlin, pp. 1092–128.

Elkoby, Z. 2016. Analysis of human-robot harvesting operations in sweet pepper greenhouses. MSc thesis. Department of Industrial Engineering and Management, Ben-Gurion University of the Negev.

Elkoby, Z., van Ooster, B. and Edan, Y. 2014. Simulation analysis of sweet pepper harvesting operations. *Advances in Production Management Systems Conference, APMS 2014 Conference*, 21–24 September, Corsica, France, pp. 441–8.

Endsley, M. R. 1988. Situation awareness global assessment technique (SAGAT). *Proceedings of the IEEE 1988 National Aerospace and Electronics Conference*.

Fong, T., Thorpe, C. and Baur, C. 2001. Collaboration, dialogue, human-robot interaction. In: R. A. Jarvis and A. Zelinsky (Eds), *Robotics Research*. Springer, Berlin-Heidelberg, Germany, pp. 255–66.

Fong, T., Nourbakhsh, I. and Dautenhahn, K. 2003. A survey of socially interactive robots. *Robotics and Autonomous Systems* 42(3–4): 143–66.

Goodrich, M. A., Michael, A. and Alan, C. 2008. Human-robot interaction: A survey. *Foundations and Trends® in Human-Computer Interaction* 1(3), pp. 203–75.

Guo, P., Dusadeeringsikul, P. and Nof, S. Y. 2018. Agricultural cyber physical system collaboration for greenhouse stress management. *Computers and Electronics in Agriculture* 150, 439–54.

Holland, S. W. and Nof, S. Y. 1999. *Handbook of Industrial Robotics: Emerging Trends and Industry Needs*. 2nd ed. John Wiley & Sons, New York, NY, 1378p.

Jaimes, A. and Sebe, N. 2007. Multimodal human–computer interaction: A survey. *Computer Vision and Image Understanding* 108(1–2): 116–34.

Jevtic, A., Doisy, G., Parmet, Y. and Edan, Y. 2015. Comparison of interaction modalities for mobile indoor robot guidance. *IEEE Transactions on Human-Machine Systems* 45(6): 653–63.

Kaber, D. B., Onal, E. and Endsley, M. R. 2000. Design of automation for telerobots and the effect on performance, operator situation awareness, and subjective workload. *Human Factors and Ergonomics in Manufacturing* 10(4): 409–30.

Murakami, N., Ito, A., Will, J. D., Steffen, M., Inoue, K., Kita, K. and Miyaura, S. 2008. Development of a teleoperation system for agricultural vehicles. *Computers and Electronics in Agriculture* 63(1): 81–8.

Olsen, D. R. and Goodrich, M. A. 2003. Metrics for evaluating human-robot interactions. In *Proceedings of PERMIS*, p. 4.

Oren, Y., Bechar, A. and Edan, Y. 2012. Performance analysis of a human-robot collaborative target recognition system. *Robotica* 30(15): 823–36.

Parasuraman, R., Sheridan, T. B. and Wickens, C. D. 2000. A model for types and levels of human interaction with automation. *IEEE Transactions on Systems, Man, and Cybernetics - Part A: Systems and Humans* 30(3): 286–97.

Rodriguez, G. and Weisbin, C. R. 2003. A new method to evaluate human-robot system performance. *Autonomous Robots* 14(2–3): 165–78.

Scholtz, J. 2003. Theory and evaluation of human robot interactions. *Proceedings of the 36th Annual Hawaii International Conference on System Sciences, 2003*. IEEE, pp. 10.

Sheridan, T. B. 1992. *Telerobotics, Automation, and Human Supervisory Control*. MIT Press, Boston, MA.

Sheridan, T. B. 1995. Teleoperation, telerobotics and telepresence: A progress report. *Control Engineering Practice* 3: 205–14.

Sheridan, T. and Verplank, W. 1978. *Human and Computer Control of Undersea Teleoperators*. Man-Machine Systems Laboratory, Department of Mechanical Engineering, MIT, Cambridge, MA.

Sim, D. Y. Y. and Loo, C. K. 2015. Extensive assessment and evaluation methodologies on assistive social robots for modelling human-robot interaction – A review. *Information Sciences* 301: 305–44.

Steinfeld, A., Fong, T., Kaber, D., Lewis, M., Scholtz, J., Schultz, A. and Goodrich, M. 2006. Common metrics for human-robot interaction. *Proceedings of the 1st ACM SIGCHI/SIGART Conference on Human-robot Interaction*, pp. 33–40.

Stentz, A., Dima, C., Wellington, C., Herman, H. and Stager, D. 2002. A system for semi-autonomous tractor operations. *Autonomous Robots* 13(1): 87–104.

Tkach, I., Edan, Y. and Bechar, A. 2009. Algorithms for dynamic switching of collaborative human-robot system in target recognition tasks. *Proceedings IFAC Symposium on Information Control Problems in Manufacturing (INCOM)*, Moscow, Russia.

van Henten, E. J., Bac, C. W., Hemming, J. and Edan, Y. 2013. Robotics in protected cultivation. *AgriControl, Proceeding of the 4th IFAC Conference on Modelling and Control in Agriculture, Horticulture and Post Harvest Industry*, 27–30 August 2013. Espoo, Finland, pp. 170–87.

Yanco, H. A. and Drury, J. 2004. Classifying human-robot interaction: An updated taxonomy. *IEEE International Conference on Systems, Man and Cybernetics* 3: 2841–6.

Yashpe, D., Bechar, A. and Edan, Y. 2009. Influence of human reaction time on performance of human-robot target recognition systems. *International Conference on Production Research*, China, July 2009.

10 Further reading

Admoni, H. J. and Shah, S. Srinivasa. 2017. Special issue on human-robot interaction. *International Journal of Robotics Research* 459–60.

Almasri, M., Tricot, N. and Lenain, R. 2018a. Toward an optimal assignment of diagnosis method to mobile robots faults. In *5th International Conference on Control, Decision and Information Technologies (CoDIT)*, pp. 835–40.

Almasri, M., Tricot, N. and Lenain, R. 2018b. Multiple model adaptive estimation for blocked wheel fault detection on mobile robots. *29th International Workshop on Principles of Diagnosis*, 27–30 August, Warsaw, Poland.

Bar, A., Edan, Y. and Alper, Y. 1996. Robotic transplanting: Adaption and simulation. ASAE Paper No. 96-3008. ASAE, St. Joseph, MI.

Bar-Gera, H., Oron-Gilad, T. and Musicant, O. 2013. In-vehicle stopping decisions advisory system for drivers approaching a traffic signal. *Transportation Research Record Series* 2365: 22–30.

Berenstein, R., Ben-Shahar, O., Shapiro, A. and Edan, Y. 2010. Grape clusters and foliage detection algorithms for autonomous selective vineyard sprayer. *Intelligent Service Robotics* 3(4): 233–43.

Berenstein, R., Hočevar, M., Godeša, T., Edan, Y. and Ben-Shahar, O. 2015. Distance-dependent multimodal image registration for agriculture tasks sensors. *Sensors (Basel)* 15: 20845–62.

Borowsky, A. and Oron-Gilad, T. 2016. The effects of automation failure and secondary task on drivers' ability to mitigate hazards in highly or semi-automated vehicles, *Advances in Transportation Studies* 1: 59–70.
Brooks, D. J. 2017. A human-centric approach to autonomous robot failures. Doctoral dissertation. University of Massachusetts Lowell.
Carlson, J. and Murphy, R. R. 2005. How UGVs physically fail in the field. *IEEE Transactions Robotics* 21(3): 423–37.
Couvent, A., Dridi, M., Tricot, N., Debain, C., Almasri, M., De Sousa, G., Chaloub, G., Izaute, M. and Coutarell, F. 2018. Impact of machine robotization on the activity of an operator in picking tasks. *1st International Conference on Human Systems Engineering and Design: Future Trends and Applications (IHSED 2018)*, 25–27 October, Reims, France.
Crestani, D. and Godary-Dejean, K. 2012. Fault tolerance in control architectures for mobile robots: Fantasy or reality? In: *CAR: Control Architectures of Robots*, May 2012, Nancy, France. 7th National Conference on Control Architectures of Robots, 2012.
Deremetz, M., Lenain, R., Thuilot, B. and Rousseau, V. 2017. Adaptive trajectory control of off-road mobile robots: A multi-model observer approach. *IEEE International Conference on Robotics and Automation (ICRA)*, pp. 4407–13.
Doisy, G. 2014. Human robot interface design for robots with learning capabilities. PhD thesis. Department of Industrial Engineering and Management, Ben-Gurion University of the Negev.
Doisy, G., Meyer, J. and Edan, Y. 2011. Toward an ecological approach to interface design for teaching robots. In: R. Groß, L. Alboul, C. Melhuish, M. Witkowski, T. J. Prescott and J. Penders (Eds), *Towards Autonomous Robotic Systems. TAROS 2011. Lecture Notes in Computer Science*, vol. 6856. Springer, Berlin, Heidelberg, pp. 420–1.
Doisy, G., Ronen, A. and Edan, Y. 2017. Comparison of three different techniques for camera and motion control of a teleoperated robot. *Applied Ergonomics* 58: 527–34. doi:0.1016/j.apergo.2016.05.001.
Driddi, M., Couvent, A., Almasri M., Chalhoub, G., De Sousa, G., Izaute, M., Tricot, N. and Coutarel, F. 2018. From acceptability assessment during design process to acceptance in real world. *1st International Conference on Human Systems Engineering and Design: Future Trends and Applications (IHSED 2018)*, 25–27 October, Reims, France (*accepted*).
Edan, Y. 1995. Design of an autonomous agricultural robot. *The International Journal of Applied Intelligence* 5(1): 41–50.
Edan, Y. and Miles, G. E. 1994. Systems engineering of agricultural robot design. *IEEE Transactions on Systems, Man and Cybernetics* 24(8): 1259–64.
Edan, Y., Miles, G. E., Flash, T., Wolf, I., Grinspun, J. and Peiper, U. M. 1996. The robotic melon harvester. *Service Robot* 2(1): 10–15.
Eizicovits, D., Edan, Y., Tabak, I. and Levy-Tzedek, S. 2018. Robotic gaming prototype for upper limb rehabilitation: Effects of age and embodiment on user preferences and movement. *Restorative Neurology & Neuroscience* 36(2): 261–74.
Eliav, A., Lavie, T., Parmet, Y., Stern, H. and Edan, Y. 2011. Mobile robot control – examining controllers and interfaces. *Applied Ergonomics* 42(6): 820–9.
Feingold-Polak R., Elishay A., Shahar Y., Stein M., Edan Y. and Levy-Tzedek S. 2018. Differences between young and old users when interacting with a humanoid robot: A qualitative usability study. *Paladyn Journal of Behavioral Robotics* 9(1): 183–92.

Giuliani, M., Mirnig, N., Stollnberger, G., Stadler, S., Buchner, R. and Tscheligi, M. 2015. Systematic analysis of video data from different human-robot interaction studies: A categorization of social signals during error situations. *Frontiers in Psychology* 6: 931.

Gompei, T. and Umemuro, H. 2015. A robot's slip of the tongue: Effect of speech error on the familiarity of a humanoid robot. *Proc. - IEEE Int. Work. Robot Hum. Interact. Commun.*, vol. 2015-November, pp. 331-6.

Guillet, A., Lenain, R., Thuilot, B. and Rousseau, V. 2017. Formation control of agricultural mobile robots: A bidirectional weighted constraints approach. *Journal of Field Robotics* 34(7): 1260–74.

Honig, S. S. and Oron-Gilad, T. 2018. Understanding and resolving failures in human-robot interaction: Literature review and model development. *Frontiers in Psychology* 9: 861.

Honig, S., Oron-Gilad, T. and Edan, Y. 2017. Multimodal communication for guiding a person following robot. *Proceedings of the 26th IEEE International Symposium on Robot and Human Interactive Communication (Ro-Man)*, 28 August–1 September 2017, pp. 1018–23.

Honig, S., Oron-Gilad T., Zaichyk, H., Fleischmann-Serna, V., Olatunji, S. and Edan, Y. 2018. Towards socially aware person-following robots. *IEEE Transactions on Cognitive and Developmental Systems* (*in press*).

Isermann, R. 2005. Model-based fault-detection and diagnosis - status and applications. *Annual Reviews in control* 29(1): 71–85.

Laprie, J. C. 1985. Dependable computing and fault-tolerance: Concepts and terminology. *Proceedings of International Symposium on Fault-Tolerant Computing*, vol. III, pp. 2–11.

Lavie, T. and Oron-Gilad, T. 2012. Perception of electronic navigation displays. *Behaviour and Information Technology* 32: 800–23.

Lee, M. K., Kiesler, S., Forlizzi, J., Srinivasa, S. and Rybski, P. 2010. Gracefully mitigating breakdowns in robotic services. *5th ACM/IEEE Int. Conf. Human-Robot Interact.*, pp. 203–10.

Lenain, R., Debain, C., Thuilot, B., Chapuis, R. and Berducat, M. 2008. Accurate and reliable navigation algorithm for off-road mobile robots. *International Conference on Intelligent Autonomous Systems (IAS).*

Norman, D. A. 1981. Categorization of action slips. *Psychology Review* 88(1): 1.

Ophir-Arbelle, R., Oron-Gilad, T., Borowsky, A. and Parmet, Y. 2012. Is more information better? How dismounted soldiers utilize video feed from unmanned vehicles - attention allocation and information extraction considerations. *Journal of Cognitive Engineering and Decision Making* 7(1): 26–48.

Oron-Gilad, T. and Oppenheim, I. 2017. Use of graphic imagery as a mean of communication between operators and unmanned systems in C3 fire tasks. In: D. Harris (Ed.), *Engineering Psychology & Cognitive Ergonomics: Performance, Emotion & Situation Awareness*. Lecture Notes in Computer Science, vol. 10275. Springer.

Oron-Gilad, T. and Parmet, Y. 2017. Close target reconnaissance: A field evaluation of dismounted soldiers utilizing video feed from an unmanned ground vehicle in patrol missions. *Journal of Cognitive Engineering and Decision Making* 11(1): 63–80.

Oron-Gilad, T., Hancock, P. A. and Helmick-Rich J. 2014. Coding warnings without interfering with dismounted soldiers' missions. *Applied Ergonomics* 45(5): 1343–52.

Parasuraman, R. and Riley, V. 1997. Humans and automation: Use, misuse, disuse, abuse. *Human Factors* 39(2): 230–53.

Porat, T., Oron-Gilad, T., Rottem-Hovev, M. and Silbiger, J. 2016. Supervising and controlling unmanned systems: A multi-phase study with subject matter experts. *Frontiers in Psychology*. doi:10.3389/fpsyg.2016.00568.

Ragni, M., Rudenko, A., Kuhnert, B. and Arras, K. O. 2016. Errare humanum est: Erroneous robots in human-robot interaction. *25th IEEE International Symposium on Robot and Human Interactive Communication (RO-MAN) 2016*, pp. 501-6.

Rahimi, M. and Karwowski, W. 1992. *Human-robot Interaction*. Taylor & Francis, Inc., Bristol, PA.

Reason, J. 1990. *Human Error*. Cambridge University Press, Cambridge, UK.

Riek, L. 2012. Wizard of Oz studies in HRI: A systematic review and new reporting guidelines. *Journal of Human-Robot Interaction* 1(1): 119-36.

Ross, R., Collier, R. and O'Hare, G. M. P. 2004. Demonstrating social error recovery with agent factory. *3rd Int. Jt. Conf. Auton. Agents Multi-agent Syst.*, pp. 1424-5.

Roy, S. and Edan, Y. 2017. Givers & Receivers perceive handover tasks differently: Implications for Human-Robot collaborative system design. *Journal of Social Robotics (accepted)*. arXiv preprint arXiv:1708.06207.

Shalev, I. and Oron-Gilad, T. 2015. What do we think we are doing: Principles of coupled self-regulation in human-robot interaction (HRI). *Frontiers in Psychology* 6: 00929.

Shapiro, A., Korkidi, E., Demri, A., Riemer, R., Edan, Y. and Ben-Shahar, O. 2009. Toward elevated agrobotics: An autonomous field robot for spraying & pollinating date palm trees. *Journal of Field Robotics* 26(6-7): 572-90.

Short, E., Hart, J., Vu, M. and Scassellati, B. 2010. No fair!! An interaction with a cheating robot. *5th ACM/IEEE International Conference on Human-Robot Interaction*, pp. 219-26.

Stanton, N. A. and Salmon, P. M. 2009. Human error taxonomies applied to driving: A generic driver error taxonomy and its implications for intelligent transport systems. *Safety Science* 47(2): 227-37.

Steinbauer, G. 2013. A survey about faults of robots used in RoboCup. In: X. Chen, P. Stone, L. E. Sucar and T. van der Zant (Eds), *RoboCup 2012: Robot Soccer World Cup XVI. RoboCup 2012. Lecture Notes in Computer Science*, vol. 7500. Springer, Berlin, Heidelberg, pp. 344-55.

Tourette, T., Deremetz, M., Naud, O., Lenain, R., Laneurit, J. and De Rudnick, V. 2018. Close coordination of mobile robots using radio beacons: A new concept aimed at smart spraying in agriculture. *IEEE International Conference on Intelligent Robots IROS and RAL*.

Vitzrabin, E. and Edan, Y. 2016. Adaptive thresholding with fusion using a RGBD sensor for red sweet pepper detection. *BioSystems Engineering* 146: 45-56.

Weinstock, A., Oron-Gilad, T. and Parmet, Y. 2012. The effect of system aesthetics on trust, cooperation, satisfaction and annoyance in an imperfect automated system. *Work: A Journal of Prevention, Assessment & Rehabilitation* 41(1): 258-65.

Xing, B. and Marwala, T. 2018. Introduction to human robot interaction. In: *Smart Maintenance for Human-Robot Interaction. Studies in Systems, Decision and Control*, vol. 129. Springer, Cambridge.

Yanco, H. A. and Drury, J. L. 2007. Rescuing interfaces: A multi-year study of human-robot interaction at the AAAI Robot Rescue Competition. *Autonomous Robots* 22: 333-52.

Yanco, H. A., Drury, J. L. and Scholtz, J. 2004. Beyond usability evaluation: Analysis of human-robot interaction at a major robotics competition. *International Journal of Human-Computer Interaction* 19: 117-49.

Chapter 5

Global positioning systems (GPS) for agriculture: an overview

John Billingsley, University of Southern Queensland, Australia

1 Introduction

The essence of a global positioning system (GPS) is the ability to define a location anywhere on Earth in terms of three coordinates. The calculation is performed in terms of Cartesian axes including one through the poles, but the result is transformed into local coordinates that include height. In fact, the system gives a fourth coordinate, the time accurate to 40 ns. Originally conceived for military purposes, GPS is finding countless civilian uses that range from finding your parked car to guiding a passenger aircraft or finding your way through a maze of roads.

The original US Navstar system provided several 'qualities' of signal. The roughest was 'coarse acquisition', originally designed to help the precision system lock on. This is the one that has been seized on for driving aids, bushwalkers and joggers, at minimal cost. A receiver is built into most cellular telephones. The precision receiver came at a much higher price, the early Beeline system for tractor guidance being marketed at a price close to $100 000.

Although GPS usually refers to the American system, the term global navigation satellite system (GNSS) embraces rival systems. An early rival was the Russian GLObal NAvigation Satellite System (GLONASS), followed by the European Galileo. China has launched its BeiDou system that already has 35 satellites, while Japan and India have their own systems. As equipment

http://dx.doi.org/10.19103/AS.2019.0056.02

manufacturers take advantage of these new constellations, the performance of low-cost receivers is improving all the time.

For farming applications, the GPS receiver is at the core of a larger system. For the guidance of planting, the signal axes must be transformed to enable straight lines to be followed at an angle that conforms to the field. For other tasks such as UAV guidance for aerial observation or for yield mapping, the requirements might not be as stringent.

2 How does the system work?

While geostationary satellites have a 24-h orbit that has a radius of some 40 000 km, the GPS satellites operate with a 12-h orbit at an altitude of about half that radius. These orbits do not lie in the equatorial plane but are inclined at 55° to it, several such orbits making a sort of woven basket pattern in terms of coverage of the Earth's surface.

For 'coarse acquisition', each satellite transmits a pseudo-random 'song' of a length of 1023 bits at a frequency of 1 MHz. The L1 carrier frequency is 1575.42 MHz. These frequencies will become more significant when we discuss accuracy and precision.

At the velocity of light, the signals will take 66 μs or more to reach the receiver. If the delay can be measured, the distance from the satellite can be calculated. Three such distances could in theory be used to define a position. Unfortunately, things are not quite so easy. Although the clocks in the satellite are very accurate indeed, an error in the inferior receiver clock will destroy the accuracy. If the signals from four satellites are compared, however, the equations for x, y, z and t can be solved, meaning that the receiver clock is corrected in the process.

All the satellites transmit on the same frequency. The separation of the signals from the received mixture is similar to the 'cocktail party' effect, where a distant conversation can be followed through the hubbub of a party. The satellite 'songs' are all different and the cross-correlation of any two songs will be less than a thousandth of the correlation of a song with itself. The receiver tests the received signal against each song, shifted over its sequence length, until it finds a substantial correlation with the 'loudest' song. To look for weaker signals it then synthesises that song and subtracts it from the received signal. Not only does the receiver have to match the timing of the song, it also has to match the phase of the carrier signal.

It then repeats the selection process until no more songs can be found. The result is a set of 'code pseudo-ranges', the delays of the individual songs themselves and of 'carrier pseudo-ranges' giving the phase-shifts of the carriers from each of the satellites.

3 Improving accuracy

Though Section 2 suggested that four satellites would be enough for a fix, that is not always the case. If the satellites instantaneously lie in a plane, the matrix solution collapses. But in general, eight or more satellites should be visible at any time, so improving the accuracy.

The code of the song is at 1 MHz, each bit being a microsecond in length. As it flies through space at the velocity of light, each bit will be of length 330 m. The receiver shifts the timing to maximise the correlation, so an error of 1% of a bit will correspond to a distance of 3.3 m. A reasonable estimate of the error of the fix might be 5 m. For the precision military system, the P-code signal is at 10 MHz, but could still be in error by half a metre. So how can the accuracy be improved to a level that is good enough for agricultural planting?

It is first necessary to understand 'ambiguity'. The song repeats after 1023 bits, so the same correlation would be found for ranges differing by intervals of some 350 km. That is a large enough ambiguity that there is little risk of choosing the wrong cycle. In contrast, the carrier wavelength is some 20 cm. The phase is matched to much better than a centimetre, but with an ambiguity so small it is difficult to choose which cycle should be matched. Once the cycle has been chosen, however, the phase is tracked accurately until the lock is lost.

The technique called 'Real Time Kinematics' (or RTK) exploits this property to gain much greater accuracy. First the P-code signal is used to estimate the location. Within an error zone of half a metre, there will be few carrier cycles to choose from. Just one combination can be selected to give sub-centimetre accuracy. This is a favourite of surveyors, but the systems are not cheap.

A simpler alternative has been to use a base station. In the early days of GPS in the 1990s, a deliberate error was added to satellite timing, increasing the error to 100 m or worse. The purpose was to forestall the use of the GPS signal by any enemy power for precision bombing, while still providing an aid to shipping. The error was called 'selective availability' and was turned off in May 2000 by order of President Bill Clinton. But before that happened, the 'Australian Maritime Safety Authority' had found ways to defeat selective availability. They mounted a receiver at a fixed location. This meant that the timing errors could be evaluated and broadcast to users for automatic application as 'Differential GPS'. Although selective availability will never again be applied, accuracy can be improved by comparing the signals of a mobile receiver with those from a fixed base station.

4 A peer-differential system

First it is necessary to consider the difference between accuracy and precision. Accuracy concerns the ability to identify the receiver's location on the planet.

If instead we have a known fixed point, such as a base station, we can consider the precision to which the vector from the base station to the mobile can be determined.

Some experimental work was conducted in the late 1990s, where two simple receivers were used. Their binary outputs included both code and carrier phases. The carrier pseudo-ranges were transmitted from the fixed receiver to be subtracted from those of the mobile receiver. The cycles were locked when the mobile receiver was at some reference point. The differences then tracked the movement of the mobile receiver. From these differences the displacement could be calculated to sub-centimetre accuracy.

It was then found that such a displacement could be tracked with very much less effort. A receiver of this sort also gave a binary velocity signal with great precision. It was clear that the velocity was calculated from the carrier pseudo-ranges, rather than from the code fix. When the three components of that velocity were integrated over a 10-km return journey, it was found that the accumulated error was less than a metre.

5 Future trends and conclusion

GNSS research and improvements will continue, advances being reflected in the subsystems that are on the market for integrating into agricultural systems. Already, receivers are available that combine signals from all the present constellations of satellites. Agricultural systems will certainly embody these off-the-shelf modules rather than developing anything specific.

With many governments pushing a move towards autonomous road vehicles, economical systems with sub-metre accuracy will become common and will be a low-cost guide for planting. Cultivating is a rather different matter, since two layers of errors are involved. First there is the error between the intended row and the one actually planted. The second error is the deviation of the vehicle doing the cultivation from the GPS target. For that reason, I am convinced that the best solution will be to combine GPS with visual row sensing. Machine vision can provide accurate row guidance when the rows are clear, with GPS taking over for precision heading following whenever necessary. GPS can then manage headland turns. All these sensory systems can be embodied in something as simple as an Android mobile phone.

6 Where to look for further information

Other chapters of this book will give details of the way that GPS technology is being exploited for agriculture. As for GPS itself, considerable information can be gleaned 'straight from the horse's mouth' at www.gps.gov, the site for official

US government information about GPS and related topics. A good introduction to the subject is Hofmann-Wellenhof, B. et al. (2001), *Global Positioning Systems: Theory and Practice*, Springer Verlag, Vienna, Austria.

A web search reveals a host of manufacturers marketing GPS systems for embedding. Dual and triple frequency receivers are already being promoted. A fresh search will be needed, since any recorded findings would rapidly become out of date.

Part 2

Applications

Chapter 6

The use of agricultural robots in crop spraying/fertilizer applications

Ron Berenstein, University of California-Berkeley, USA

1 Introduction

Application of nutrients, fungicides and pesticides is one of the most important processes in agricultural production and can have a significant impact on crop yield, quality and ultimately profitability (Singh et al., 2005). It is estimated that approximately 30–35% of crop losses can be prevented when harmful insects and diseases are eliminated by applying pesticides (Cho and Ki, 1999). Although pesticides are needed in modern agriculture, they are poisonous and dangerous for humans (Dasgupta et al., 2007; Rogan and Chen, 2005) and the environment (Pimentel and Lehman, 1993; Reus et al., 2002).

The current common approach for pesticide application is mechanized non-selective spraying in which a human drives a tractor connected with a mechanized sprayer that sprays the crops continuously. The sprayer can be mounted (Fig. 1a) or towed by the tractor (Fig. 1b). Another method of pesticide application includes a human operator selectively spraying targets using a backpack sprayer. However, this type of spraying is rarely in use due to the long operational times, the exposure of the human to the hazardous pesticide material and human fatigue. Despite the use of pesticide protection equipment (e.g. central filtration system and personal head mask for the mechanized and manual spraying methods, respectively) the human is still exposed to hazardous pesticides that can cause negative health issues (Swan et al., 2003).

http://dx.doi.org/10.19103/AS.2019.0056.10

Figure 1 Tractor sprayers. (a) The sprayer is mounted directly on the tractor (also known as three-point sprayer), (b) trailed sprayer where the sprayer is being towed by the tractor.

Agricultural robots have been developed for many operations, such as field cultivation, planting, spraying, pruning and selective harvesting (Edan et al., 2009; Nishiwaki et al., 2004; Oberti et al., 2016). Robotic spraying technology can benefit from these latest developments in agri-related robots.

Robotic technology can provide a way to reduce the quantity of pesticide applied, improve its sustainability and reduce its environmental impact (Slaughter et al., 2008). A target-specific robotic sprayer can reduce the amount of pesticides applied in modern agriculture and potentially remove or minimize the human presence during the pesticide spraying process (Lee et al., 1999). Studies show that up to 60% of pesticide use can be reduced when fruit and vegetables are sprayed specifically (Elkabetz et al., 1998; Gil et al., 2007; Goudy et al., 2001).

A review on autonomous robotic weed control systems (Slaughter et al., 2008) describes the current status of the four core technologies (guidance, detection and identification, precision in-row weed control and mapping) required for the successful development of a general-purpose robotic system for weed control and sprayer robot. Although several complete robotic weed control systems have demonstrated the potential of technology in the field, additional research and development are needed to fully realize this potential. The two core tasks for an agricultural spraying robot are sensing - for target detection - and 'robotics' - for the spray execution (Song et al., 2015).

Automation in the weed control process has received increased attention from the scientific community and is already in use in the agricultural industry (Slaughter et al., 2008). Since mechanical removal is a difficult process, the plantations use herbicides to remove unwanted plants (Lameski et al., 2018). Slaughter et al. (2008) presented a review describing the current status of the four core technologies (guidance, detection and identification, precision in-row weed control and mapping) required for the successful development of a general-purpose robotic system for weed control. Since then, several publications showed advances in robotic weeding (Vuong et al., 2017; McCool et al., 2018; Pérez-Ruíz et al., 2014). A commercially available automated weeder has become available for the farm (BlueRiver[1]).

2 Challenges in current robotic sprayers

There are four core technologies required for developing a general-purpose robotic sprayer (Slaughter et al., 2008): guidance, detection and identification, precision in-row weed control and mapping. The developments must deal with the three sources of variation in a crop environment: objects, environment and crops (Bac et al., 2014).

2.1 Guidance and mapping

Accurately travelling along crop rows is a crucial proficiency agricultural robots must have. While the most common solution to guide vehicles along pre-planned routes is to use global navigation satellite systems (GNSS), the high cost of using high-precision GNSS and the lack of availability in all agricultural environments (Vázquez-Arellano et al., 2016) has led to investigations on how to guide the robot using computer vision (English et al., 2014). In the last few decades, vision-based guidance systems and algorithms were investigated in depth (Li et al., 2009; Reid et al., 2000). Research and development included guidance using RGB images, point cloud (depth images) and the use of multispectral sensors (Table 1).

Global positioning systems (GPS) are the key tool used for positioning and maintaining maps for precision agricultural tasks and are used for yield mapping and variable chemical applications. Real-time kinematics (RTK) GPS which has improved accuracy can be used for automated vehicle guidance (Slaughter et al., 2008). Bak and Jakobsen (2004) developed a robotic platform for mapping of weed populations in the field. Their work included the development of the robotic platform, the platform controller and experiments of the platform manoeuvring in the field. Peteinatos et al. (2014) developed ground-based sensors carried by a robot in the field to detect weed and weed infestation level.

2.2 Target detection

One of the main barriers to pesticide reduction using robotic sprayers is the target detection process. Despite R&D performed along the past decade

1 http://www.bluerivertechnology.com.

Table 1 Agricultural robotic guidance

Application	Sensor	Reference
Robot guidance along crops rows - open field	RGB	English et al. (2014)
Driving along row crops - open field	RGB	Xue and Xu (2010)
Row navigation - open field	RGB	Chang and Song (2017)
Autonomous speed sprayer guidance using vision and fuzzy logic	RGB	Cho and Ki (1999)
Crop rows detection for agricultural robots navigation	RGB	Jiang et al. (2010)
3D imaging systems for agricultural applications	3D	Vázquez-Arellano et al. (2016)
Path planning and obstacle avoidance - row crops	Stereo vision	Rovira-Más et al. (2006)
Autonomous navigation - maize field	LIDAR	Hiremath et al. (2014)
Robot navigation and plant detection	Ultrasonic	Harper and McKerrow (2001)
Mapping and navigation - open field	Sonar	Toda et al. (1999)
Vegetation detection for mobile robot navigation	Multispectral	Bradley et al. (2004)
Vehicle steering control using overhead guide	N/A	Gat et al. (2016)
High-precision straight-line navigation - greenhouse	Six axis compass	Chang et al. (2016)
Robot navigation in greenhouses	Ultrasonic (low and mid-range), digital compass, radar	González et al. (2009)

Table 2 Detection and identification

Application	Sensor	Reference
Real-time apple detection - apples	RGB	Bulanon et al. (2004)
Grape clusters and foliage detection	RGB	Berenstein et al. (2010)
Precision weed control - cotton	RGB	Lamm et al. (2002)
Human-robot collaboration for target detection - vineyards	RGB	Berenstein and Edan (2012a)
Fusion of visible and thermal images for fruit detection	Thermal and RGB	Bulanon et al. (2009)
Adaptive thresholding for pepper detection	RGBD	Vitzrabin and Edan (2016)
Orange fruit detection algorithms	RGB	Hannan et al. (2010)
Fruit detection system using deep neural networks	RGB and NIR	Sa et al. (2016)
Weed detection (*Avena sterilis*)	RGB	Tellaeche et al. (2008)

(Table 2) detection rates are low with a maximum of a 90% hit rate. The reasons for these low rates are because of the complicated field conditions caused by the changes in illumination (due to changes in the sun direction, clouds), shading, the highly variable plant characteristics (size, colour, texture, shape, location) and the occlusion of targets (caused by leaves, branches). Berenstein et al. (2010) showed a 90% hit rate for grape clusters (green type) detection while using a colour CCD sensor and applying morphological manipulation on the RGB channels. Work by Diago et al. (2012) showed 95% hit rate for detecting grape clusters (red type) but since the background was an artificial white screen (to avoid confounding effects from the background vegetation) the actual hit rate performance in real-world conditions will probably decrease. A recent review of this topic is Luo et al. (2016).

2.3 Control

The use of spraying nozzles in modern industry is widespread for different applications such as cleaning (Canny, 1986), coating (Sharifi et al., 2002) and painting (Breiman et al., 1984). Manufacturers offer a wide range of nozzles with manually adjustable spraying angles and even automatic spraying systems that can control the flow rate (e.g. Spraying Systems Co, PulsaJet, AutoJet). Due to the nature of products and applications in the industrial domain, the nozzle spraying angle is preset manually according to the designated target, which is well defined. In the agricultural domain, the targets have inherent high variability in size (e.g. watermelon, lettuce) and shape (e.g. grape clusters, cherry tomatoes, eggplant, kiwi, strawberry) (Kapach et al., 2012) that requires adjusting the spraying coverage to the specific target.

Extensive research has been performed over the past two decades on spraying robots, mainly for the automotive industry (Berenstein et al., 2015), with focus on path planning of the robotic arm and achieving uniform paint thickness layers (Diao et al., 2009; Sahir Arikan and Balkan, 2000; From et al., 2011; Conner et al., 2005). Agricultural spraying robots (Table 3) have been developed mostly for weed control and plant protection applications (Pergher and Petris, 2008; Singh et al., 2005; Slaughter et al., 2008; Steward et al., 2002; Mandow et al., 1996; Zhao et al., 2016; Gazquez et al., 2016; Guan et al., 2015). One of the main goals of the agriculture engineering research, and in particular the precision agriculture community, is to reduce the use of pesticides while preventing crop losses due to pests (Pérez-Ruiz et al., 2015).

3 Case study: robotic sprayers in vineyards

In the following case study, we discuss the design and implementation of a smart spraying framework that includes a fully functional mobile robot positioned in the field equipped with a smart sprayer and a remote human supervisor assisting the robot with the target detection task. The case study deals with the smart implementation of pesticides towards grape clusters aiming to direct spray towards only the grape clusters.

Table 3 Weed and spray methods

Application	Control method	Reference
Site-specific weed management	Direct-injection sprayer	Goudy et al. (2001)
Non-chemical weed control	Electrical discharge	Blasco et al. (2002)
Weed vision-based perception	Mechanical weed control	Strand and Baerveldt (2002)
Plant protection with a variable rate application	Spraying, ultrasonic sensors	Gil et al. (2007)
Spraying robot - grape production	Chemical spraying	Ogawa et al. (2006)
Individual weed treatment using a robotic arm	Chemical spraying	Jeon et al. (2005)
Unmanned aerial vehicle for spray application	Chemical spraying	Huang et al. (2008)
Pesticide dose adjustment in vineyard spraying	Chemical spraying	Pergher and Petris (2008)
Precision spraying methods	Coloured water spraying	Berenstein and Edan (2012c)
Human-robot collaboration for vineyard spraying	Coloured water spraying	Berenstein and Edan (2017)
Teleoperated robotic sprayer	Water spraying	Adamides et al. (2017)
Automatic adjustable device for precision spraying	Coloured water spraying	Berenstein and Edan (2018)
Specification development of a robotic system for pesticide spraying in greenhouse	Chemical spraying	Komasilovs et al. (2013)
Information from spraying, harvesting and grading operation robot	Chemical spraying	Arima et al. (2003)

The potential reduction of pesticide use is inversely proportional to the per cent of grape clusters in the image. An analysis of the relation between the per cent of grape clusters in real field images to the saving potential is shown in Fig. 2. The saving potential increases with a smaller number of grape clusters in the image. Fewer grape clusters are common early in the season and can be caused by a gap between the grapevines.

The human and robot work collaboratively to detect targets in a sequential mode. The framework places the human at a remote location equipped with a target-marking device (e.g. a stationary computer, laptop, tablet, PDA or smartphone) and uses the human's excellent perception skills to mark targets on images captured by the robot in the field.

3.1 Automatic adjustable spraying device

An adjustable spraying device (ASD) was designed and built as an experimental tool in order to implement the one target-one shoot (OTOS) spraying method

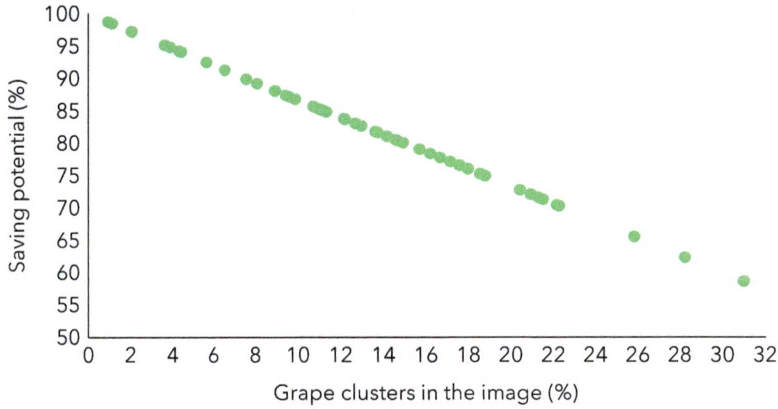

Figure 2 Evaluation results of the relation between saving potential and grape clusters in the image. Source: adapted from Berenstein et al. (2010).

(Berenstein and Edan, 2012c), applying single spray for each target by adjusting the spraying diameter according to the target. Two other alternatives were evaluated, fixed nozzle spacing and optimal spray spacing, both with fixed spray diameter (i.e. several sprays to completely cover the target). Analysis of the spraying costs (Fig. 3) reveals the advantage of selecting the OTOS spraying method.

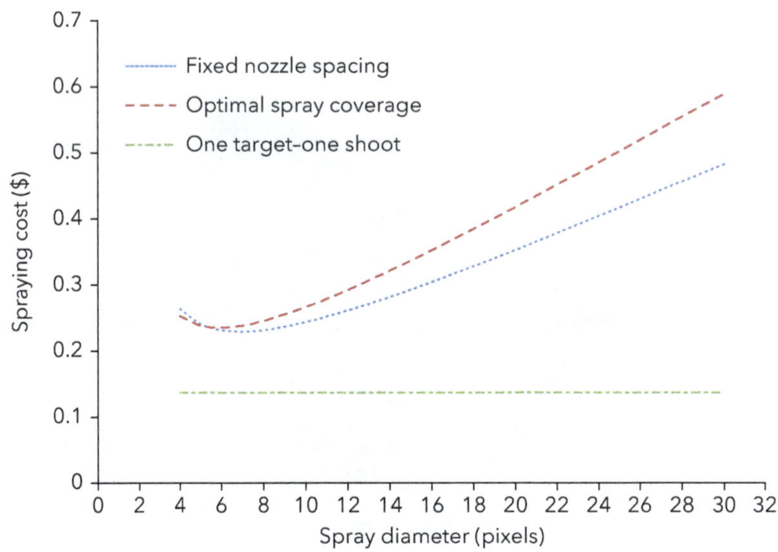

Figure 3 Economic function results.

The device is mounted on to a mobile robotic sprayer that supplies pressurized pesticide. The operational concept of the ASD is as follows:

1 Direct the nozzle to face the crop (perpendicular to the crop).
2 Capture an image using the ASD camera.
3 Find the targets positions and diameters.
4 For each target perform the following routine:
   a Direct the ASD towards the target centre,
   b Adjust the nozzle diameter to equal the closing circle diameter of the target, and
   c Open the sprayer electric valve for a specific pre-defined time.

3.1.1 ASD design and characteristics evaluation

The ASD is presented in Fig. 4. The ASD base is constructed from three aluminium parts, two pressure plates that mount the spraying nozzle and the two line beam lasers, and a shoulder. The shoulder is connected to the pressure plate with four screws and its height can be adjusted.

The ASD was based on a commercial spraying nozzle (AYHSS 16) using the recommended spraying pressure of 20 (bar). The spraying nozzle is constructed from two parts, the nozzle base and the nozzle cup. The nozzle base is mounted on to the pressure plates. The pressurized pesticide hose is connected to the nozzle base and the flow is controlled using an electric valve (on/off). The spraying diameter can be controlled by rotating the nozzle cap over the nozzle base. This nozzle was chosen since it is in common use among farmers that adjust the spraying diameter prior to the spraying task.

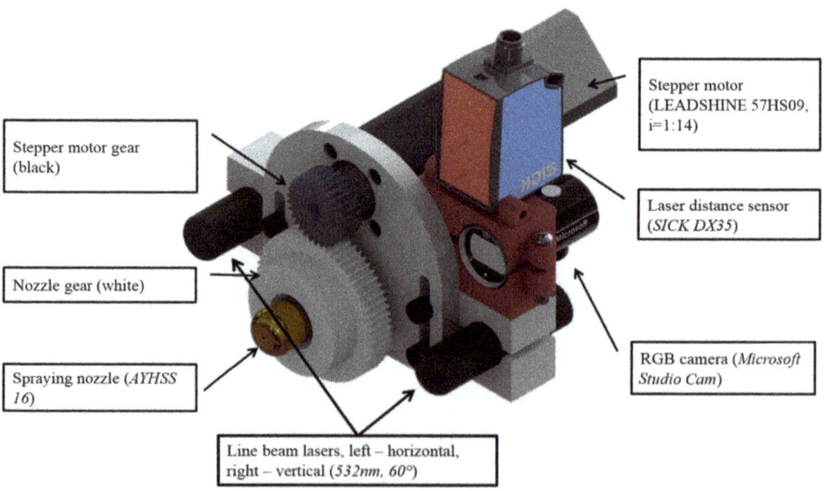

Figure 4 Spraying device. (a) Isometric view - CAD, (b) front view, (c) side view.

A stepper motor, mounted on the shoulder, is used to control the spraying diameter. The stepper motor is connected to the nozzle cap using two tangent gears, one connected to the stepper motor and the other connected to the spraying nozzle cap (Fig. 4). Rotational feedback of the stepper motor is acquired using a rotational potentiometer (ten rounds, 1 KΩ) connected to the stepper motor gear. An Arduino (uno) board closes the stepper motor position loop using feedback from the potentiometer and the desired circular position.

Other peripheral sensors are mounted on to the ASD; a laser distance sensor (SICK DX35) for measuring the distance between the device and the target, a colour camera (Microsoft LifeCam Studio) for capturing images from the field and later used for automatic target detection and two line beam marking lasers (532 nm, 50 mW, 60°) positioned horizontally and vertically for marking a cross (+) over the target. The entire device is mounted on a Pan Tilt Unit (PTU) (FLIR D46-17) able to rotate horizontally ±180° and vertically +31° up to −80°.

A PC is connected to an Arduino board, laser distance sensor, colour camera, PTU and the electric valve controlling the pesticide flow. The main software for managing the ASD was based on Microsoft Visual Studio (C#). The software collects data from the ASD sensors and controls the ASD orientation by adjusting the PTU, the ASD nozzle by rotating the stepper motor and the electric valve opening/closure, according to the collected data.

The spray diameter was evaluated in order to find the spray diameter (spray cone) for varying nozzle apertures. Using this relation between the nozzle aperture and the spray diameter, the spray diameter can be adjusted according to the target size.

The experimental setup (Fig. 5a and b) included the ASD facing the target base with a target attached. The target base was constructed from steel net and was mounted vertically on a manually controlled conveyor in front of the ASD (Fig. 5b). The target used for evaluation was a white paper sheet, 0.5 m wide, which was stretched top to bottom and fixed to the target base (Fig. 5b shows

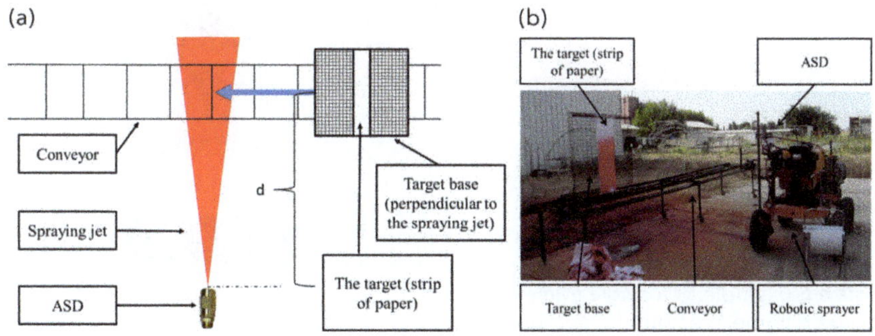

Figure 5 The configuration of an experiment for spray diameter evaluation. (a) Experiment scheme, (b) field view of the experiment.

the target fixed to the target base after spraying). In order to view the spray deposition and post-analyse the position of the spray, a red water-soluble food dye (Florma red 696) was used as pesticide replacement.

Each spray repetition included the following steps: (1) attaching a new target to the target base, (2) setting the nozzle aperture to the desired value, (3) opening the spray flow, (4) starting the conveyor movement towards the spray jet and (5) after the entire target base has crossed the spraying jet, the spray flow is closed and the conveyor stops. During the experiment, the robot was kept at a static position to ensure the ASD was static.

Image acquisition software was designed to capture a movie using the ASD camera along the spray process. After each spray repetition, the captured movie was saved for post-analysis. Each movie was manually scanned by a human expert to extract a single frame containing the target in mid-frame. The extracted frame was analysed manually for the spray boundaries (Fig. 6). Since the spray is cone-shaped, the spray diameter was evaluated by measuring the distance (in pixel units) between the upper and lower spray boundaries.

The experiments were performed at three distances between the ASD and the target (0.5, 1 and 1.5 m). For each distance, the nozzle angular position was between 175 and 210 with increments of 5 (units in potentiometer Ω). Three measurements were conducted for each distance-aperture combination. All experiments were performed at dawn ensuring no-wind conditions (this was confirmed by measuring the wind speed using Skywatch Xplorer 1).

The experimental results shown in Fig. 7 reveal the correlation between the nozzle aperture and the spray diameter for three measured distances. The

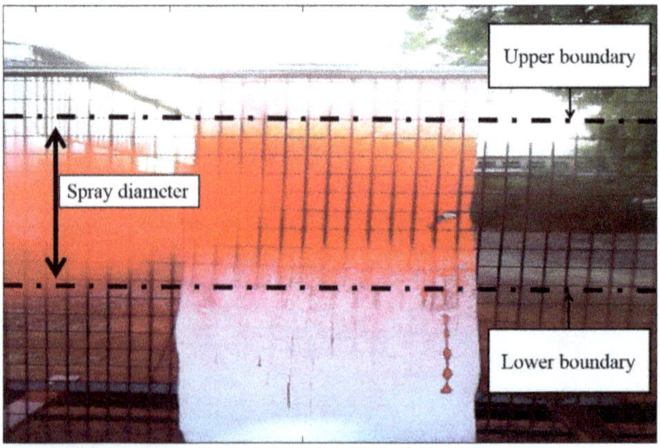

Figure 6 Example of a single frame extracted from captured spraying movie. Using the captured frame, the boundaries (upper and lower) and the spray diameter of the sprayed target were extracted.

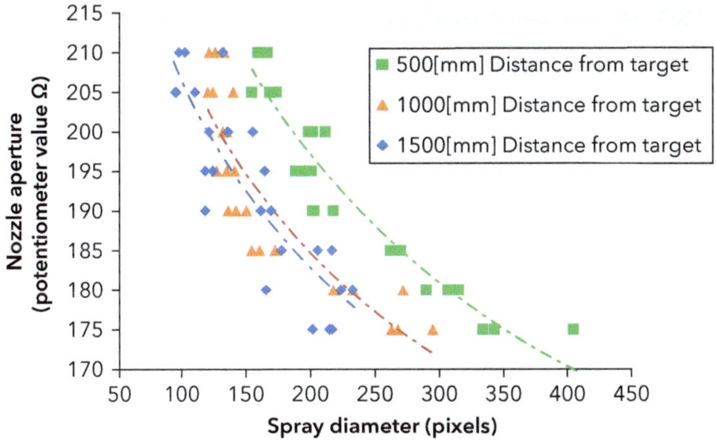

Figure 7 Experimental results of the spray diameter for three measured distances.

measured spray diameter increases as the distance increases. In theory, the three curves are supposed to unite since both the camera's field of view and the spraying cone have a linear trajectory. The spray dispersion is probably caused by the spray jet turbulence and air drag that affects the spray dispersion.

The spray diameter increases with the increase in distance between the nozzle and the target. This is because in the experiment the spray diameter is measured using the digital camera, which is located at the same distance as the spraying nozzle (Fig. 4, the spraying nozzle and the camera are located together), and is expressed using pixel units. Therefore, the measured spray diameter does not increase with the increase in distance but remains approximately constant due to the digital camera perspective.

Table 4 presents the curve fitting parameters for Fig. 7, where NA is the nozzle aperture and SD is the spray diameter.

By using the resulting curves for the different distances, the nozzle aperture can be calculated after extracting the target diameter. The spraying distance in most commercial vineyards is between 500 and 1500 mm. In order to correlate between the spraying distance and the nozzle aperture, an interpolation of the distance and the nozzle aperture can be applied.

Table 4 Experimental results summary

Distance	Trend line (power)	R^2
500	$NA = 600.22 \cdot SD^{-0.210}$	0.911
1000	$NA = 490.97 \cdot SD^{-0.184}$	0.782
1500	$NA = 467.12 \cdot SD^{-0.177}$	0.761

3.1.2 ASD performance evaluation

An experiment was conducted in order to evaluate the performance of the ASD while implementing the results of the previous experiment (Table 4). Currently, the robotic sprayer is designed to perform the spraying task in step mode (Fig. 8): the robot travels a single step along the vineyard row, stops, captures images from the field, sprays the targets and moves another step forward. The experiment is based on the same work procedure. The robot movement speed was constant. One of the secondary goals of this experiment was to provide insights regarding the overall work procedure of the complete spraying system which will include the robot equipped with an ASD.

During the experiment, the ASD was attached to the robotic sprayer and was operated similarly to the planned robotic procedure which moved in step mode along the vineyard row (Fig. 8). During this experiment, the robotic sprayer was programmed to track a straight line placed at a 1.6 m distance from the target base (red plastic strip 50 mm width) (Fig. 9a). The robot was programmed to travel 1.6 m at each step. The ASD is mounted perpendicular to the robot's travel direction and faces the target's base (Fig. 9a). The target's base is a polyethylene net (50 mesh), 11 m long, stretched between two anchoring poles and positioned parallel to the line at the floor. The targets are attached to the target's base and the centre of the target is positioned 1.55 m high. In order to ensure a single target per image, the targets were positioned at intervals of 1.6 m, similar to the robot's travel distance.

The targets are blue polyethylene round circles with varying diameters (300, 250, 230, 210, 190, 170 and 150 mm). To simplify the detection and classification of the targets, a red circle was attached to the centre of the main target. The diameter of the red circle was one-third of the blue circle diameter.

Artificial targets were used to enable accurate target detection. The targets consisted of a round, blue, polyethylene target with different diameters (150, 170, 190, 210, 230, 250 and 300 mm). A round, red, polyethylene target was mounted on to the centre of the blue target. The diameter of the red target was one-third of the blue corresponding target (Figs. 9b and 10).

The target detection algorithm was based on colour thresholding and was implemented using MATLAB software equipped with the image processing toolbox. The detailed description of the target detection algorithm is presented in Berenstein and Edan (2018).

Following the detection process, the programme extracts the coordinates of the detected target's centre and the minimum closing circle diameter in pixel units. These measures are used to control the sprayer (i.e. direct the PTU towards the target centre and adjust the spraying diameter according to the closing circle diameter).

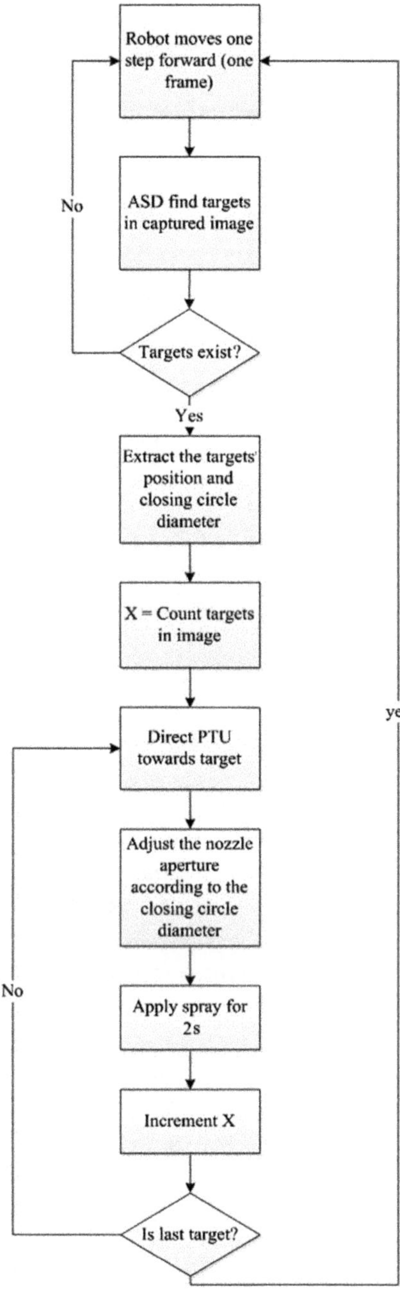

Figure 8 Robotic sprayer work procedure. The following experimental procedure was based on this figure procedure including the steps of directing the PTU towards the target core, adjusting the spraying nozzle and the actual spraying.

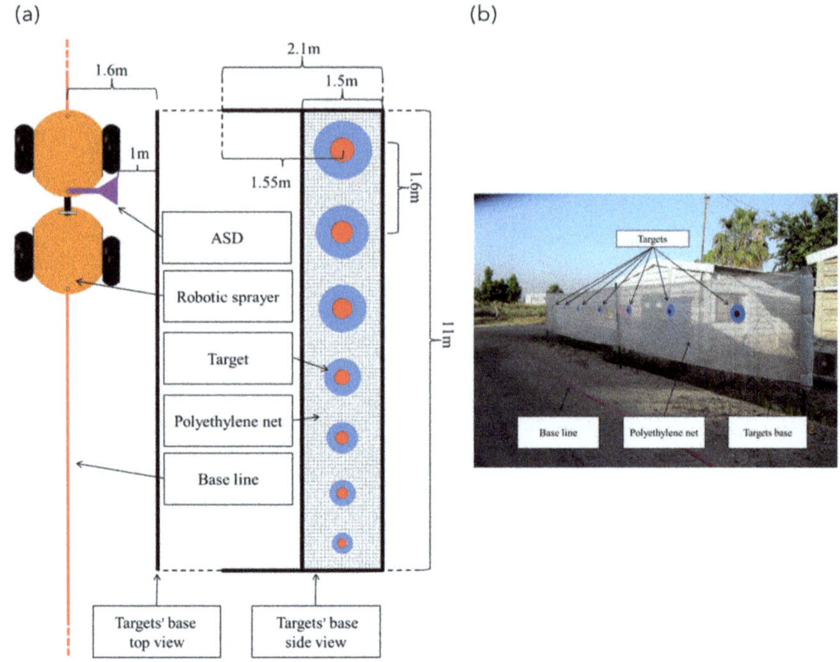

Figure 9 Experiment configuration. (a) Experiment scheme, (b) field view of the experiment.

The target detection algorithm, with all of its steps and unique values, was developed specifically for the artificial targets that were used in the experiments and it is not the core of this work. Needless to say that in order to use the suggested ASD, a specific target detection algorithm must be developed for the specific crop.

Similar to the previous experiment, a red water-soluble food dye (Floma red 696) was used as pesticide replacement to simplify the detection of the spray deposition.

The sprayed area was evaluated by both manually measuring the sprayed area's diameter immediately after each spray and image processing of images captured immediately after each spray (Fig. 10).

The experiment included 12 repetitions of the robot travelling along the line on the floor and spraying the seven targets attached to the target base. Each target was sprayed for 2 s. All experiments were conducted early morning. Measured wind speed was zero in all experiments (measured using Skywatch Xplorer 1).

The results described here use the ASD in automatic mode: the ASD automatically directs the PTU towards the target centre and adjusts the spray diameter according to the closing circle diameter of the detected artificial target.

Figure 10 Image captured immediately after spraying.

A visual inspection revealed that each target was fully covered by the spray. Experimental results are summarized in Fig. 11. The spray flows under gravitation force (Fig. 10) increasing the spray spot size which complicates the spray diameter analysis and was eliminated from the spray diameter evaluation.

The results of the automatically adjustable spray diameter show a constant increase of the sprayed diameter with the increase in target size; however, the ratio between the sprayed diameter and the target size decreases. This ratio can be addressed as the false detection ratio, and according to Fig. 11, this

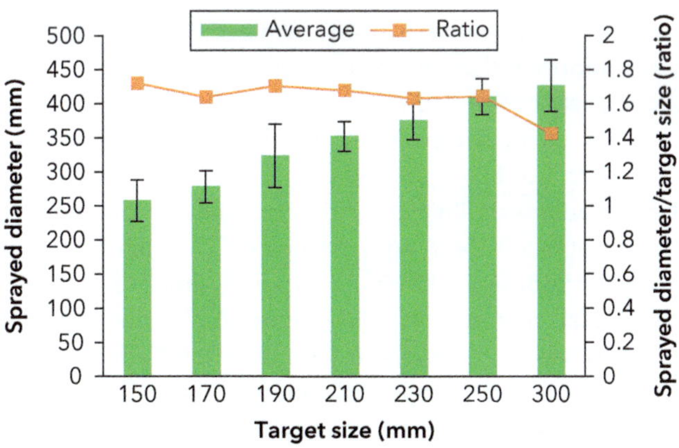

Figure 11 Experimental results. Each column represents the average sprayed diameters of 12 sprays (robot repetitions). The resulting standard deviation are shown in each column. The orange line (secondary axis on the right) measures the ratio between the sprayed diameter and the target size.

ratio decreases with the increase of the target size. The 150 mm diameter target was sprayed with coverage diameter of ~250 mm, whereas a 300 mm diameter target was sprayed with coverage of ~425 mm.

3.2 Agri-robotic platforms: description and experiments

A robotic platform was designed and built to serve as a research tool for investigating methods and devices designated for the agricultural domain in general and specifically for vineyard operations (Berenstein, 2016). The robot was designed to include all the necessary equipment, hardware and software required to accomplish autonomous and semi-autonomous (human-assisted) field tasks such as navigation along the vineyard row and spraying accurately towards the target area using the ASD (Section 3.1) (Berenstein and Edan, 2017).

3.2.1 Platform description

The robotic chassis (Fig. 12a) is assembled from two identical platforms that are interconnected using a two degrees of freedom (DOF) universal joint (cardan joint). The first DOF is used to improve the turning radius and the second DOF allows the platform designers to neglect the need for a complicated suspension system. Although the robot is capable of turning using differential steering, allowing a relative angle between the platforms contributes to a smaller turning radius and minimizes side slip of the wheels, resulting in reduced wear of the vehicle and less ground trace. The platform payload is designed for 300 kg. A modular approach is taken with four identical wheel modules. Each wheel module consists of an ATV wheel (0.5 m diameter), wheel shoulder that connects the wheel to the platform and a 24 V-480 W electric motor. The electric motor is fixed to the platform and connected to the wheel using chain wheels. Using incremental encoders connected to each wheel, the wheel position and speed can be controlled using a developed kinematic model (Berenstein, 2016; Zaidner and Shapiro, 2016).

The robot is equipped with an electrical box that is mounted on to the front platform and contains a PC with an i7 processor, 7″ touch screen, two electric motor controllers (Roboteq AX3500) and some small peripheral aids (e.g. Arduino boards, step motor controller). Other equipment is mounted on the platform including two colour cameras (Microsoft LifeCam Studio) (one facing forward for navigation and the second facing sideways for target detection), two car batteries 12 V 110 A/h, power generator 2500 W (Geko 2801) and commercial sprayer (200 L tank with internal combustion engine connected to a liquid pressure pump). A gamepad controller (Microsoft Xbox 360) is wirelessly connected to the robot platform and is used for manual manoeuvring of the robotic platform.

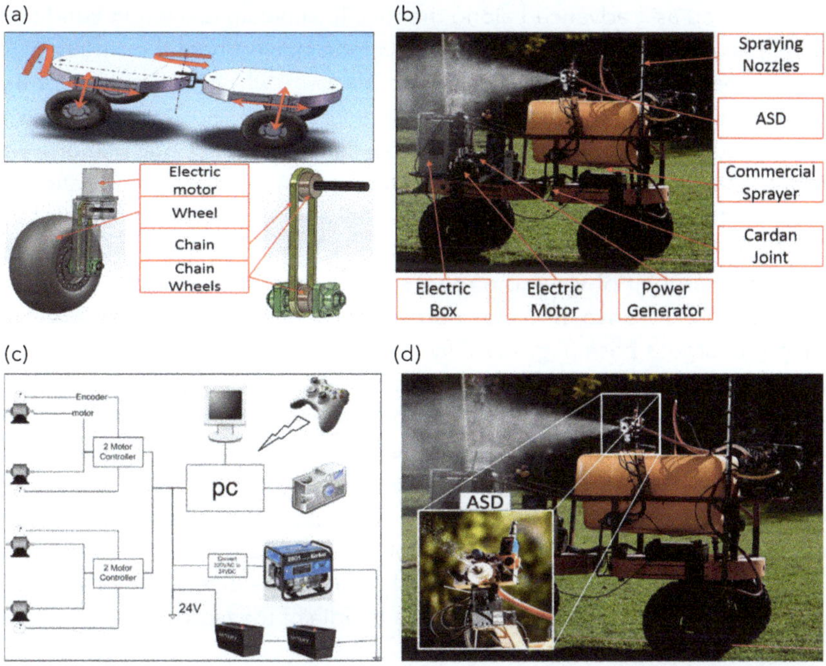

Figure 12 Robotic sprayer. (a) CAD drawing of two identical platforms interconnected using a two-DOF universal joint and wheel unit, (b) complete robotic sprayer with all of the main components and peripheral accessories attached, (c) robotic sprayer electric power scheme, (d) a focused ASD image within the overall robotic sprayer (lower right).

The robot uses a fan-less spraying design in order to minimize the spraying drift and to achieve highly accurate spraying. The focus of the current experiments was to evaluate the detection and control aspects of the sprayer. It must be noted that it is important to follow up with agronomy experiments to validate the sprayer efficiency.

3.2.2 An integrative site-specific sprayer experiment

An experiment was designed to evaluate the human-robot collaboration framework for the site-specific target spraying task, focusing on the integrative performance of the three main components of the collaboration framework that were previously tested and evaluated separately: human marking methods (Berenstein and Edan, 2012a), levels of human-robot collaboration (Berenstein and Edan, 2012b) and the spraying device (Berenstein and Edan, 2018). For simplification and better control and evaluation, artificial targets were set in an artificial outdoor environment. The human-robot task was to spray the targets as accurately as possible, within a limited time frame that corresponded to the sprayer speed as it advanced along the row. To simulate real-world conditions the human operator was located at Ben-Gurion University of the Negev, Be'er Sheva, Israel, 90 km south of the robotic platform, which was located at Beit Dagan, Israel.

To focus on the target detection/spraying tasks, the robotic platform was programmed to autonomously follow a red line on the floor (red plastic strip 50 mm width) that was fixed at a 1.6 m distance in parallel to the target's base (Fig. 13a). During each step, the robot travels 1 m to completely change the current frame point of view. The robot's travel speed was set to 0.2 m/s. The ASD was mounted on to the robot, perpendicular to the robot's travel direction, facing the target's base (Fig. 13). Fifty targets were randomly spread along an 18 m long path and were set at least 20 cm apart, imitating grape clusters.

Artificial targets were constructed from blue polyethylene plastic and were hand cut according to four shape patterns (Berenstein and Edan, 2017). In order to be as close as possible to commercial field conditions, the experiment included pre-defined TP and FP rates. Two types of targets were used: 38 targets that can be detected by the robotic sprayer (targets with a red circle in the middle of the target) and 12 targets that cannot be detected by the robotic sprayer (targets with a yellow circle in the middle of the target).

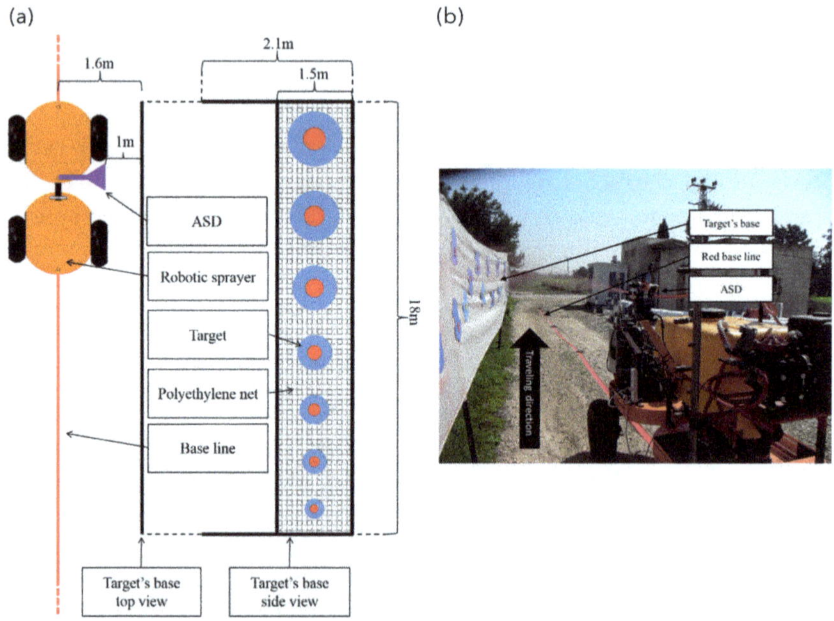

Figure 13 Robotic platform following red strip. (a) Experimental scheme including the robotic platform and target base and (b) a photo of the robotic platform during the experiment.

An artificial target detection algorithm was developed specifically for detecting the artificial targets and the colour of the circle in the middle (red for robot detectable and yellow for targets not detected). The target detection algorithm was based on simple colour thresholding and was implemented using MATLAB software equipped with the image processing toolbox. The algorithm was based on isolating the blue target in the image (the background of the image is the target base, which is white in colour) and identifying the colour of the circle in mid-target.

Since artificial targets were used in the experiment, the artificial target detection algorithm can reach a 100% TP rate and close to zero FP. In order to match vineyard field conditions, an FP area was added to the detected target surroundings. The pre-defined FPs were added using the MATLAB image processing tool. The mathematical morphology operation dilation was used to expand the computer detected target. Since each of the captured images is unique in the sense of different numbers of targets, target orientation and position, the added FP area is different for each image. The FP rate was set between 10% and 20%, corresponding to field FP results evaluated in Berenstein et al. (2010) in field conditions with an average of 17.3% (with a standard deviation of 5.5).

Using the ASD, the targets were sprayed with red water-soluble food dye (Florma red 696). Each target was sprayed for 1 s and immediately after the spraying operation stopped, an image of the spray was captured and saved.

The communication between the robot and the remote human was based on the TCP-IP protocol. The robot obtained internet access using a smartphone HotSpot (4G LTE with random switching to 3G). The remote human–computer was connected to a high-speed academic network (Ben-Gurion University of the Negev internet) with a maximum rate of 1 Gbit/s.

The human task was to mark the target area using one of the marking methods and one of the suggested human–robot collaborations described above. The human used a desk computer equipped with a 21″ screen. Each user was trained before the experiment with 20 images according to the training rate evaluated in Berenstein and Edan (2012a).

The spray quality was evaluated using four methods:

1 Marking comparison: comparison between the targets that exist in the image and the marked areas. The performance measures were TP and FP rates.
2 Spraying comparison: comparison between the targets that exist in the image and the theoretically sprayed areas. The performance measures were TP and FP rates.
3 Qualitative evaluation: analysis of the sprayed target (Fig. 14). Each sprayed target image was presented to an expert and was graded on a 1–5 scale (Fig. 14). The performance measure was the TP rate.

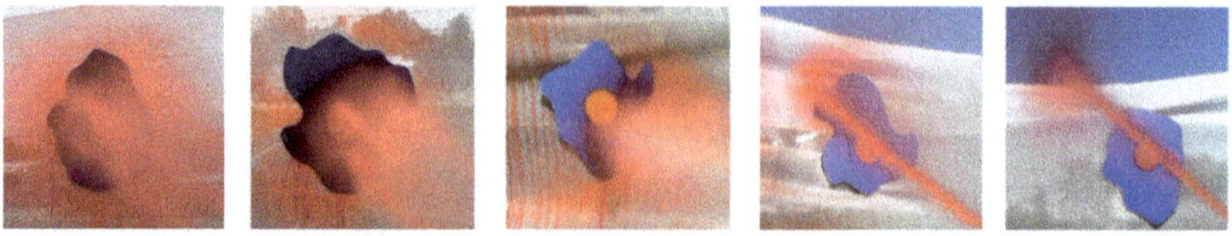

5 - outstanding 4 - very satisfactory 3 - satisfactory 2 - unsatisfactory 1 - poor

Figure 14 Target spraying evaluation scale 5 (outstanding)→1 (poor).

4 Spraying material estimation: an estimation of the quantity of spraying material used was conducted based on the ASD development results (Berenstein and Edan, 2018). The estimation compares the quantity of liquid that was used in each of the spraying experiments above compared to a simulated experiment of the robot continuously driving along the targets' base with three open nozzles (similar to the traditional spraying method; Fig. 1b).

Twenty male and female students aged 25–40 participated in the experiments and were divided into two groups, one for each marking method (CDC and free hand). Each participant practised the three collaboration levels. For each collaboration level, the robot travelled a single time along the target base with steps of 1 m. The image switching time was set to 12 s.

For the fourth collaboration level (autonomous) the robot performed ten repetitions in each of which the robot travelled along the target's base with 1 m intervals, captured the target's frame, analysed the captured frame using the artificial target detection algorithm and sprayed towards each of the detected targets.

4 Conclusion

The experimental results of the target marking and collaboration levels indicate that the desired hit and FA values that can be selected by the farmer, and the best marking method and image switching time can be chosen. The highest hit rate was measured while using the first collaboration method (fully manual with no robot assistance). However, this was achieved only for the long switching time. This collaboration level also yielded high FA. When using the faster switching time, the best collaboration level is level 3. The lowest FA was measured while using collaboration level 2 with the free hand marking method for both image switching times.

The ASD and spraying method show the ability to perform the spraying task efficiently and economically. Pesticide application is reduced by spraying individual targets by directing the spraying device towards the centre of the target using a PTU and setting the diameter of the spraying according to the shape and size of the target (according to the closing circle diameter of the target). The suggested ASD can be incorporated for different agricultural crops and for a variety of commercial applications.

The overall spraying duration for a single target was 11 s. This duration included general software commands, communication between main software and peripherals (MATLAB, Arduino), machine vision, PTU repositioning, spraying nozzle aperture adjustment, spraying and capture of image post-spraying. It also included some software pauses located at critical points of the software. These pauses were used to control the experiment and to verify that the ASD is functioning as designed. The accumulated time of the pauses was 8 s and spray time was 2 s. By eliminating the software pauses, the spraying time for a single target can be reduced to 3 s including the 2 s spraying time. Further time reduction can be achieved by optimizing the machine vision algorithms and the overall ASD control software.

An exercise was conducted to evaluate the possible pesticide savings when using the ASD in comparison with a traditional spraying technique, with the robot travelling at constant speed with open nozzles. The estimation was based on the number of targets (7.89) per frame presented in Berenstein and Edan (2012c) and on a robot travelling speed of 0.33 m/s (speed needed to spray targets for 1 s while using spraying nozzles with spray diameter of 0.33 m). The estimation assumes spraying one side of a single row in a commercial vineyard with a row length of 100 m. The expected pesticide use while using the ASD was 26.27 L and when using traditional spraying techniques (three nozzles constantly open) the expected pesticide use was 48.45 L. The estimation shows that pesticide use can be reduced by up to 45% when using the ASD.

An experiment to evaluate the elements of the collaborative human–robot framework working in sync was designed, implemented and evaluated. The experiment proves the feasibility of human–robot collaboration for the complex task of targeted spraying considering both TP and FP rates. The collaborative spraying system reduces the quantity of sprayed material by 50%, which has both economic and environmental impacts.

Results obtained can be used to implement a human–robot operational system by deciding on the best target-marking method and collaboration level according to the selected criterion. For example, if the TP rate is prioritized to ensure maximum application, full manual collaboration should be employed with a CDC marking method. To achieve the lowest FP rate (to minimize waste of material), collaboration level 2 should be employed using the free hand marking method. This is important since in the spraying task the selected criterion can change along the season depending on the conditions of pests, and environmental and growing conditions. For example, when there is a high risk of dangerous pests it is more important to ensure high coverage of targets (maximize TP rate) than wasted material (FPs). When risks are low the farmer prefers to save spray material as much as possible (minimize FPs).

With minor adaptations, the human-robot collaborative framework can be easily used with other agricultural applications such as fruit picking, yield and disease monitoring and field exploration. For full robot operation, crop-specific target detection (e.g. Berenstein et al., 2010) and navigation algorithms (e.g. Rovira-Más et al., 2015) must be integrated. The framework can also be used in other commercial applications that require complex target detection such as border control and hazardous material environment.

5 Future trends

The modern agricultural industry is facing several challenges such as worldwide population growth and ageing (Gerland et al., 2014), climate changes affecting crops and migration (both rural-to-urban and political). Using field robots in the agricultural industry may help humanity to cope with these challenges.

As described in Section 2, several agricultural robots were developed and tested over the past three decades; however, there are several challenges that must be solved before deploying these robots in a commercial environment. BlueRiver is the first to present a smart commercial agricultural-selective sprayer. The BlueRiver device is carried by a tractor in the field driven by a human. By using deep-learning techniques and deploying GPUs at the edge, they were able to develop a reliable weed detection system. Developing an autonomous spraying robot brings new challenges such as accuracy of the navigation and spraying, the robustness of the system to weather, light, thermal and other external factors in the field. Other limitations are the safety and legal aspects. An autonomous robot operating in the field must be safe to humans and animals that might enter along the robot's operations. As the robot may be considered as an agent acting on the behalf of others, legal responsibility for the actions of a robot falls on the individual who grants the robot permission to act on their behalf (Asaro, 2007).

Future research can also concentrate on improving the human-remote perception by adding advanced sensors (e.g. stereo vision, 3D cameras, a combination of RGB and LIDAR). It must be noted that if sensors are added, efficient design methods must be employed to display the sensed information so as to maximize information display while minimizing distraction. Hence this is an area recommended for future research. However, evaluating the robotic sprayer performance in real-world conditions is necessary.

An advanced remote human interface can also be developed in future work. We suggest that such an interface be implemented on a web platform to allow human control from different devices, such as smartphones, handheld computers, tablets and laptops. Another subject for future work related to the interface design can be the evaluation of different pointing devices such as touch screen, 3D mouse and a digital pen, and their effect on the human marking performance similar to what has been recently investigated by Adamides (2016) for teleoperation tasks.

In order to maximize the farmer's benefits, the overall spraying process must be economically optimized. Both the robotic sprayer and the human operator have operational costs (e.g. pesticide material, human salary, robot power consumption); the sprayed pesticides also influence the economics of the system (e.g. a sprayer missing the fruit can draw pests and damage the crop, and over-spraying can damage the foliage and exceed pesticide regulations).

6 Where to look for further information

Agricultural engineering journals and conferences are the best source for past and up-to-date information. One of the leading sources is the *Journal of Field Robotics* that presents the state-of-the-art research on agricultural robotics in general.

Other sources of information can be found in the leading agricultural and engineering journals and conferences:

- International Conference on Robotics and Automation (ICRA)
- *Biosystems Engineering*
- *IEEE Transactions on Automation Science and Engineering (T-ASE)*
- *IEEE Transactions on Robotics*
- AgEng Conference

7 References

Adamides, G. 2016. User interfaces for human-robot interaction: Application on a semi-autonomous agricultural robot sprayer. Ph.D. thesis. Open University of Cyprus.

Adamides, G., Katsanos, C., Constantinou, I., Christou, G., Xenos, M., Hadzilacos, T. and Edan, Y. 2017. Design and development of a semi-autonomous agricultural vineyard sprayer: Human-robot interaction aspects. *Journal of Field Robotics* 34(8), 1407–26. doi:10.1002/rob.21721.

Arima, S., Shibusawa, S., Kondo, N. and Yamashita, J. 2003. Traceability based on multi-operation robot: Information from spraying, harvesting and grading operation robot. *Proceedings 2003 IEEE/ASME International Conference on Advanced Intelligent Mechatronics* (AIM 2003), 20–24 July 2003, pp. 1204–9, vol. 2.

Asaro, P. M. 2007. Robots and responsibility from a legal perspective. *Proceedings of the IEEE International Conference on Robotics and Automation*, Rome, pp. 20–4.

Bac, C. W., Henten, E. J., Hemming, J. and Edan, Y. 2014. Harvesting robots for high-value crops: State-of-the-art review and challenges ahead. *Journal of Field Robotics* 31(6), 888–911. doi:10.1002/rob.21525.

Bak, T. and Jakobsen, H. 2004. Agricultural robotic platform with four wheel steering for weed detection. *Biosystems Engineering* 87(2), 125–36. doi:10.1016/j.biosystemseng.2003.10.009.

Berenstein, R. 2016. A human-robot cooperative vineyard selective sprayer. Ph.D. thesis, Ben-Gurion University of the Negev.

Berenstein, R. and Edan, Y. 2012a. Evaluation of marking techniques for a human-robot selective vineyard sprayer. *International Conference of Agricultural Engineering (CIGR-AgEng)*, Valencia, Spain (Unpublished), C-1090.

Berenstein, R. and Edan, Y. 2012b. Human-robot cooperative precision spraying: Collaboration levels and optimization function. *Symposiums on Robot Control (SYROCO)*, Dubrovnik, Croatia (Unpublished).

Berenstein, R. and Edan, Y. 2012c. Robotic precision spraying methods. *ASABE Annual International Meeting*, Dallas, Texas.

Berenstein, R. and Edan, Y. 2017. Human-robot collaborative site-specific sprayer. *Journal of Field Robotics* 34(8), 1519–30. doi:10.1002/rob.21730.

Berenstein, R. and Edan, Y. 2018. Automatic adjustable spraying device for site-specific agricultural application. *IEEE Transactions on Automation Science and Engineering* 15(2), 641–50. doi:10.1109/TASE.2017.2656143.

Berenstein, R., Shahar, O. B., Shapiro, A. and Edan, Y. 2010. Grape clusters and foliage detection algorithms for autonomous selective vineyard sprayer. *Intelligent Service Robotics* 3(4), 233–43. doi:10.1007/s11370-010-0078-z.

Berenstein, R., Hočevar, M., Godeša, T., Edan, Y. and Ben-Shahar, O. 2015. Distance-dependent multimodal image registration for agriculture tasks. *Sensors* 15(8), 20845–62. doi:10.3390/s150820845.

Blasco, J., Aleixos, N., Roger, J. M., Rabatel, G. and Molto, E. 2002. AE–Automation and emerging technologies: Robotic weed control using machine vision. *Biosystems Engineering* 83(2), 149–57. doi:10.1006/bioe.2002.0109.

Bradley, D., Thayer, S., Stentz, A. and Rander, P. 2004. Vegetation detection for mobile robot navigation. Tech. Rep. CMU-RI-TR-04-12. Pittsburgh, PA: Robotics Institute, Carnegie Mellon University.

Breiman, L., Friedman, J., Stone, C. J. and Olshen, R. A. 1984. *Classification and Regression Trees*. CRC Press.

Bulanon, D. M., Kataoka, T., Okamoto, H. and Hata, S. 2004. Development of a real-time machine vision system for the apple harvesting robot. *SICE 2004 Annual Conference*, 4–6 August 2004, pp. 595–8, vol. 1.

Bulanon, D. M., Burks, T. F. and Alchanatis, V. 2009. Image fusion of visible and thermal images for fruit detection. *Biosystems Engineering* 103(1), 12–22. doi:10.1016/j.biosystemseng.2009.02.009.

Canny, J. 1986. A computational approach to edge detection. *IEEE Transactions on Pattern Analysis and Machine Intelligence* 8(6), 679–98. doi:10.1109/TPAMI.1986.4767851.

Chang, C.-L. and Song, G.-B. 2017. Vision-based guidance control scheme for autonomous navigation of 4WDS agricultural vehicle in a narrow space environment. *2017 ASABE Annual International Meeting*. St. Joseph, MI: ASABE.

Chang, C.-L., Song, G.-B. and Lin, K.-M. 2016. Two-stage guidance control scheme for high-precision straight-line navigation of a four-wheeled planting robot in a greenhouse. *Transactions of the American Society of Agricultural and Biological Engineers* 59(5), 1193–204. doi:10.13031/trans.59.11801.

Cho, S. I. and Ki, N. H. 1999. Autonomous speed sprayer guidance using machine vision and fuzzy logic. *Transactions of the American Society of Agricultural Engineers* 42, 1137–43.

Conner, D. C., Greenfield, A., Atkar, P. N., Rizzi, A. A. and Choset, H. 2005. Paint deposition modeling for trajectory planning on automotive surfaces. *IEEE Transactions on Automation Science and Engineering* 2(4), 381-92. doi:10.1109/TASE.2005.851631.

Dasgupta, S., Meisner, C., Wheeler, D., Xuyen, K. and Thi Lam, N. 2007. Pesticide poisoning of farm workers-implications of blood test results from Vietnam. *International Journal of Hygiene and Environmental Health* 210(2), 121-32. doi:10.1016/j.ijheh.2006.08.006.

Diago, M., Correa, C., Millan, B., Valero, C., Barreiro, P. and Tardaguila, J. 2012. Grapevine yield and leaf area estimation using supervised classification methodology on RGB images taken under field conditions. *Sensors* 12, 1698-706.

Diao, X. D., Zeng, S. X. and Tam, V. W. Y. 2009. Development of an optimal trajectory model for spray painting on a free surface. *Computers and Industrial Engineering* 57(1), 209-16. doi:10.1016/j.cie.2008.11.010.

Edan, Y., Kondo, N. and Shufeng, H. 2009. Automation in agriculture. In: Nof, S. Y. (Ed.), *Handbook of Automation*. Berlin: Springer Verlag.

Elkabetz, P., Edan, Y., Grinstein, A. and Pasternak, H. 1998. Simulation model for evaluation of site−specific sprayer design. *ASAE Annual International Meeting*, Orlando, FL.

English, A., Ross, P., Ball, D. and Corke, P. 2014. Vision based guidance for robot navigation in agriculture. *2014 IEEE International Conference on Robotics and Automation (ICRA)*, 31 May-7 June 2014, pp. 1693-8.

From, P. J., Gunnar, J. and Gravdahl, J. T. 2011. Optimal paint gun orientation in spray paint applications-experimental results. *IEEE Transactions on Automation Science and Engineering* 8(2), 438-42. doi:10.1109/TASE.2010.2089450.

Gat, G., Gan-Mor, S. and Degani, A. 2016. Stable and robust vehicle steering control using an overhead guide in greenhouse tasks. *Computers and Electronics in Agriculture* 121, 234-44. doi:10.1016/j.compag.2015.12.019.

Gazquez, J. A., Castellano, N. N. and Manzano-Agugliaro, F. 2016. Intelligent low cost telecontrol system for agricultural vehicles in harmful environments. *Journal of Cleaner Production* 113, 204-15. doi:10.1016/j.jclepro.2015.11.015.

Gerland, P., Raftery, A. E., Ševčíková, H., Li, N., Gu, D., Spoorenberg, T., Alkema, L., Fosdick, B. K., Chunn, J., Lalic, N., et al. 2014. World population stabilization unlikely this century. *Science* 346(6206), 234-7. doi:10.1126/science.1257469.

Gil, E., Escol, A., Rosell, J. R., Planas, S. and Val, L. 2007. Variable rate application of plant protection products in vineyard using ultrasonic sensors. *Crop Protection* 26(8), 1287-97. doi:10.1016/j.cropro.2006.11.003.

González, R. S., Rodríguez, F., Sánchez-Hermosilla, J. and Donaire, J. G. 2009. Navigation techniques for mobile robots in greenhouses. *Applied Engineering in Agriculture* 25(2), 153-65. doi:10.13031/2013.26324.

Goudy, H. J., Bennett, K. A., Brown, R. B. and Tardif, F. J. 2001. Evaluation of site-specific weed management using a direct-injection sprayer. *Weed Science* 49(3), 359-66. doi:10.1614/0043-1745(2001)049[0359:EOSSWM]2.0.CO;2.

Guan, Y., Chen, D., He, K., Liu, Y. and Li, L. 2015. Review on research and application of variable rate spray in agriculture. *IEEE 10th Conference on Industrial Electronics and Applications (ICIEA)*. IEEE, pp. 1575-80.

Hannan, M., Burks, T. and Bulanon, D. M. 2010. A machine vision algorithm combining adaptive segmentation and shape analysis for orange fruit detection. *Agricultural Engineering International: CIGR Journal*.

Harper, N. and McKerrow, P. 2001. Recognising plants with ultrasonic sensing for mobile robot navigation. *Robotics and Autonomous Systems* 34(2–3), 71–82. doi:10.1016/S0921-8890(00)00112-3.

Hiremath, S. A., Van Der Heijden, G. W. A. M., Van Evert, F. K., Stein, A. and Ter Braak, C. J. F. 2014. Laser range finder model for autonomous navigation of a robot in a maize field using a particle filter. *Computers and Electronics in Agriculture* 100, 41–50. doi:10.1016/j.compag.2013.10.005.

Huang, Y., Hoffmann, C., Fritz, B. and Lan, Y. 2008. Development of an unmanned aerial vehicle-based spray system for highly accurate site-specific application. *ASAE Annual International Meeting*, Rhode Island.

Jeon, H. Y., Tian, L. F. and Grift, T. 2005. Development of an individual weed treatment system using a robotic arm. *ASABE Annual International Meeting*, Tampa, FL.

Jiang, G., Zhao, C. and Si, Y. 2010. A machine vision based crop rows detection for agricultural robots. *2010 International Conference on Wavelet Analysis and Pattern Recognition*, 11–14 July 2010, pp. 114–18.

Kapach, K., Barnea, E., Mairon, R., Edan, Y. and Ben-Shahar, O. B. 2012. Computer vision for fruit harvesting robots-state of the art and challenges ahead. *International Journal of Computational Vision and Robotics* 3(1/2), 4–34. doi:10.1504/IJCVR.2012.046419.

Komasilovs, V., Stalidzans, E., Osadcuks, V. and Mednis, M. 2013. Specification development of robotic system for pesticide spraying in greenhouse. *2013 IEEE 14th International Symposium on Computational Intelligence and Informatics (CINTI)*, 19–21 November 2013, pp. 453–7.

Lameski, P., Zdravevski, E. and Kulakov, A. 2018. Review of automated weed control approaches: An environmental impact perspective. In: Kalajdziski, S. and Ackovska, N. (Eds), *ICT Innovations 2018. Engineering and Life Sciences. ICT 2018. Communications in Computer and Information Science*, vol. 940. Cham: Springer, pp. 132–47.

Lamm, R., Slaughter, D. and Giles, D. 2002. Precision weed control system for cotton. *Transactions of the American Society of Agricultural Engineers* 45, 231–8.

Lee, W. S., Slaughter, D. C. and Giles, D. K. 1999. Robotic weed control system for tomatoes. *Precision Agriculture* 1(1), 95–113. doi:10.1023/A:1009977903204.

Li, M., Imou, K., Wakabayashi, K. and Yokoyama, S. 2009. Review of research on agricultural vehicle autonomous guidance. *International Journal of Agricultural and Biological Engineering* 2, 1–16.

Luo, L., Tang, Y., Zou, Z., Wang, C., Zhang, P. and Feng, W. 2016. Robust grape cluster detection in a vineyard by combining the AdaBoost framework and multiple color components. *Sensors* 16, 2098–312.

Mandow, A., Gomez-De-Gabriel, J. M., Martinez, J. L., Munoz, V. F., Ollero, A. and Garcia-Cerezo, A. 1996. The autonomous mobile robot Aurora for greenhouse operation. *IEEE Robotics and Automation Magazine* 3(4), 18–28. doi:10.1109/100.556479.

McCool, C. S., Beattie, J., Firn, J., Lehnert, C., Kulk, J., Bawden, O., Russell, R. and Perez, T. 2018. Efficacy of mechanical weeding tools: A study into alternative weed management strategies enabled by robotics. *IEEE Robotics and Automation Letters* 3, 1184–90. doi:10.1109/LRA.2018.2794619.

Nishiwaki, K., Amaha, K. and Otani, R. 2004. Development of nozzle positioning system for precision sprayer. *Automation Technology for Off-Road Equipment, Kyoto, Japan*. Kyoto, Japan.

Oberti, R., Marchi, M., Tirelli, P., Calcante, A., Iriti, M., Tona, E., Hočevar, M., Baur, J., Pfaff, J., Schütz, C., et al. 2016. Selective spraying of grapevines for disease control using a modular agricultural robot. *Biosystems Engineering* 146, 203–15. doi:10.1016/j.biosystemseng.2015.12.004.

Ogawa, Y., Kondo, N., Monta, M. and Shibusawa, S. 2006. *Spraying Robot for Grape Production: Springer Tracts in Advanced Robotics*. Berlin: Springer.

Pérez-Ruíz, M., Slaughter, D. C., Fathallah, F. A., Gliever, C. J. and Miller, B. J. 2014. Co-robotic intra-row weed control system. *Biosystems Engineering* 126, 45–55. doi:10.1016/j.biosystemseng.2014.07.009.

Pérez-Ruiz, M., Gonzalez-De-Santos, P., Ribeiro, A., Fernández-Quintanilla, C., Peruzzi, A., Vieri, M., Tomic, S. and Agüera, J. 2015. Highlights and preliminary results for autonomous crop protection. *Computers and Electronics in Agriculture* 110, 150–61. doi:10.1016/j.compag.2014.11.010.

Pergher, G. and Petris, R. 2008. Pesticide dose adjustment in vineyard spraying and potential for dose reduction. *Agricultural Engineering International: The CIGR EJournal* X.

Peteinatos, G. G., Weis, M., Andújar, D., Rueda Ayala, V. and Gerhards, R. 2014. Potential use of ground-based sensor technologies for weed detection. *Pest Management Science* 70(2), 190–9. doi:10.1002/ps.3677.

Pimentel, D. and Lehman, H. 1993. *The Pesticide Question: Environment, Economics, and Ethics*. London: Chapman & Hall.

Reid, J. F., Zhang, Q., Noguchi, N. and Dickson, M. 2000. Agricultural automatic guidance research in North America. *Computers and Electronics in Agriculture* 25(1–2), 155–67. doi:10.1016/S0168-1699(99)00061-7.

Reus, J., Leendertse, P., Bockstaller, C., Fomsgaard, I., Gutsche, V., Lewis, K., Nilsson, C., Pussemier, L., Trevisan, M., Van Der Werf, H., et al. 2002. Comparison and evaluation of eight pesticide environmental risk indicators developed in Europe and recommendations for future use. *Agriculture, Ecosystems and Environment* 90(2), 177–87. doi:10.1016/S0167-8809(01)00197-9.

Rogan, W. J. and Chen, A. 2005. Health risks and benefits of bis (4-chlorophenyl)-1, 1,1-trichloroethane (DDT). *The Lancet* 366(9487), 763–73. doi:10.1016/S0140-6736(05)67182-6.

Rovira-Más, F., Reid, F. J. and Zhang, Q. 2006. Stereovision data processing with 3D density maps for agricultural vehicles. *Transactions of the American Society of Agricultural and Biological Engineers* 49, 1213–22.

Rovira-Más, F., Millot, C. and Sáiz-Rubio, V. 2015. Navigation strategies for a vineyard robot. *ASABE Annual International Meeting*, Orleans, LA.

Sa, I., Ge, Z., Dayoub, F., Upcroft, B., Perez, T. and McCool, C. 2016. DeepFruits: A fruit detection system using deep neural networks. *Sensors* 16(8), 1222. doi:10.3390/s16081222.

Sahir Arikan, M. A. and Balkan, T. 2000. Process modeling, simulation, and paint thickness measurement for robotic spray painting. *Journal of Robotic Systems* 17(9), 479–94. doi:10.1002/1097-4563(200009)17:9<479::AID-ROB3>3.0.CO;2-L.

Sharifi, M., Fathy, M. and Mahmoudi, M. T. 2002. A classified and comparative study of edge detection algorithms. *International Conference Proceedings on Information Technology: Coding and Computing*. IEEE, pp. 117–20.

Singh, S., Burks, T. F. and Lee, W. S. 2005. Autonomous robotic vehicle development for greenhouse spraying. *Transactions of the American Society of Agricultural Engineers* 48, 2355–61.

Slaughter, D. C., Giles, D. K. and Downey, D. 2008. Autonomous robotic weed control systems: A review. *Computers and Electronics in Agriculture* 61(1), 63–78. doi:10.1016/j.compag.2007.05.008.

Song, Y., Sun, H., Li, M. and Zhang, Q. 2015. Technology application of smart spray in agriculture: A review. *Intelligent Automation and Soft Computing* 21(3), 319–33. doi: 10.1080/10798587.2015.1015781.

Steward, B. L., Tian, L. F. and Tang, L. 2002. Distance-based control system for machine vision-based selective spraying. *Transactions of the American Society of Agricultural Engineers* 45, 1255.

Strand, B. and Baerveldt, A. J. 2002. An agricultural mobile robot with vision-based perception for mechanical weed control. *Autonomous Robots* 13(1), 21–35. doi:10.1023/A:1015674004201.

Swan, S. H., Kruse, R. L., Liu, F., Barr, D. B., Drobnis, E. Z., Redmon, J. B., Wang, C., Brazil, C. and Overstreet, J. W. 2003. Semen quality in relation to biomarkers of pesticide exposure. *Environmental Health Perspectives* 111(12), 1478–84. doi:10.1289/ehp.6417.

Tellaeche, A., Burgosartizzu, X. P., Pajares, G., Ribeiro, A. and Fernández-Quintanilla, C. 2008. A new vision-based approach to differential spraying in precision agriculture. *Computers and Electronics in Agriculture* 60(2), 144–55. doi:10.1016/j.compag.2007.07.008.

Toda, M., Kitani, O., Okamoto, T. and Torii, T. 1999. Navigation method for a mobile robot via sonar-based crop row mapping and fuzzy logic control. *Journal of Agricultural Engineering Research* 72(4), 299–309. doi:10.1006/jaer.1998.0371.

Vázquez-Arellano, M., Griepentrog, H. W., Reiser, D. and Paraforos, D. S. 2016. 3-D imaging systems for agricultural applications–A review. *Sensors* 16(5), 618. doi:10.3390/s16050618.

Vitzrabin, E. and Edan, Y. 2016. Adaptive thresholding with fusion using a RGBD sensor for red sweet-pepper detection. *Biosystems Engineering* 146, 45–56. doi:10.1016/j.biosystemseng.2015.12.002.

Vuong, V. L., Slaughter, D. C., Nguyen, T. T., Fennimore, S. A. and Giles, D. K. 2017. An automated system for crop signaling and robotic weed control in processing tomatoes. *2017 ASABE Annual International Meeting*. St. Joseph, MI: ASABE.

Xue, J. and Xu, L. 2010. Autonomous agricultural robot and its row guidance. *2010 International Conference on Measuring Technology and Mechatronics Automation*, 13–14 March 2010, pp. 725–9.

Zaidner, G. and Shapiro, A. 2016. A novel data fusion algorithm for low-cost localisation and navigation of autonomous vineyard sprayer robots. *Biosystems Engineering* 146, 133–48. doi:10.1016/j.biosystemseng.2016.05.002.

Zhao, D. J., Zhao, Y., Wang, X. L. and Zhang, B. 2016. Theoretical design and first test in laboratory of a composite visual servo-based target spray robotic system. *Journal of Robotics* 2016, 1–11. doi:10.1155/2016/1801434.

Chapter 7

The use of intelligent/autonomous systems in crop irrigation

Stefano Carpin, University of California-Merced, USA; Ken Goldberg, University of California-Berkeley, USA; Stavros Vougioukas, University of California-Davis, USA; Ron Berenstein, University of California-Berkeley, USA; and Josh Viers, University of California-Merced, USA

1 Introduction

Agriculture is the major consumer of managed freshwater worldwide. Some studies have estimated that more than 80% of managed freshwater in the United States is used for agriculture, and worldwide trend is around 70% (Schaible and Aillery, 2017). Climate change, combined with the need to feed an increasing population with decreasing arable land, requires to radically rethink the way water is delivered to crops to increase efficiency and minimize wasted water (Wallace, 2000). Precision irrigation systems using drip emitters have been used for decades but only offer a partial solution. In particular, the problem of regulating the amount of water delivered to an individual plant is still outstanding, especially for large-scale operations. This problem is evident in wine grapes and particularly acute in California.

In recent years, California has endured one of its most severe droughts (Wang et al., 2017). While these cyclic phenomena are known to periodically repeat, because of climate change they are in the future expected to increase in frequency, intensity, and length (Rahmstorf and Coumou, 2011). California produces more than 80% of US wine (WineInstitute.org, 2018), and the wine sector is growing in quantity and quality. Vines are best grown with a strategy

http://dx.doi.org/10.19103/AS.2019.0056.11

known as *stress irrigation* or *deficit irrigation,* whereby the amount of water delivered to vines is limited to improve grape quality and curb vegetative growth. Given the intrinsic soil heterogeneity found in large-scale farming, commonly used strategies using uniform irrigation over large blocks are suboptimal and lead to non-uniform growth. In particular, when in doubt, ranch managers tend to over-irrigate, with the doubly negative effect of consuming more water and possibly reducing the quality of the final product. One of the leading challenges in this domain is that if one looks at an irrigation infrastructure as a control system (McCarthy et al., 2013), we find an abundance of sensor data, but a scarcity of fine grain actuation mechanisms (see Fig. 1).

Funded by the US Department of Agriculture as part of the National Robotics Initiative led by the National Science Foundation, robot-assisted precision irrigation delivery (RAPID) is a multiyear, multicampus research project aiming at overcoming these limitations by developing a novel irrigation system relying on a combination of remote sensing and mobile robots to carefully tune the amount of water delivered to each vine. The basic idea is to retrofit existing irrigation systems with passive, low-cost drip emitters that can be individually regulated to adjust the amount of water delivered to each vine. Due to the sheer scale of the ranches commonly found in intensive grape farming, manually adjusting them is an approach that does not scale. To this end, we are designing a fleet of robots that will periodically traverse the tree rows and make the necessary adjustments to increase or decrease water flow

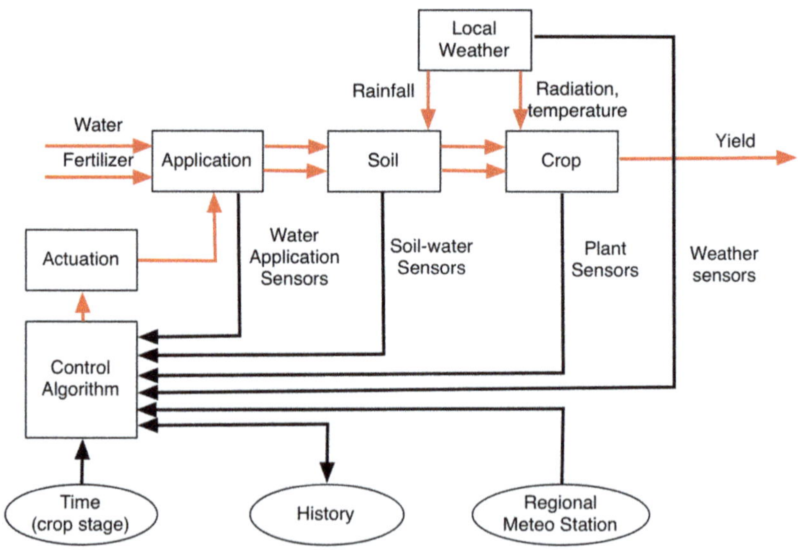

Figure 1 Irrigation system as a control system. Source: adapted from McCarthy et al. (2013).

to match desired conditions. Decisions about the necessary adjustment will be made based on an inference system combining, through learning, weather data, remote sensing, and soil moisture measures. Note that electromechanical systems or IoT-based solutions are currently not considered viable on a large scale by end users, because of their cost and the necessity to withstand extremely harsh environmental conditions for extensive periods of time.

The rest of this chapter is organized as follows. In Section 2, we discuss selected works in the area of estimation and robotics. Section 3 provides an overview of the RAPID system and then in Section 4 we provide preliminary results for estimation and routing. We conclude the chapter with Section 5, where conclusions and opportunities for future work are discussed.

2 Related work

2.1 Soil and field modeling

The RAPID project incorporates modeling water levels available at a per-plant basis, and evaluating sensing modalities for updating those models. The problem of modeling how subsurface water propagates in space and over time has been extensively investigated and a variety of models and software packages are available, both commercially and as open source. These methods aim at solving a problem known as *soil water balance calculation* and compute the water balance as the difference between inputs (precipitation and irrigation) and outputs (evapotranspiration, runoff, and drainage). The problem of spatially varying moisture measurement and simulation has been extensively studied (Clement et al., 1994; Romano, 2014; Tian et al., 2011) using models based on finite differences, nonlinear differential equations, and partial differential equations. Temporal variability has been considered as well, for example in Choi and Jacobs (2007), Ojha et al. (2014), and Vereecken et al. (2014). Methods specifically aiming at modeling subsurface moisture with drip irrigation have been developed (Kandelous and Šimůnek, 2010) and experimentally shown to be quite accurate.

Building upon these models, several simulation packages are available for modeling surface, subsurface, and groundwater flow. Models for water flow through soil typically involve solving three-dimensional partial differential equations, for example the commonly used Richards equation for variably saturated soil water flow (Lafolie et al., 1989; Or and Coehlo, 1996; van Dam and Feddes, 2000; Tian et al., 2011). Software packages such as HYDRUS 2D/3D (Hydrus, 2018) or LandLab (Adams et al., 2014) have been used for modeling flow and designing drip irrigation systems (Deb et al., 2013; Rahimi et al., 2004). However, such models have not been implemented to design irrigation systems at the per-plant scale, and have not been used for model-based closed-loop regulation. Furthermore, such models rely on many different parameters, including physical parameters related to soil or topographical properties, as well as being able to estimate initial and boundary conditions for moisture levels.

2.2 Environmental sensing

Different sensing modalities for estimating plant water availability across vineyards have received a lot of attention, and there is a rich history regarding soil moisture measurement techniques (Dane and Topp, 2002; Robinson et al., 2008; Smith and Mullins, 2000). Recently, airborne thermal imagery has been used to assess the spatial variability of water stress, an indicator of soil moisture availability across vineyards (Baluja et al., 2012; Bellvert et al., 2014), as well as soil moisture probes based on soil electroconductivity. Wireless sensor networks (WSNs) have been proposed for environmental monitoring (Pon et al., 2005; Rahimi et al., 2004), and applications in agriculture (Rehman et al., 2014). A WSN composed of 135 soil moisture and 27 temperature sensors was deployed in an apple tree orchard of about 5000 m^2 (Majone et al., 2013). The network is in charge of estimating soil moisture, but does not include an actuation subsystem capable of adjusting the application of water. A similar system was proposed in Navarro-Hellín et al. (2015) where it was demonstrated that current sensing technology is mature to determine soil moisture levels. In Gutierrez Jaguey et al. (2015) an irrigation sensor was developed using a smartphone camera, and a water pump was then activated based on the results provided by the image-processing algorithm. This system assumes that a human will manually take a picture of its crops and therefore does not scale when large extensions are considered. Moreover, the pump activates a whole irrigation line and does not allow a per-plant adjustment.

2.3 Irrigation scheduling

While irrigation schedules based on moisture measurements and water balance calculation are extensively used, in plant sciences there is increasing interest to switch toward plant-based methods, that is to determine the amount of water based on data related to plant physiology and water stress (Jones, 2004; Romero et al., 2010). However, to date, the question of which quantities should be measured and how these measures could be turned into a decision about how much water should be delivered to the plant is an active area of study. This is a research problem pertaining to the domain of plant science and orthogonal to the one tackled in this project. With RAPID we develop technology that enables individual adjustment of water emitters so that every plant can receive a different amount of water. In our control algorithm, the amount to be delivered is based on soil water measurements and soil water balance calculations; our control algorithm is, however, amenable to systems where emitter states are controlled based on plant water stress levels.

2.4 Mobile robotics

From the robotics and control theory communities, there is a large body of literature concerning the use of robots and mobile sensor networks for data collection over stationary or time varying distributions (Bullo et al., 2009; Martinez et al., 2007; Schwager et al., 2009; Smith et al., 2012). Among the many approaches proposed in literature, the Voronoi-based coverage control algorithms introduced by Bullo et al. have gained notable traction (Bullo et al., 2009; Martinez et al., 2007). These approaches have in general only been studied in theory and have seen limited applications in the field (Pierson et al., 2015).

A related problem of interest is how to combine together multiple maps or models providing partial information about a domain of interest. This work is in the general area of map merging (Aragüés et al., 2011; Aragüés et al., 2012; Erinc and Carpin, 2014; Saeedi et al., 2016) and different solutions are available depending on the structure of the underlying model, that is occupancy grids, topological maps, appearance-based maps, and more. Moisture maps essentially represent a scalar field, and the problem of combining them together can then be seen as a special instance of image stitching. Methods aiming at combining together multiple occupancy grid maps can be used, for example given in Birk and Carpin (2006), Carpin (2008), and Saeedi et al. (2012). Recently, these techniques have indeed been used to stitch together multiple images for agricultural applications (Li and Isler, 2016) where images were obtained from unmanned aerial vehicles (UAVs) flying at low elevations.

2.5 Robot task allocation

Robot and multi-robot task allocation (MRTA) is a mature, active area of research drawing insights from multiple domains, such as economy, operations research, and more (Korsah et al., 2013; Khamis et al., 2015). When a cooperative system is considered with multiple tasks to be assigned and multiple heterogeneous agents available to execute them, the question 'Which robot should execute which task?' (Gerkey, 2003) naturally emerges, and MRTA algorithms aim at answering such question optimizing some objective function accounting for the tasks being solved and the agents. Many MRTA algorithms can handle heterogeneous agents and could then in principle be used to solve task allocation problems where some of the agents are robots and some are humans.

With increased interest in the development of co-robot systems, the problem of scheduling human-robot teams became a question of independent value (Clare et al., 2012; Gombolay and Shah, 2014; Murphy, 2015). In this case, the main focus is to leverage the peculiar strengths of humans and robots and to implement an allocation of tasks that does not alienate humans. Note that in general this is a different problem than human-robot interaction, because the scheduling algorithm may generate an allocation of tasks such that humans and robots do not interact at all. In the context of RAPID, we face a situation where the set of tasks to be allocated is homogeneous, that is it consists of a set of spatially distributed emitters to be adjusted. Hence, the problem of picking a set of tasks engaging for humans (Gombolay et al., 2015) is not relevant because all tasks are the same. What is instead important is assigning tasks so that humans can improve their productivity without being overwhelmed.

3 Overview of RAPID

Figure 2 shows how introducing RAPID would refine the control system presented in Fig. 1.

RAPID implements both a new estimation algorithm to determine how much water is to be delivered, as well as the actuation system. Both components rely on existing infrastructure and methods to gather data in the field, and consequently novel sensors or sensing systems are not being developed as part of this effort. Figure 3 illustrates in more detail the overall architecture of

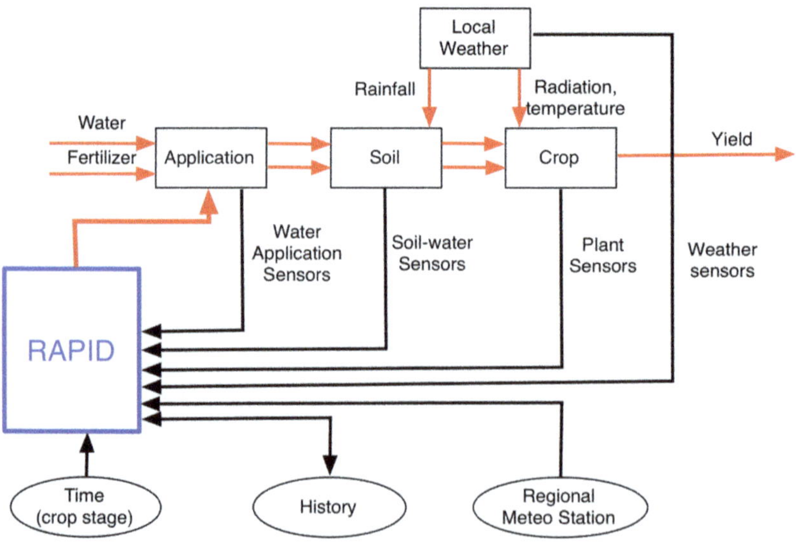

Figure 2 Revised irrigation control system with the introduction of RAPID.

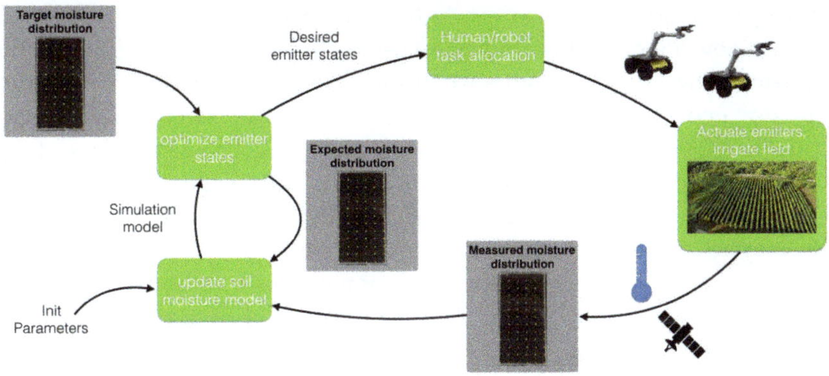

Figure 3 RAPID architecture.

the RAPID system we are developing, and evidences the feedback loop. As part of this project, we are currently developing three different efforts to develop three subsystems that will interact to implement the envisioned solution.

The first effort consists in developing a learning-based system to infer soil moisture levels from aerial images (Tseng et al., 2018). Based on deep convolutional neural networks (CNN), this module aims at learning a mapping between soil moisture data and aerial images.

The second thread consists in the co-design of a variable rate emitter and a robotic gripper that can perform the desired adjustments (Gealy et al., 2016; Berenstein et al., 2018). Mounted on a robotic arm placed on a mobile robot, or on a handheld device carried by a human, the gripper can latch on a variable rate emitter and autonomously perform the desired adjustment. In fact, while we anticipate an extensive use of robots to address the scalability aspect, preliminary discussions with farmers and growers have evidenced that a human component will also be important, as the presence of a human expert in the vineyard has multiple benefits in terms of quality assessment and the overall process monitoring.

The last component consists in a suite of planning algorithms to distribute the load between robots operating in the vineyard (Thayer et al., 2018a,b). An essential part underlying these three efforts is the experimental validation in the field in collaboration with our commercial partners. In this chapter, due to space limitations, we mostly discuss the first and third task, that is inference and routing algorithms.

Our underlying assumption, corroborated by multiple interactions with vine growers, is that in most instances uniform growth is desired over a given block. Starting from this premise, the overall information flow of the RAPID system is as follows. Through the collection of aerial images, the trained CNN infers the moisture level in the soil of a given vineyard. This soil moisture map is then presented to a human expert (e.g., a ranch manager) who, based on domain knowledge, can then indicate a desired (*target)* soil moisture level. The mismatch between the inferred soil moisture level and the desired soil moisture translates into actionable information about how each variable rate emitter should be adjusted to increase or decrease the amount of water released and match the target distribution. This information is provided to the planner that then schedules one or more robots to make the desired adjustments throughout the vineyard. Once the mobile robot reaches an emitter to be adjusted, the robotic arm mounted on the robot performs the required change.

4 Preliminary results

4.1 Soil moisture inference

Critical to the RAPID system is the knowledge of soil moisture to determine how emitters should be adjusted. Due to the sheer size of the vineyards managed by large-scale growers, measurements based on stationary sensors or mobile probes would provide only coarse scale data. To overcome these limitations, one possible alternative is using a data-driven approach to infer soil moisture conditions from aerial images. This approach is potentially advantageous because thanks to recent advances in UAV-related technologies, it has become relatively easy and inexpensive to acquire images covering large areas. Aerial images could be color images, hyperspectral, or thermal, just to name a few. In fact, there exist various previous works attempting to solve this very problem using a variety of machine-learning algorithms, such as support vector machines (SVM), Bayesian Neural Networks, and multilayer perceptrons (Hassan-Esfahani et al., 2014, 2015, 2017; Zaman et al., 2012; Ahmad et al., 2010; Bachour et al., 2014). Estimation methods based on deep learning and related approaches (Goodfellow et al., 2016) have enormous potential but to the best of our knowledge have not yet been applied to this specific problem. One of the main limitations of these techniques is the necessity to acquire large-scale datasets for training. In a domain such as the one considered, this could be prohibitively time consuming. To sidestep this entry-level obstacle and assess the potential of deep learning in this domain, we developed a simulator synthesizing artificial aerial images of a vineyard subject to different irrigation regimes (Tseng et al., 2018). The simulator is based on a discretized version of the Richards equation. In particular, the volumetric water content at time t is approximated as

$$s_t = s_{t-1} + \Delta s_t$$

where the variation Δs_t is defined as

$$\Delta s_t = -\left(r - a_t\right) - U\left(t\right)$$

In this last equation, r is the soil drainage, a_t is the irrigation rate, and $U(t)$ is the plant uptake that is assumed constant in our simulation. The quantity $d = r + U$ is called *dissipation rate* and is what the algorithm tries to learn from the images generated by the simulator. More specifically, the simulator generates 320 × 200 RGB images where the growth of nm plants spaced in a regular grid with m row and n columns is determined based on a vector $\bar{d}$ of soil moisture dissipation rates and a vector $\bar{a}$ of applied irrigation rates (see Fig. 4).

Both vectors have nm real components and are input parameters to the simulator. In all experiments presented in the following, $m = 20$ and $n = 10$. Starting from an initial uniform soil moisture distribution, the simulator propagates the distribution of moisture and adds leaves to a tree if the soil moisture value is positive. If the soil moisture is zero, then no leaves are added and the color of the existing leaves is altered by adding a shade of yellow. With this method, 1200 images were generated and then randomly split into 1000 training images and 200 validation images.

Starting from these images, the objective is to learn the dissipation rate. More formally, let x_i be the ith synthetic image and $\bar{d_i}$ be the corresponding dissipation rate. The objective is to determine the parameters θ of a model H_θ that predicts the dissipation rate $\bar{d_i}$ associated with the image x_i. Given an aerial image x, for a given choice of parameters θ' the predicted soil moisture dissipation will be $H_{\theta'}(x)$. As is common in machine-learning literature, for a given choice of the vector of parameters θ we define a loss function L as

$$L\left(\bar{d}, H_\theta(x)\right) = \frac{1}{p}\sum_{j=1}^{p}\left(\bar{d_j} - H_\theta(x_j)\right)^2$$

The optimal set of parameters would therefore be the vector θ^* solving the following minimization problem

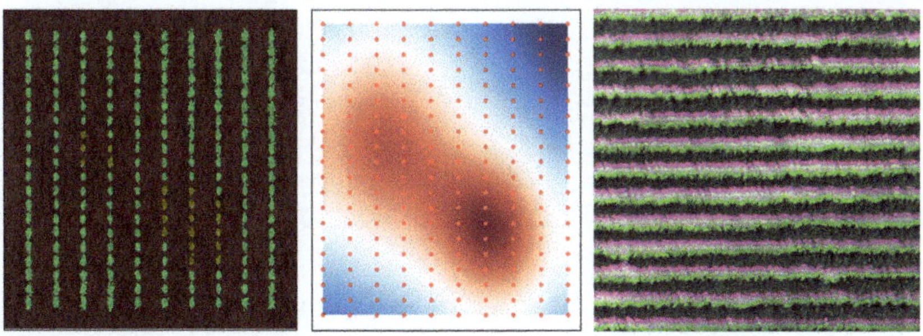

Figure 4 The left side shows one of the synthetic aerial images generated by the simulator, while the center figure shows the corresponding drainage image used for training. On the right, an actual cropped aerial image of a vineyard collected during our field experiments.

$$\theta^* = \text{argmin}_\theta \sum_{j=1}^{N_{\text{train}}} L\left(\overline{d_j}, H_\theta\left(x_j\right)\right)$$

To test the potential of deep CNN in this domain, we consider two different architectures. The first, dubbed deep Convolutional Neural Networks Uncorrelated Plants (CNNUP), is fed with a tree cropped out from the aerial image and produces the dissipation rate for the single tree (see Fig. 5).

The second, dubbed Deep Convolutional Neural Networks Correlated Field (CNNCF), instead receives as input the entire image and predicts $nm = 200$ dissipation rates, that is one for each tree in the input image (see Fig. 6).

These architectures were implemented using TensorFlow and the Adam algorithm was used for optimization (see Tseng et al. (2018) for more implementation details).

To assess the potential of CNN in this domain, we compared the two networks against five alternative approaches:

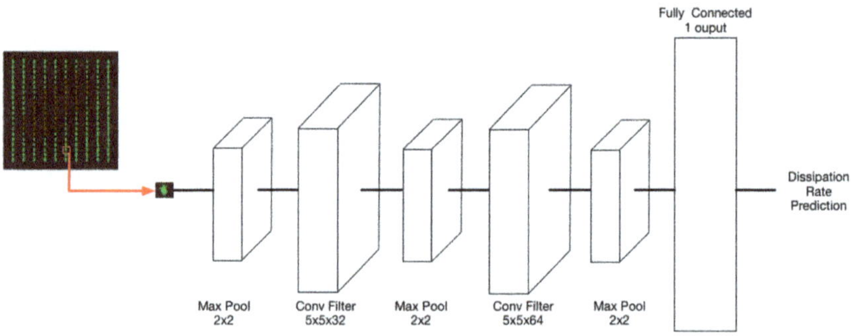

Figure 5 The CNNUP architecture takes as input a single tree cropped from an image with 200 trees and predicts the dissipation rate for the input tree.

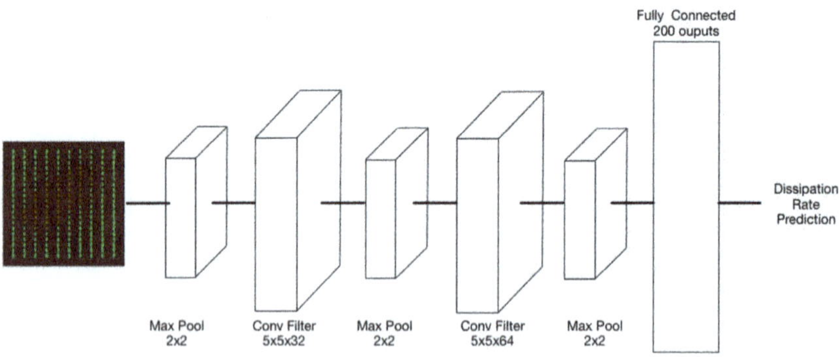

Figure 6 The CNNCF architecture takes as input an entire image with 200 trees and produces as output the soil moisture prediction for the whole set of trees.

1 a constant baseline always producing a fixed value *c* for every input;
2 a set of 200 SVM using a linear kernel, each predicting the moisture dissipation rate for one vine;
3 an uncorrelated method based on random forests using the RGB values to make predictions about moisture dissipation rate for one vine (RFUP);
4 a correlated method based on random forests where the whole image is used to make 200 predictions (RFCF); and
5 a two-layer feed-forward neural network (NN) inspired by the work presented in Hassan-Esfahani et al. (2017).

In all cases, 1000 examples were used for training and 200 examples were used in the validation set. To assess the merit of each approach and contrast them with each other, we consider the absolute error in the dissipation rate prediction $\left|\overline{d_j} - H(x)_j\right|$, that is the absolute difference between the predicted and actual dissipation rates. Table 1 shows mean, median, as well as the 25th and 75th percentile for this performance measure. Training times are also provided and refer to a system equipped with an Intel Core i7-6850k and three Nvidia Titan X Pascal GPUs.

From the table, we can see that the CNNCF architecture outperforms all other methods in terms of the considered error metric. As expected, its training time exceeds the time needed to train the two-layer NN, but the gap is moderate and the absolute time is contained. In particular, given that this refers to the offline training stage, this difference is acceptable from a practical standpoint. The overall objective of this preliminary study is to explore possible network architectures to identify promising approaches and to determine whether this solving method is practical at all. Results suggest that this is possible and efforts are ongoing to extend and adapt these deep architectures to deal with real data collected in the field.

Table 1 Results obtained comparing the proposed convolutional neural network architectures and five alternative machine-learning algorithms

Method	25th percentile	Mean	Median	75th percentile	Training time (s)
Constant	0.082	0.177	0.162	0.246	-
SVM	0.015	0.038	0.030	0.053	4141
RFUP	0.030	0.071	0.060	0.101	578
RFCF	0.047	0.109	0.094	0.152	655
NN	0.037	0.086	0.075	0.119	243
CNNUP	0.021	0.054	0.044	0.077	538
CNNCF	0.012	0.034	0.027	0.047	603

4.2 Routing algorithms in vineyards

Once a uniform target moisture level for the block has been determined, either by a human expert or an algorithm, it will be necessary to adjust variable rate emitters to match it. The discrepancy between the current soil moisture level and the desired moisture is an indicator of how much the emitters should be adjusted. Therefore, for each emitter e we define a quantity $R(e) = |m(e) - T|$ as the difference between the moisture measured at the emitter $m(e)$ and the desired target moisture T. The quantity $R(e)$ indicates by how much the water flow at each emitter should be adjusted. In particular, if $R(e) = 0$ for a certain emitter, no adjustment is needed because the current soil moisture level matches the desired value. Large values of $R(e)$ indicate significant mismatches, either due to over-irrigation or under-irrigation.

In general, in large vineyards a mobile robot, or a person, will not be able to visit and adjust all emitters at once, due to the necessity to periodically recharge the batteries (or rest, in the case of a person). In the following, we indicate this limitation as *travel budget*, and we will indicate it with the letter B. Intuitively, B captures the constraint that a robot can only travel a finite distance between two successive charges. Therefore, an optimization problem emerges, that is selecting the subset of spatially distributed emitters that can be adjusted given the travel budget B. In particular, it appears intuitive to prioritize adjusting emitters with a high mismatch value $R(e)$.

This problem is related to a classic optimization problem known as *orienteering* (Golden et al., 1987). More precisely, the orienteering problem is defined as follows. Let B be a positive budget and G a graph where every edge has a non-negative weight and every vertex has a value. The orienteering problem asks to determine a path in G maximizing the sum of the values of the visited vertices and whose cost is no larger than B. It shall be noted that if a vertex is visited multiple times, the reward is collected only once, but the cost of an edge is incurred every time it is traversed. In our problem setting, this corresponds to the case where a robot visits the same emitter more than once. In such a case, the emitter would be adjusted only once, and accordingly the reward would be collected only once. It was shown in Golden et al. (1987) that orienteering is an NP-hard optimization problem, and it was more recently shown that it is also APX-hard (Blum et al., 2007).

Numerous variations have been developed through the years. One that is particularly relevant to our discussion is the team orienteering problem (TOP) where one is given M agents, all subject to the same budget B, and the objective is to determine M paths maximizing the sum of the rewards collected by the M agents (Chao et al., 1996). Note that in TOP if a vertex is visited multiple times, then its reward is collected only by the first robot. Because of its intrinsic complexity and the difficulty in developing suitable approximation algorithms, heuristic approaches have been the most common choice for solving large instances of the orienteering problem (Gunavan et al., 2016). General-purpose heuristics working for a large variety of graphs have, however, remained elusive.

The connection between orienteering and our vineyard routing problem is as follows. For a given vineyard, we build a graph where every emitter is associated with a vertex, and edges are added between vertices to model the motion constraints typical of a vineyard. One such graph is shown in Fig. 7. The connections between the vertices model the constraints that in a vineyard a robot can move along tree rows (horizontal edges), but row switching is allowed only at either end of the rows (see also Fig. 8). A graph of this type will be called *irrigation graph (IG)* in the following, and in particular we write *IG(m,n)* for an irrigation graph with *m* rows and *n* emitters along each row.

For each vertex, the reward is then set to *R*(*e*), as defined above. For each edge, the cost is set as the distance between the vertices it connects. Graphs

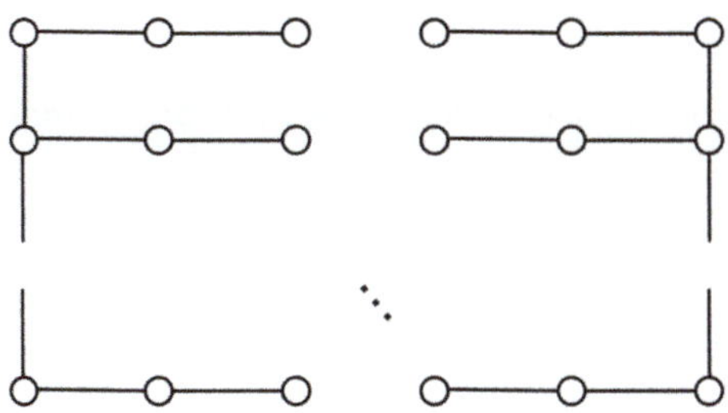

Figure 7 An irrigation graph modeling the spatial layout of emitters in a vineyard.

Figure 8 A robot moving in a vineyard can change row only at either end because of the irrigation lines running between vines.

such as the one shown in Fig. 7 are a particular instance of *BP3,* that is bipartite, planar graphs with vertices of degree at most three. A special instance is obtained when the cost of each edge is constant, for example to model the fact that emitters are regularly spaced in a vineyard. In Thayer et al. (2018a), we defined the following problem.

> *Constant Cost Orienteering on BP3 Problem (CCOBP3P)*: let $G = (V, E)$ a *BP3* graph, $r : V \rightarrow \mathbb{R}$ a value function, and $c : E \rightarrow \mathbb{R}$ a constant cost function defined over the edges. Moreover, let v_1, v_n be two vertices in V and T_{MAX} a positive constant. Find a path from v_1 to v_n of maximum value with cost no larger than T_{MAX}.

In Thayer et al. (2018a), we have shown that CCOBP3P is NP-hard and therefore we developed domain-specific heuristics both for the single robot case and for the multi-robot case (Thayer et al., 2018b). We refer the reader to the aforementioned paper of the detailed algorithmic formulation and we here sketch the underlying ideas. In all cases, we assume that $v_1 = v_n$, that is at the end the robot must return to the start location. For the single robot case, we have developed two heuristics called *greedy* row and *greedy partial-row* (GPR). The greedy row heuristic on an *IG(m,n)* works as follows. Preliminarily, for each of the *m* rows we compute the quantity

$$R_i = \sum_{j=1}^{n} R(i,j)$$

that is, the overall reward collected if the robot fully traverses the i-th row in the graph.[1] Moreover, to each row we associate a *feasible* flag, indicating whether with the remaining budget the robot can move from the current position to the row, traverse it, and go to the final position. This flag is incrementally updated, as the robot route is updated and more of its allocated budget T_{MAX} is consumed. The overall route is then built as follows. Considering the current robot location *v* and remaining budget, the robot updates the feasible flag for all rows, and for each feasible row it computes the normalized cost

$$R_i^{'} = \frac{R_i}{\text{cost}(v,i)}$$

that is, the ratio between the reward of the *i-th* row and the cost to reach it, and traverse it. Among all feasible rows, the path is then extended by appending the row maximizing the normalized cost and the residual budget is updated

1 Since emitters are placed in a grid layout, *R(i,j)* is the reward associated with the emitter placed at position *(i,j)* on the grid.

subtracting the budget spent to extend the route. This process is iterated until no more feasible rows are left. At that point, the robot returns to the starting location. Note that by construction this is always possible because a row is appended to the tour only if it is feasible, that is if there is enough residual budget to reach the final location. Therefore, a valid solution is always found.

Observing that soil moisture displays a 'local' structure, that is wet and dry locations tend to be clustered, we designed a different heuristic called GPR. The intuition is that if there is a region with high rewards at either end of the rows, it may be more advantageous to enter the row, adjust some of its emitters, and then come back from the same side from which it was entered, and then move to an adjacent row. The trade-off is that the robot will spend some budget without getting any reward when traveling back along the row, but this may enable to move to an adjacent row with high rewards without having to first traverse an entire row. This would not be possible in the previous heuristic where once one row is entered it must be fully traversed. This idea can be implemented following an approach similar to the previous one, that is the route is iteratively extended by greedily selecting feasible rows or partial rows. The definition of a feasible row or partial row is exactly the same, that is there has to be sufficient residual budget to cover the partial row and then proceed to the final vertex v_n. However, given that in this case a row can be partially entered by either side, we compute two rewards, that is $L(i,j)$ and $R(i,j)$. $L(i,j)$ is the reward obtained if the robot enters the *i-th* row from the left, proceeds until the *j-th* emitter, and then goes back. The definition of $R(i, j)$ is the same but considers starting from the right. Such costs are then scaled by the travel cost as in the full row heuristic, and the path is then extended greedily (see Thayer et al. (2018a) for more details). Both these heuristics have computational complexity polynomial in the number of rows m and columns n. In particular, the former is $O(m^2)$ while the latter is $O(m^2n)$.

Besides the single robot heuristics, we also developed three heuristics for the TOP problem, as it is foreseeable that multiple robots will be developed in a vineyard to service it more efficiently (Thayer et al, 2018b). In the following, we indicate with M the number of robots. The first approach, dubbed *sectioning*, splits the vineyard into M sections (each consisting of a certain number of rows) and assigns a robot to each section. Within each section, the robot then runs the GPR heuristic. Sections are defined aiming at having the same sum of rewards and not the same area, so that each robot can potentially make the same impact. With this approach, all robots stay separate and cannot collide with each other. The second heuristic, called *SeriesGPR*, runs M times the GRP heuristic in sequence. At every iteration, the path for a single robot is planned, and the map is updated at every iteration to zero out the rewards associated with the emitters already serviced. With the SeriesGPR algorithm, all robots can visit any area in the vineyard. Finally, the *ParallelGPR* approach plans the path for all robots at the same time. This means that at every iteration, the algorithm

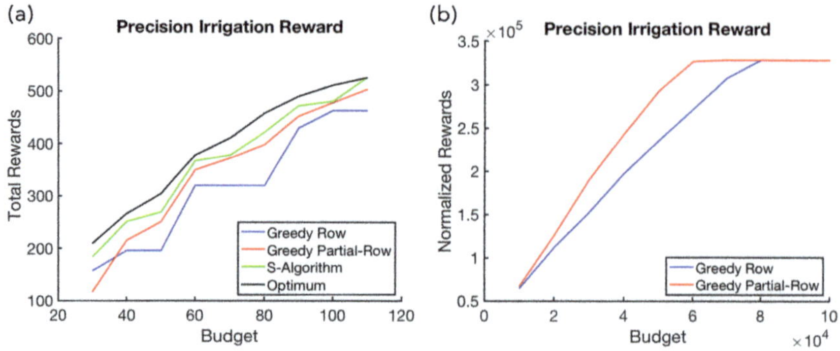

Figure 9 Performance of the single robot routing heuristics.

determines for each robot what is the best next move (using the GPR heuristic) and the path is extended for that robot. The approach is iterated until all robots have expended their budget.

Extensive results are presented in Thayer et al. (2018a,b), and we here report just a small representative set. The two single robot heuristic approaches we developed have been numerically compared with a general-purpose heuristic for orienteering called the *S-algorithm* (Tsiligirides, 1984). For the TOP case, instead, we compared our findings with the guided local search heuristic formerly proposed in the literature (Vansteenwegen et al., 2009).

Figure 9 shows the performance of the single robot heuristics. The left figure shows the results for a small vineyard where the optimal solution can be found with an integer programming approach. The figure shows that the GPR heuristic and the S-algorithm are relatively close to the optimal solution and show comparable performance. However, the S-algorithm does not scale well and cannot efficiently solve instances with hundreds of thousands of vines. In the right figure, we then compare the performance of the two heuristics we proposed for the case of a full-sized vineyard. The chart confirms that the GPR is the most competitive one. In Fig. 10, instead, we compare the performance of the various heuristics as a function of the budget for a case with six agents. In this case, the ParallelGPR algorithm emerges as the most competitive and the result is representative of other cases, too.

5 Future trends and conclusion

We are currently in the second year of a project expected to last for 4 years. Therefore, while many of the individual components have been prototyped, system integration of the whole RAPID system is still ongoing. Due to limited space, we described just selected contributions but refer the reader to cited references for additional results, for example regarding the design of the gripper.

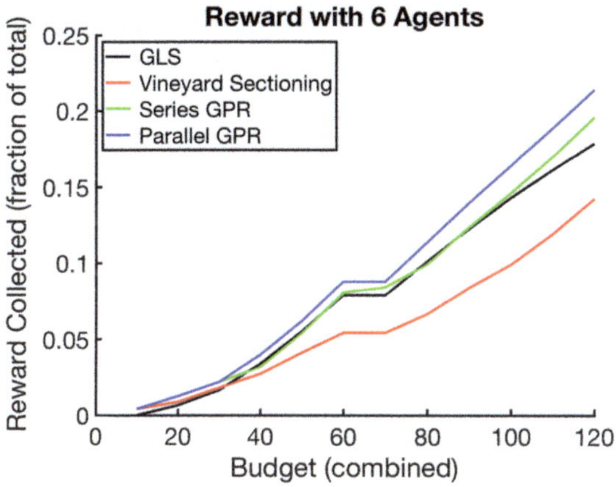

Figure 10 Performance of the heuristics for the multiagent case.

Despite the fact that we are still in an early stage of the project, there are some conclusions that can be drawn. With regard to estimation, deep CNN seem to be a promising method to infer soil moisture content from aerial images and they appear to have the potential to outperform methods proposed in the past. With regard to routing in vineyards, the peculiar structure of the environment calls for the development of domain-specific heuristics capable of scaling to very large instances. Our preliminary results have identified methods superior to the state-of-the-art and scalable.

In terms of future work, besides the necessary integration of all components, we identify the following avenues for further investigation. First, it will be necessary to test the CNN architecture on data collected in the field, and possibly later its structure to better model the data we gathered. For routing, we will incorporate uncertainty in the rewards, since those are obtained through an estimation process that is necessarily uncertain. Uncertainty shall therefore be considered when deciding where the robot should go next. Finally, extensive field evaluation will be needed to assess the viability of the whole system and ultimately quantify water savings.

6 Acknowledgements

This material is based upon work that is supported by the USDA-NIFA under award number 2017-67021-25925 (National Robotics Initiative). Any opinions, findings, conclusions, or recommendations expressed in this publication are those of the authors and do not necessarily reflect the view of the US Department of Agriculture or the US government.

We gratefully acknowledge Luis Sanchez and Nick Dokoozlian from E&J Gallo Winery for having granted access to their vineyards for data collection, and for the valuable information provided during this project. We also thank the many lab members and collaborators who contributed to this project: Thomas Thayer, Jose Manuel Gonzalez and Carlos Diaz Alvarenga, Christine Breckenridge, Jonathan Garache, David Tseng, David Wang, Carolyn Chen, Lauren Miller, William Song, and Juan Aparicio Ojea.

7 Where to look for further information

There is growing interest in the use of robotics and automation for precision agriculture. Funding agencies in the United States, Europe, and Asia have specialized initiatives to support this research. For example, the US Department of Agriculture currently supports programs such as:

- Food and Agriculture Cyberinformatics and Tools: https://nifa.usda.gov/program/fact
- National Robotics Initiative: https://nifa.usda.gov/funding-opportunity/national-robotics-initiative-realization-co-robots-acting-direct-support

General overviews about recent developments in this domain can be found in https://arxiv.org/pdf/1806.06762.pdf and Bergman et al. (2016).

Research centers working in this area include:

- The National Robotics Engineering Center at Carnegie Mellon University: https://www.nrec.ri.cmu.edu/nrec/solutions/agriculture/index.html
- The Center for Precision and Automated Agricultural Systems (CPAAS) at Washington State University: http://cpaas.wsu.edu/ (joint venture with the University of Technology Sydney in Australia).
- The Australian Centre for Field Robotics at the University of Sydney (ACFR): https://sydney.edu.au/engineering/our-research/robotics-and-intelligent-systems/australian-centre-for-field-robotics/agriculture-and-the-environment.html

8 References

Adams, J. M., Nudurupati, S. S., Gasparini, N. M., Hobley, D., Hutton, E. W. H., Tucker, G. E. and Istanbulluoglu, E. 2014. LandLab: sustainable software development in practice. In: *Proceedings of the Second Workshop on Sustainable Software for Science: Practice and Experiences*.

Ahmad, S., Kalra, A. and Stephen, H. 2010. Estimating soil moisture using remote sensing data: a machine learning approach. *Advances in Water Resources* 33(1), 69–80. doi:10.1016/j.advwatres.2009.10.008.

Aragüés, R., Cortés, J. and Sagüés, C. 2011. Distributed consensus algorithms for merging feature-based maps with limited communication. *Robotics and Autonomous Systems* 59(3–4), 163–80. doi:10.1016/j.robot.2011.01.002.

Aragüés, R., Cortés, J. and Sagüés, C. 2012. Distributed consensus on robot networks for dynamically merging feature-based maps. *IEEE Transactions on Robotics* 28(4), 840–54. doi:10.1109/TRO.2012.2192012.

Bachour, R., Walker, W. R., Ticlavilca, A. M., McKee, M. and Maslova, I. 2014. Estimation of spatially distributed evapotranspiration using remote sensing and a relevance vector machine. *Journal of Irrigation and Drainage Engineering* 140(8), 04014029. doi:10.1061/(ASCE)IR.1943-4774.0000754.

Baluja, J., Diago, M. P., Balda, P., Zorer, R., Meggio, F., Morales, F. and Tardaguila, J. 2012. Assessment of vineyard water status variability by thermal and multispectral imagery using an unmanned aerial vehicle (UAV). *Irrigation Science* 30(6), 511–22. doi:10.1007/s00271-012-0382-9.

Bellvert, J., Zarco-Tejada, P. J., Girona, J. and Fereres, E. 2014. Mapping crop water stress index in a 'pinot-noir' vineyard: comparing ground measurements with thermal remote sensing imagery from an unmanned aerial vehicle. *Precision Agriculture* 15(4), 361–76. doi:10.1007/s11119-013-9334-5.

Berenstein, R., Fox, R., McKinley, S., Carpin, S. and Goldberg, K. 2018. Robustly adjusting indoor drip irrigation emitters with the Toyota HSR robot. In: *Proceedings of the IEEE International Conference on Robotics and Automation*, pp. 2236–43.

Bergman, M., Billingsley, J., Reid, J. and van Henten, E. 2016. Robotics in agriculture and forestry. In: Siciliano, B. and Khatib, O. (Eds), *Springer Handbook of Robotics*. Springer, pp. 1463–92.

Birk, A. and Carpin, S. 2006. Merging occupancy grids from multiple robots. *Proceedings of the IEEE* 94(7), 1384–97. doi:10.1109/JPROC.2006.876965.

Blum, A., Chawla, S., Karger, D. R., Lane, T., Meyerson, A. and Minkoff, M. 2007. Approximation algorithms for orienteering and discounted-reward TSP. *SIAM Journal on Computing* 37(2), 653–70. doi:10.1137/050645464.

Bullo, F., Cortes, J. and Martinez, S. 2009. *Distributed Control of Robotic Networks*. Princeton University Press, Princeton, NJ.

Carpin, S. 2008. Fast and accurate map merging for multi-robot systems. *Autonomous Robots* 25(3), 305–16. doi:10.1007/s10514-008-9097-4.

Chao, I., Golden, B. L. and Wasil, E. A. 1996. The team orienteering problem. *European Journal of Operational Research* 88(3), 464–74. doi:10.1016/0377-2217(94)00289-4.

Choi, M. and Jacobs, J. M. 2007. Soil moisture variability of root zone profiles within SMEX02 remote sensing footprints. *Advances in Water Resources* 30(4), 883–96. doi:10.1016/j.advwatres.2006.07.007.

Clare, A. S., Cummings, M. L., How, J. P., Whitten, A. K. and Toupet, O. 2012. Operator object function guidance for a real-time unmanned vehicle scheduling algorithm. *Journal of Aerospace Computing, Information and Communication* 9(4), 161–73. doi:10.2514/1.I010019.

Clement, T. P., Wise, W. R. and Molz, F. J. 1994. A physically based, two-dimensional, finite-difference algorithm for modeling variably saturated flow. *Journal of Hydrology* 161(1–4), 71–90. doi:10.1016/0022-1694(94)90121-X.

Dane, J. H. and Topp, G. C. (Eds). 2002. *Methods of Soil Analysis. Part 4. Physical Methods*. Soil Society of America, Madison, WI.

Deb, S. K., Shukla, M. K., Šimunek, J. and Mexal, J. G. 2013. Evaluation of spatial and temporal root water uptake patterns of a flood-irrigated pecan tree using the HYDRUS (2D/3D) model. *Journal of Irrigation and Drainage Engineering* 139(8), 599–611. doi:10.1061/(ASCE)IR.1943-4774.0000611.

Erinc, G. and Carpin, S. 2014. Anytime merging of appearance-based maps. *Autonomous Robots* 36(3), 241–56. doi:10.1007/s10514-013-9352-1.

Gealy, D. V., McKinley, S., Guo, G., Miller, L., Vougioukas, S., Viers, J., Carpin, S. and Goldberg, K. 2016. DATE: a handheld co-robotic device for automated tuning of emitters to enable precision irrigation. In: *Proceedings of the IEEE International Conference on Automation Science and Engineering*, pp. 922–7.

Gerkey, B. P. 2003. On multi-robot task allocation. PhD thesis. University of Southern California.

Golden, B. L., Levy, L. and Vohra, R. 1987. The orienteering problem. *Naval Research Logistics* 34(3), 307–18. doi:10.1002/1520-6750(198706)34:3<307::AID-NAV3220340302>3.0.CO;2-D.

Gombolay, M. and Shah, J. 2014. Challenges in collaborative scheduling of human-robot teams. In: *Proceedings of the Association for the Advancement of Artificial Intelligence Symposium*, pp. 73–5.

Gombolay, M. C., Gutierrez, R. A., Clarke, S. G., Sturla, G. F. and Shah, J. A. 2015. Decision-making authority, team efficiency and human worker satisfaction in mixed human–robot teams. *Autonomous Robots* 39(3), 293–312. doi:10.1007/s10514-015-9457-9.

Goodfellow, I., Bengio, Y. and Courville, A. 2016. *Deep Learning*. MIT Press, Cambridge, MA.

Gunavan, A., Chuin Lin, H. C. and Vansteenwegen, P. 2016. Orienteering problem: a survey of recent variants, solution approaches and applications. *European Journal of Operational Research* 255(2), 315–32. doi:10.1016/j.ejor.2016.04.059.

Gutierrez Jaguey, J., Villa-Medina, J. F., Lopez-Guzman, A. and Porta-Gandara, M. A. 2015. Smartphone irrigation sensor. *IEEE Sensors Journal* 15(9), 5122–7. doi:10.1109/JSEN.2015.2435516.

Hassan-Esfahani, L., Torres-Rua, A., Ticlavilca, A. M., Jensen, A. and McKee, M. 2014. Topsoil moisture estimation for precision agriculture using unmanned aerial vehicle multispectral imagery. In: *Proceedings of the IEEE International Geoscience and Remote Sensing Symposium*, pp. 3263–6.

Hassan-Esfahani, L., Torres-Rua, A., Jensen, A. and McKee, M. 2015. Assessment of surface soil moisture using high-resolution multi-spectral imagery and artificial neural networks. *Remote Sensing* 7(3), 2627–46. doi:10.3390/rs70302627.

Hassan-Esfahani, L., Torres-Rua, A., Jensen, A. and McKee, M. 2017. Spatial root zone soil water content estimation in agricultural lands using Bayesian-based artificial neural networks and high-resolution visual, NIR, and thermal imagery. *Irrigation and Drainage* 66(2), 273–88. doi:10.1002/ird.2098.

Hydrus. 2018. Hydrus 2D/3D (online). Available at: https://www.pc-progress.com/en/Default.aspx?hydrus-3d (accessed October 2018).

Jones, H. G. 2004. Irrigation scheduling: advantages and pitfalls of plant-based methods. *Journal of Experimental Botany* 55(407), 2427–36. doi:10.1093/jxb/erh213.

Kandelous, M. M. and Šimůnek, J. 2010. Comparison of numerical, analytical, and empirical models to estimate wetting patterns for surface and subsurface drip irrigation. *Irrigation Science* 28(5), 435–44. doi:10.1007/s00271-009-0205-9.

Khamis, A., Hussein, A. and Elmogy, A. 2015. Multi-robot task allocation: a review of the state-of-the-art. In: Koubâa, A. and Martínez-de Dios, J. (Eds), *Cooperative Robots and Sensor Networks*. Springer, pp. 31–51.

Korsah, G. A., Stentz, A. and Dias, M. B. 2013. A comprehensive taxonomy for multi-robot task allocation. *The International Journal of Robotics Research* 32(12), 1495–512. doi:10.1177/0278364913496484.

Lafolie, F., Guennelon, R. and Van Genuchten, M. T. 1989. Analysis of water flow under trickle irrigation: I. Theory and numerical solution. *Soil Science Society of America Journal* 53(5), 1310–8. doi:10.2136/sssaj1989.03615995005300050002x.

Li, Z. and Isler, V. 2016. Large scale image mosaic construction for agricultural applications. Technical Report 16-002. University of Minnesota.

Majone, B., Viani, F., Filippi, E., Bellin, A., Massa, A., Toller, G., Robol, F. and Salucci, M. 2013. Wireless sensor network deployment for monitoring soil moisture dynamics at the field scale. *Procedia Environmental Sciences* 19, 426–35. doi:10.1016/j.proenv.2013.06.049.

Martinez, S., Cortes, J. and Bullo, F. 2007. Motion coordination with distributed information. *IEEE Control Systems Magazine* 27(4), 75–88. doi:10.1109/MCS.2007.384124.

McCarthy, A. C., Hancock, N. H. and Raine, S. R. 2013. Advanced process control of irrigation: the current state and an analysis to aid future development. *Irrigation Science* 31(3), 183–92. doi:10.1007/s00271-011-0313-1.

Murphy, R. R. 2015. Meta-analysis of autonomy at the DARPA robotics challenge trials. *Journal of Field Robotics* 32(2), 189–91. doi:10.1002/rob.21578.

Navarro-Hellín, H., Torres-Sánchez, R., Soto-Valles, F., Albaladejo-Pérez, C., López-Riquelme, J. A. and Domingo-Miguel, R. 2015. A wireless sensors architecture for efficient irrigation water management. *Agricultural Water Management* 151, 64–74. doi:10.1016/j.agwat.2014.10.022.

Ojha, R., Morbidelli, R., Saltalippi, C., Flammini, A. and Govindaraju, R. S. 2014. Scaling of surface soil moisture over heterogeneous fields subjected to a single rainfall event. *Journal of Hydrology* 516, 21–36. doi:10.1016/j.jhydrol.2014.01.057.

Or, D. and Coelho, F. E. 1996. Soil water dynamics under drip irrigation: transient flow and uptake models. *Transactions of the ASAE* 39(6), 2017–25.

Pierson, A., Figueiredo, L. C., Pimenta, L. C. A. and Schwager, M. 2015. Adapting to performance variations in multi-robot coverage. In: *Proceedings of the IEEE International Conference on Robotics and Automation*, pp. 415–20.

Pon, R., Kansal, A., Liu, D., Rahimi, M., Kaiser, W. J., Pottie, G. J., Srivastava, M., Sukhatme, G. and Estrin, D. 2005. Networked infomechanical systems (NIMS): next generation sensor networks for environmental monitoring. In: *IEEE MTT-S International Microwave Symposium Digest*, pp. 373–6.

Rahimi, M., Pon, R., Kaiser, W. J., Sukhatme, G., Estrin, D. and Srivastava, M. 2004. Adaptive sampling for environmental robotics. In: *Proceedings of the IEEE International Conference on Robotics and Automation*, pp. 3537–44.

Rahmstorf, S. and Coumou, D. 2011. Increase of extreme events in a warming world. *Proceedings of the National Academy of Sciences* 108(44), 17905–9. doi:10.1073/pnas.1101766108.

Rehman, A., Zafar Abbasi, A., Islam, N. and Shaikh, Z. A. 2014. A review of wireless sensors and networks' applications in agriculture. *Computer Standards and Interfaces* 36(2), 263–70. doi:10.1016/j.csi.2011.03.004.

Robinson, D. A., Campbell, C. S., Hopmans, J. W., Hornbuckle, B. K., Jones, S. B., Knight, R., Ogden, F., Selker, J. and Wendroth, O. 2008. Soil moisture measurement for ecological and hydrological watershed-scale observatories: a review. *Vadose Zone Journal* 7(1), 358–89. doi:10.2136/vzj2007.0143.

Romano, N. 2014. Soil moisture at local scale: measurements and simulations. *Journal of Hydrology* 516(4), 6–20. doi:10.1016/j.jhydrol.2014.01.026.

Romero, P., Fernandez-Fernandez, J. I. and Martinez-Cutillas, A. 2010. Physiological thresholds for efficient regulated deficit-irrigation management in winegrapes grown under semiarid conditions. *American Journal of Enology and Viticulture* 61(3), 300–12.

Saeedi, S., Paull, L., Trentini, M., Seto, M. and Li, H. 2012. Map merging using Hough peak matching. In: *Proceedings of the IEEE/RSJ International Conference on Intelligent Robots and Systems*, pp. 4683–8.

Saeedi, S., Trentini, M., Seto, M. and Li, H. 2016. Multiple-robot simultaneous localization and mapping: a review. *Journal of Field Robotics* 33(1), 3–46. doi:10.1002/rob.21620.

Schaible, G. D. and Aillery, M. P. 2017. Challenges for US irrigated agriculture in the face of emerging demands and climate change. In: Ziolkowska, J. and Peterson, J. M. (Eds), *Competition for Water Resources*. Elsevier, pp. 44–79.

Schwager, M., Rus, D. and Slotine, J. J. 2009. Decentralized, adaptive coverage control for networked robots. *The International Journal of Robotics Research* 28(3), 357–75. doi:10.1177/0278364908100177.

Smith, K. A. and Mullins, C. E. (Eds). 2000. *Soil and Environmental Analysis: Physical Methods*. Marcel Dekker, New York, NY.

Smith, S. L., Schwager, M. and Rus, D. 2012. Persistent robotic tasks: monitoring and sweeping in changing environments. *IEEE Transactions on Robotics* 28(2), 410–26. doi:10.1109/TRO.2011.2174493.

Thayer, T., Vougioukas, S., Goldberg, K. and Carpin, S. 2018a. Routing algorithms for robot assisted precision irrigation. In: *Proceedings of the IEEE International Conference on Robotics and Automation*, pp. 2221–8.

Thayer, T., Vougioukas, S., Goldberg, K. and Carpin, S. 2018b. Multi-robot routing algorithms for robots operating in vineyards. In: *Proceedings of the IEEE International Conference on Automation Science and Engineering*, pp. 7–14.

Tian, F., Gao, L. and Hu, H. 2011. A two-dimensional Richards equation solver based on CVODE for variably saturated soil water movement. *Science China Technological Sciences* 54(12), 3251–64. doi:10.1007/s11431-011-4566-y.

Tseng, D., Wang, D., Chen, C., Miller, L., Song, E., Viers, J., Vougioukas, S., Carpin, S., Aparicio Ojea, J. and Goldberg, K. 2018. Towards automating precision irrigation: deep learning to infer local soil moisture conditions from synthetic aerial agricultural images. In: *Proceedings of the IEEE International Conference on Automation Science and Engineering*, pp. 284–91.

Tsiligirides, T. 1984. Heuristic methods applied to orienteering. *Journal of the Operational Research Society* 35(9), 797–809. doi:10.1057/jors.1984.162.

van Dam, J. C. and Feddes, R. A. 2000. Numerical simulation of infiltration, evaporation and shallow groundwater levels with the Richards equation. *Journal of Hydrology* 233(1-4), 72–85. doi:10.1016/S0022-1694(00)00227-4.

Vansteenwegen, P., Souffriau, W., Berghe, G. V. and Oudheusden, D. V. 2009. A guided local search metaheuristic for the team orienteering problem. *European Journal of Operational Research* 196(1), 118–27. doi:10.1016/j.ejor.2008.02.037.

Vereecken, H., Huisman, J. A., Pachepsky, Y., Montzka, C., van der Kruk, J., Bogena, H., Weihermüller, L., Herbst, M., Martinez, G. and Vanderborght, J. 2014. On the spatio-temporal dynamics of soil moisture at the field scale. *Journal of Hydrology* 516(4), 76–96. doi:10.1016/j.jhydrol.2013.11.061.

Wallace, J. S. 2000. Increasing agricultural water use efficiency to meet future food production. *Agriculture, Ecosystems and Environment* 82(1–3), 105–19. doi:10.1016/S0167-8809(00)00220-6.

Wang, S.-Y., Yoon, J.-H., Gillies, R. R. and Hsu, H.-H. 2017. The California drought. In: Wang, S.-Y., Yoon, J.-H., Funk, C. C. and Gillies, R. R. (Eds). *Climate Extremes: Patterns and Mechanisms*. Wiley, pp. 223–35.

WineInstitute.org. 2018. The wine institute website (online). Available at: https://www.wineinstitute.org/resources/statistics (accessed October 2018).

Zaman, B., McKee, M. and Neale, C. M. U. 2012. Fusion of remotely sensed data for soil moisture estimation using relevance vector and support vector machines. *International Journal of Remote Sensing* 33(20), 6516–52. doi:10.1080/01431161.2012.690540.

Chapter 8

The use of agricultural robots in weed management and control

Brian Steward, Jingyao Gai, and Lie Tang, Iowa State University, USA

1 Introduction

Weeds are a curse to agricultural production and lead to diminished crop yield and quality. Weed management and control are thus essential to the production of high-yielding and high-quality crops. Advancements in weed control technology have had a huge impact on agricultural productivity. Meeting the food and fiber demands of the world's growing population will only be possible with highly productive agricultural systems in which weed management is a critical component.

Any effective weed control technology needs to be both robust and adaptable. Robust weed control technology will successfully control weeds in spite of variability in the field conditions. Adaptable weed control technology has the capacity to change its strategy in the context of evolving weed populations, genetics, and climatic conditions. Conventional weed control practices, such as chemical weed control or mechanical cultivation, tend to be robust with the input of agricultural producer experience and management decisions. However, they may be less adaptable because they are often applied with large machines, typically large boom sprayers. While these machines have a substantial amount of controller technology to vary several application parameters, the changes are made on a larger scale that is consistent with the size of the sprayers and boom widths.

http://dx.doi.org/10.19103/AS.2019.0056.13

Agricultural robots have great potential to deliver weed control technologies that are much more adaptable even down to the plant scale. They potentially could direct chemical or cultivation tools to directly target weed plants. Agricultural robots can have these characteristics because they bring recent advances in artificial intelligence (AI) to bear on the control of weeds in crop fields. However, bringing AI and robotics technology to weed control has several challenges that may limit robotic weed control robustness, at least with the current state of technology.

Weeds are plants that are misplaced or undesirable to the purpose of the crop plants in the field. Crop plants are being cultivated because of the economic value to the producer. Thus, any plant can be a weed, like volunteer corn growing in a soybean field, if it is not serving the purposes of the producer's management scheme. Thus plants are weeds based on their location and competition relative to the crop plants. So robotic weed control is an ill-posed problem until the agricultural producer's intentions for a field are made known to the robot, which will then identify and make decisions about which plants are the weeds that need to be controlled.

Another challenge is that while crop plants are mechanically planted in a structured manner that is compatible with agricultural machinery, weed plants emerge and grow in patterns that are consistent with their ecology. Thus weed plants exist in random patterns in a field. The species of plants that make up the collection of weeds in the field varies. This variability changes at different scales as well. Weed plant emergence changes on a meter-by-meter basis, but also at larger scales: field, farm, county, state, region, and climatic zone.

Weed management strategies take this variability into account and employ weed control techniques that are general and robust enough to be efficacious to control the weeds. In the development of robotic weed control methods, there are obvious challenges to success. These challenges include informing the robot which plants need to be controlled and determining the distinguishing features of those plants. To control weed plants, their growth needs to be retarded or stopped, while not injuring the nearby crop plants. Implementing the control is particularly difficult when the weeds are in close proximity to the crop plants.

The Institute of Electrical and Electronics Engineers (IEEE) provides the following definition for a robot: 'A robot is an autonomous machine capable of sensing its environment, carrying out computations to make decisions, and performing actions in the real world' (IEEE, 2019). Autonomy requires some degree of machine intelligence which involves achieving a particular goal in the context of uncertainty and variability (Jarvis and Grant, 2014; Rzevski, 2003). So while the advances in technology in the area of 'robotic weeding' have been substantial, current automated weeding systems generally lack autonomy so the use of the term robotic is questionable.

Han et al. (2015) developed a multilayer design framework for intelligent machines and field robots consisting of four technology layers (Fig. 1). The layers tend to build on each other, starting with the machine architecture layer which contains both the hardware and software architecture required for the robot's function. Next, the machine awareness layer consists of the perception, localization, and monitoring technology needed for the robot to be aware of its own systems and environment. Robots act upon the world, and thus need to have control systems for the actuators in the machine control layer. For autonomy, being able to achieve goals under uncertainty, the machine planning and supervision technologies associated with the top machine behavior layer are required.

From a review of the literature, technologies that are classified as robotic weeders have typically encompassed technology in the lower three layers such as hardware architecture, perception, and localization and navigation, and implement control. Technologies in the machine behavior layer along with condition monitoring technology are largely absent. Thus it is difficult to say, while in no way intending to diminish the technical work reported in the literature, that we have robotic weeding technology today. However, the developments in the area are progressing rapidly.

Along a similar line of thought, Merfield (2016) compellingly argues that the robotic weeders documented in the literature or commercially developed are not truly robotic weeders, but are 'essentially self-guiding vehicles carrying weeding tools.' His argument is based on the observation that the use of mechanical weeding tools in practice is complex. This complexity is due to several factors such as soil properties varying based on soil type and environmental factors such as soil moisture content, variability in weed plants and crop plants, and the response of these plants to the actions of the mechanical weeding tools.

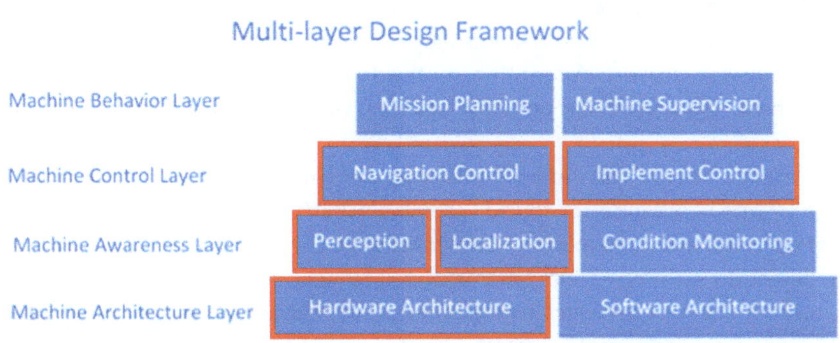

Figure 1 A multilayer design framework for intelligent agricultural machines and field robots. The red boxes indicate technologies that have been typically reported in the robotic weed control literature. Source: adapted from Han et al. (2015).

For agricultural robots to successfully control weeds, several primary challenges must be overcome. First, both crop and weed plants must be accurately perceived in the crop field. The perception system must include plant detection, weed and crop plant classification and plant localization. Secondly, the mechanisms for controlling the weeds must be developed in the context of the weed control strategy that best fits the cultural practices of the production system. Third, based on the information from the perception system, the mechanism must be directed to act at the weed plant locations and away from the crop plants. Fourth, all of these technologies must be integrated together into a weeding system.

This chapter will focus on key work in the development of robotic weeders including weed perception systems and weed control mechanisms. An example of an automated weeding system will be described, and readers will be pointed to where they can learn more.

2 Addressing the challenges of robotic weed control

Much work has been done to address the challenges of robotic weed control. In the subsections below, we will review the literature to provide a perspective on how the challenges have been addressed. The first section will focus on perception systems which can detect and classify weed plants from crop plants. The second section will focus on weed control mechanisms.

2.1 Weed and crop plant perception

After decades of research and development, numerous methods have been developed for weed and crop plant perception. The main challenges in crop and weed plant perception are vegetation detection, classification of weed plants and crop plants, and plant localization. Machine vision is the most widely used technique. Studies vary in terms of crop or weed species, complexity of visual scene (from indoor-controlled environments to commercial fields) and the sensors used. Perception methods can be categorized in terms of the vehicle platforms carrying the sensors, and the plant features used in processing.

Satellite-, aerial- (including unmanned aerial vehicles, UAVs), and ground-based (with unmanned ground vehicles (UGV), such as field robots, or commercially available off-highway vehicles) vehicle platforms are common for detecting and monitoring plants. Satellite- and aerial-based sensing were commonly used for large-scale field monitoring in applications such as variable-rate herbicide spraying (Lan et al., 2010; Torres-Sánchez et al., 2013). These platforms have a lower spatial resolution, and the working time is affected by the weather and air conditions (Moran et al., 1997). Ground vehicle-based sensing and low-altitude aerial-based sensing can acquire higher spatial resolution plant imagery enabling accurate detection of crop rows and plant localization for applications such as real-time, in-row weed control (Hassanein and El-Sheimy, 2018; Li and Tang, 2018). Ground vehicle-based methods, however, must meet requirements such as having clearance over the crop, matching the crop row spacing and being able to traverse the field under a range of soil conditions (Hague et al., 2000).

Spectral reflectance and biological morphology characteristics were two categories of features commonly used in these methods (Slaughter et al., 2008). In the subsections below, weed plant perception methods are reviewed in terms of the type of features used.

2.1.1 Spectral reflectance characteristics

Spectral reflectance characteristics were first investigated, based on the observation that the soils and plants have different spectral reflectance properties (Kyllo, 2003). Soil reflectance is typically low across the visible and near-infrared (NIR) regions with a gradually increasing slope going from visible to NIR and IR regions (Fig. 2). There will be some variations in the reflectance spectrum across soil types. Vegetation reflectance has a very distinctive shape with low reflectance in the visible spectrum, but with increased reflectance in the green region. At the red region, entering into the NIR region, there is a

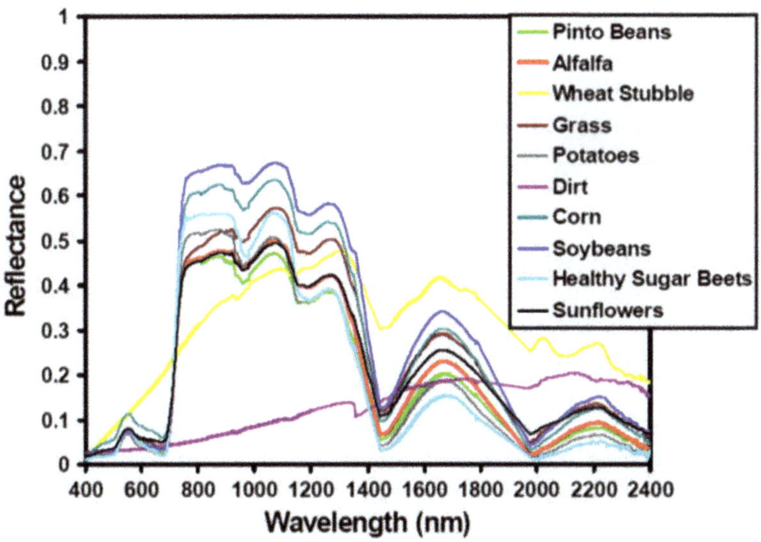

Figure 2 Spectral signatures of different plants and soil. Source: adapted from Kyllo (2003).

dramatic increase in reflectance up into the 50% range. This dramatic increase in plant reflectance is called the red edge and forms the basis for some of the vegetative indices described below.

Starting in the 1970s, spectral reflectance differences between crop, weeds, and soil were explicitly studied (Moran et al., 1997). Algorithms were developed for satellite-, aerial-, and ground-based sensing platforms. Most of the algorithms used images as input signals acquired with cameras that contained charge-coupled device (CCD) or complementary metal-oxide-semiconductor (CMOS) photosensor arrays. Techniques such as machine learning have been used to classify individual pixels into crop, weeds, or soil background categories based on the reflectance intensity values of each pixel. Photosensor arrays are sensitive to visible and NIR light, so by placing light filters on the sensors, light of specific wavelengths can be passed to the sensor. Color cameras, for example, have filters so that the intensity of red, green, and blue light is measured and communicated in different channels. Cameras that measure NIR light are also used as well as combinations of measurements of color and NIR.

Based on the sensor availability, algorithms using visible color reflectance differences were first investigated, using data acquired from color cameras. Various color indices such as the excess green index (ExG) or the hue channel of the hue-saturation-value (HSV) color space were found effective in enhancing vegetation pixels and distinguishing certain species of crop and weeds (Andújar et al., 2012; Foglia and Reina, 2006; Philipp and Rath, 2002; Tang et al., 2000). Since the contrast between plants and soil is strong at NIR wavebands (700–1400 nm), algorithms using NIR were investigated for discriminating plants from soil background. Algorithms using single-channel NIR cameras were developed to segment plants from soil (Brivot and Marchant, 1996; Chang et al., 2004). Additionally, the normalized difference vegetation index (NDVI), which is a function of both NIR and visible color reflectance, was investigated for vegetation pixel segmentation (Gerhards and Christensen, 2003; López-Granados et al., 2006; Torres-Sánchez et al., 2013). Studies were also reported about the use of photodetectors with NIR bands for weed detection in some commercialized selective chemical application systems such as the WeedSeeker (Andújar et al., 2011).

While methods using color and infrared reflectance are reliable for vegetation pixel segmentation, the classification of different plant species was challenging using such a limited number of channels, because of the similarity in light reflectance characteristics shared by most agronomic vegetation (Slaughter et al., 2008). Thus, the reflectance characteristics of a larger quantity of narrower wavebands have been investigated for distinguishing crop and weed plants. Specifically, algorithms using hyperspectral imaging (HSI) were investigated, in which the high data dimension was exploited to achieve higher correct classification rates comparing with only wide color and NIR bands (López-Granados, 2011; López-Granados et al., 2006; Zhang et al., 2012).

In general, one of the greatest potential advantages of the reflectance-based techniques introduced above is the robustness to partial occlusion compared to methods analyzing shapes. Additionally, these methods are normally less computationally intensive than shape-based methods. However, the greatest limitation of the reflectance-based methods is the shortage of reliable information. Reflectance information acquired under imperfectly controlled illumination (affected by the sunlight, especially) is subject to change or may even cause algorithms to fail with many factors, such as weather (e.g. sunny or cloudy) and time (e.g. morning and noon-time). Research is still needed to investigate the stability of classifiers using reflectance information in different field conditions and illumination conditions (Peteinatos et al., 2014). At the same time, introducing plant morphological characteristics has more potential to improve plant classification performance.

2.1.2 Biological morphology characteristics

Another category of weed plant perception methods focuses on biological morphology characteristics, such as the shapes of plant canopy and leaves in the projected two-dimensional (2D) image plane or in three-dimensional (3D) space, to discriminate plants of different species. These methods were mostly designed for applications such as selective spraying and intra-row weeding, in which crop or weeds need to be accurately detected and localized. Ground- and low-altitude aerial-based sensing are two main sensing platforms, on which adequate image spatial resolution can be provided. Techniques such as machine learning are applied to the extracted features to classify species of plants. Since the shapes of plant canopies and leaves are complex and varied, the most challenging task is the investigation of effective and robust descriptors to differentiate different crop species in images.

Historically, most computer vision-based weed perception systems have used light reflectance images as input. Segmentation is usually performed at the beginning to extract vegetation pixels in images. After segmentation, features representing the morphology of plant leaves and plant canopies for plant discrimination are extracted. Common morphological features include length, width, perimeter dimensions, roundness, circularity and convexity of plant leaves or plant canopy. They were reported effective in crop/weed identification (Dyrmann et al., 2018; Tang and Tian, 2008; Wu et al., 2007). General image feature descriptors, such as features from accelerated segment test (FAST), scale-invariant feature transform (SIFT) features, histogram of gradient (HOG), local binary pattern (LBP), and Gabor wavelet transformation, were also found effective in plant detection and discrimination, and robust to illumination variations (Bawden et al., 2017; dos Santos Ferreira et al., 2017; Tang et al., 2003).

However, using light reflectance sensors alone to obtain high discrimination accuracy is challenging unless the light is controlled and the plants are sparse. Since most spectral cameras are passive receivers of reflected light, they are dependent on the quality of the reflected light received. The spectral reflectance similarity of vegetation pixels can lead to difficulties in separating leaves or plants with occlusions, and shadow effects or saturation effects caused by uncontrolled illumination may affect the segmentation and feature extraction performance. There is also loss of structural information because the image scene is projected onto a 2D image plane. Incorrect plant identification results may be obtained using features extracted with such conditions.

3D shape features extracted from 3D point clouds were found promising in addressing some of the problems in plant identification associated with reflectance-based sensors alone. These features are more robust to external illuminance changes and shadow effects than those extracted from color images. Plant height, which is a discriminating parameter between crop and weeds at early crop growth stages, was found effective and beneficial in crop/weed segmentation and classification (Piron et al., 2011).

In the literature, stereovision, light detection and ranging (LiDAR) and time-of-flight (TOF) sensors were three types of sensors commonly used (Weiss et al., 2010). These different sensors have different trade-offs when applied to this perception problem.

Stereovision measures distance between the same real-world point in images acquired in multiple (often two) perspectives. Stereovision is biologically inspired by human binocular vision. To measure the distance between corresponding real-world image points in different images, correspondence algorithms are needed to find corresponding points. Stereovision has advantages of high image resolution, available color information and detailed textural information (Kise et al., 2005), but it is still affected by illumination and has a high computational cost to find correspondences in images (Tippetts et al., 2016). Jin and Tang (2009) demonstrated that 3D structural features extracted from stereovision images were promising in detecting individual corn plants of V2–V3 growth stages and estimating their center locations.

Active range measurement sensors such as semiconductor-based time-of-flight (TOF) and LiDAR sensors measure distance based on the time difference between transmission and reception of modulated light signals. The active sensing mode makes these range sensors more robust under varying outdoor lighting conditions. However, most LiDAR sensors have lower data sampling rates and resolution than TOF cameras, and TOF cameras are more sensitive to sunlight, since infrared light also exists in the ambient sunlight spectrum. In the example of the Deepfield BoniRob robot (Weiss and Biber, 2011), a LiDAR sensor was used to effectively detect and localize outdoor maize plants. Li and Tang (2018) used a TOF camera-based perception system for crop plant detection. 3D features such as point density, curvature, and normal vectors were extracted from the 3D point cloud to detect broccoli and green bean leaves.

Since the morphological features extracted from 2D images and 3D point clouds are complementary, the fusion of color and depth images have been investigated for crop and weed plant discrimination. Common sensors are calibrated color and depth sensor pairs (Herrera et al., 2012), or commercial RGB-D cameras such as Kinect (Microsoft, Redmond, Wash), which directly output registered RGB color and depth images. Algorithms were developed using the fusion of color and depth in different agricultural applications. Kusumam et al. (2017) developed an algorithm to detect mature broccoli heads using an RGB-D sensor for robotic harvesting and obtained a high detection rate. Andújar et al. (2016) employed a Kinect v2 sensor to estimate weed densities in corn fields, in which features such as plant height and canopy volume were used for the classification of crop and weeds. Gai et al. (2019) developed an algorithm to detect broccoli and lettuce crop plants using a Kinect v2 sensor for automated weeding and demonstrated the benefits of fusing color and depth information for plant segmentation.

Compared with algorithms using reflectance features, algorithms based on biological morphology features (especially the shape features) have a higher accuracy in discrimination of plant species with obvious shape differences. These algorithms need data with adequate spatial resolution, and they are more complex than reflectance feature-based algorithms and are thus more computationally intensive. The algorithms are more robust to illuminance changes. The disturbance from sunlight such as shadow effects and image saturation in outdoor applications can be alleviated by providing artificially controlled light sources. But occlusion, leaf damage or other visual 'defects' are the major challenges to algorithms using shape features (Slaughter et al., 2008).

2.2 Weed control mechanisms

While there are many approaches to weed control, most weed control machines typically employ either mechanical or chemical weed control methods. These approaches have been used in conventional mechanized agriculture for many years, and recently, they have been coupled with automation technology to either reduce inputs or exert more precise control of weed plants. While other weed control mechanisms exist, such as flaming, hot water or steam, or high voltage (Blasco et al., 2002), the adoption of these technologies has been low, and little research has been reported to automate them. Thus, this section will focus only on chemical and mechanical weed control mechanisms.

2.2.1 Selective chemical application

Selective spraying systems, sometimes called spot spraying systems, turn nozzles on or off based on what is at nozzle locations. In the case of weed control, once a selective chemical application system perceives the existence of weed plants, herbicide can be applied to the area surrounding the weed locations and not applied in areas where no weed plants exist. This selective chemical application strategy selectively switches nozzles on and off based on presence or absence of weeds and has been investigated for some time in the context of conventional spray application systems. There are some limited examples of commercialized products on conventional spraying platforms such as the WeedSeeker (Trimble, Sunnyvale, California), which uses photodetectors and an active light source to detect vegetation between crop rows. Machine vision-based selective patch spraying has been investigated for row crops that could deliver herbicide onto areas where weeds are present while travelling at typical operational speeds (Tang et al., 2000a; Steward et al., 2002). More recently, selective spraying system are emerging, often in the context of agricultural robotics.

There are also examples of work to increase the spatial resolution of the nozzle with the idea that the spray pattern is intended to be directed to individual plants. Lee et al. (1999), for example, developed and tested a prototype pulsed-jet, micro-dosing actuator capable of applying microliter herbicide dose rates to small target areas as small as 9 mm x 12.7 mm. This small target area was achieved with eight micro-spray nozzles each consisting of five hypodermics each forming a micro-spray nozzle array. Eight micro-spray nozzles were spaced horizontally at a 1.27-cm interval and were located at a 10.16-cm nozzle height (Lee et al., 1999). Similarly, Zhang et al. (2009) reported a system that applied heated oil to weeds using a micro-spray manifold in 6.3 mm by 12.5 mm rectangular areas. This high-spatial-resolution approach to selective chemical application should substantially reduce the volume of applied chemicals to the field over conventional low-resolution nozzles. It may also enable the use of nonselective herbicides to control weeds in the context of high-value vegetable crops.

Other examples of selective spraying include Zaman et al. (2011) who developed a selective sprayer which sensed tall weeds in blueberries using an ultrasonic sensor and selectively applied herbicide to the weeds. Wiedemann et al. (2002) described the development of a selective sprayer designed to control mesquite in road right-of-ways. Infrared light was transmitted horizontally to a receiver. When this beam was interrupted, the sprayer sensed the presence of a mesquite plant and correspondingly applied herbicide to the area nearby and containing the plant. Both of these examples demonstrate the concept of using basic sensing technology to perceive a highly distinguishable feature of weed plants to determine their presence. Additionally, there are examples of selective spraying for other pests (Shen et al., 2017; Zhu et al., 2017).

2.2.2 Robotic selective application

While selective herbicide application systems show tremendous promise in reducing the volume of chemicals applied (e.g. Lee et al., 1999), they by themselves should probably not be considered to be robotic, rather they are automated control systems that respond to perceived weeds in the field. They generally have not included much machine behavior layer technology to deal with uncertain situations. However, more recently, selective spraying technology has been coupled with smaller, automatically guided vehicle technology with hardware and control technology to more precisely apply herbicide. For example, the ecoRobotix spraying robot is autonomously guided through crop rows using GPS and machine vision sensors to follow crop rows. This robot detects weed plants, and then uses two spray nozzles on delta robot arms to position two nozzles over weed plants and selectively apply herbicide directly to detected weed plants (ecoRobotix, 2019; Fennimore and Cutulle, 2019; Fig. 3). The Australian Centre for Field Robotics' Ladybird robot is similar to the ecoRobotix robot, but used a six-axis robot arm (model UR5, Universal Robots, Odense, Denmark) to move a spray nozzle end effector to the weed plants (Bogue, 2016; Fig. 4).

2.2.3 Mechanical weeders

For situations where chemical weed control is not consistent with the producer's management practices, mechanical cultivation or tillage is often used. There is

Figure 3 The ecoRobotix robot senses weeds in the field and then moves spray nozzle to the weed to achieve goal of directly delivering herbicide to each weed plant. Source: ecoRobotix.

Figure 4 Ladybird robot by University of Sydney's Australian Centre for Field Robotics features a novel design covered with solar panels (left) and a selective spraying system in which the spray nozzle is moved close to weed plants by a six-axis robotic arm (right).

a wide variety of mechanical tool designs for mechanical cultivation that can be used with both interrow and intra-row weeds (Bowman, 1997; Bond et al., 2003). Generally, they rely on three main physical techniques for controlling weeds which include burying, cutting, and uprooting the weed plants (Bin Ahmad, 2012). Each of these techniques interferes with the growth of weed plants by killing them or slowing their growth so that the crop plants can overtake the weed plants achieving greater canopy closure and reducing the light interception by the weed plants. Mechanical cultivation is used widely in organic production systems and has been investigated with automated control.

A review of the research and commercial literature found that the majority (75%) of robotic weeders used mechanical over chemical weed control. Of the mechanical types of solutions, there were generally two main classes of approaches: (1) passive cultivation tools being automatically guided through crop rows, and (2) active control of weeding tools in the row and sometimes between the row.

Examples of automatically guided cultivation tools are Naïo Technologies Oz weeder and Dino robot (Naïo Technologies, 2019a,b). The Oz weeder is electrically powered with four electric motors, one in each wheel. It comes in a small package that is 60 cm high and 40 cm wide, weighing 110 kg. Because of its size, it fits between rows of vegetable crops and carries a variety of cultivation tools including several types of harrows or brush or spring weeders for intra-row weeding. The Dino weeding robot (Naïo Technologies, 2019b) is a larger platform with a 130 cm height and adjustable wheelbase from 140 to 180 cm. It is able to span 120 cm of row width, and guides different passive cultivation tools through multiple rows. The Kongskilde Robotti has a similar platform spanning multiple rows and is automatically guided to cultivate the crop rows that it spans (Bawden et al., 2014).

The systems that apply active control of mechanical tillage tools are typically using local scene information from the perception system to actively move the tools into and out of the crop row depending on where crop plants are located. Following this basic strategy, different systems have taken different approaches largely based on their mechanism design.

One common approach is to use a horizontal knife or blade hoe tool that is positioned shallowly below the soil. The blade is moved into the crop row in the spaces between the crop plants cutting and burying weed plants. An early implementation of this strategy was the Sarl Radis weeder developed in France (Van Der Weide et al., 2008; Cloutier et al., 2007). This automated weeder detected reflected light from the crop plants, and accordingly moved the hoes out of the row when a crop plant was detected. Originally developed for transplanted crops, it had highest performance when the weeds were substantially smaller than the crop plants. Similar realizations of this concept include the Robovator, Frank Poulsen Engineering ApS., Denmark; IC-Weeder, Steketee, the Netherlands; and Remoweed, Costruzioni Meccaniche Ferrari, Italy (Peruzzi et al., 2017).

Another approach to intra-row weeding is to use a rotating mechanism that is moved in and out of the crop row. Astrand and Baerveldt (2002) used a vertically oriented tool described as 'a rotating wheel that is rotated perpendicular to the row line.' The wheel was raised and lowered pneumatically in the presence and absence of crop plants. Gobor (2013) investigated the kinematics of a similarly oriented hoe mechanism that consisted of 'duck foot' hoes. The distance from the axis of rotation out to the hoes could be varied as well as the rotational speed of the system for flexibility in moving and positioning the tools below the surface near each crop plant. A prototype mechanism was tested under laboratory conditions in a soil bin, and the distance between the crop plant and the disturbed soil was measured. The location of disturbed soil was greater than 25 mm from the crop plants and typically less than 70 mm.

O'Dogherty et al. (2007) investigated the kinematics of a horizontally oriented disk that was pulled shallowly under the soil for intra-row weeding. The circular disk had a 130° section or notch removed from it. In the presence of a crop plant, the disk was rotated so that the notch was positioned to the location of the crop plant leaving it undisturbed while the soil between plants was disturbed. Field trials of the disk tillage tool with a computer vision system for crop plant detection found that crop damage levels were low and the system reduced intra-row weeds in transplanted cabbage (Tillett et al., 2008). This technology was commercialized under the trade name Robocrop (Garford Farm Machinery Ltd, England; Fig. 5).

There are other examples of novel intra-row weeding tools. A cycloid hoe concept was developed and reported on by Nørremark et al. (2008, 2012). The cycloid hoe consisted of eight tines that rotated around a vertical axis (Fig. 6). The hoe mechanism was designed so that individual tines could go into the row in the absence of a crop plant or be retracted when a crop plant was present.

Figure 5 Robocrop automated mechanical weeding system (left) that uses a disk weeding tool for intra-row weeding. The disk is rotated so that the notch in the disk goes around crop plants (right; Source: Tillett and Hague Technology Ltd).

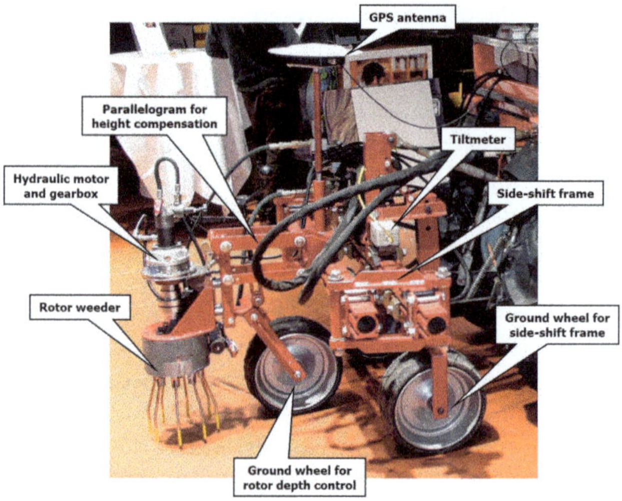

Figure 6 Cycloid hoe mechanical weeding mechanism in which the tines are directly on paths that avoid crop plants when doing intra-row weeding. Source: adapted from Griepentrog et al. (2006).

The system was tested with artificial plants and georeferenced crop plants and was found to be reliable in performing intra-row weeding without interfering with the crop plants. As an example of another mechanism, Langsenkamp (2014) described the development of a weed stamping tool. To injure weed plants, the tool is inserted 47 mm into the soil while centered on individual weed plants. The stamping tool was mounted on a delta robot arm that positioned the tool over detected weed plants. In field tests, 85% of the plants that were stamped died. Johari Jiken (2016) investigated the effectiveness of a rotating tine mechanism in disturbing simulated weed plants with tine depth and rotation speed as experimental factors.

These many clever approaches to mechanically control weeds in and between the crop rows represent substantial technical advances to this problem. Clearly a combination of crop and/or weed plant recognition, actuator control and automatic vehicle guidance are needed for a weeding robot (Slaughter et al., 2008). However, to obtain the autonomy expected by an automated weeding system that is considered robotic, higher levels of machine behavior decision-making is required, such as selection of weeding tool for the combination of crop, weeds, soil type, and soil condition (Merfield, 2016). In addition, mechanical weeding requires operator knowledge to make adjustments to the cultivation tool (Bond et al., 2003), so for a robot to be autonomous, it would need to also make cultivation tool adjustments. Additionally, weeding robots need to make decisions relative to weed control efficacy, but there is a gap here as the efficacy studies associated with weeding robots tend to be fairly limited.

3 Case study

Gai et al. (2019) described the development of a robotic weeder using a new weeding actuator design for mechanical weeding for row crops and multiple crop species. The actuator was designed to be used as a tractor implement mounted on a toolbar, but it could be integrated into a field robot as well. The weeder employs rotating vertical tines as the weeding tool for cutting, uprooting and burying weeds (Fig. 7). Each rotating tine group is positioned relative to the crop row using closed-loop control of servo-motor-driven pivoting arms. The rotating tines move in and out of the crop row based on the presence or absence of crop plants. The perception system employed an RGB-D sensor (Kinect v2, Microsoft, Redmond, Wash) to detect and localize crop plants using color and shape features in real time. After detecting and

Figure 7 The actuators (left) were designed to be used as a tractor implement (right), employing rotating vertical tines as the weeding tool for effectively cutting, uprooting and burying weeds. Source: adapted from Gai et al. (2019).

localizing crop plants, the tines were controlled to move close to the crop row to remove weed plants regardless of their species while avoiding crop plant disturbance and damage.

The perception system of this weeder used data from both color and range sensors and fused these data together for a high-performing crop plant perception system. The image processing pipeline consists of six steps including data preprocessing, vegetation pixel segmentation, plant extraction, feature extraction, feature-based localization refinement and crop plant classification (Fig. 8). In the preprocessing step, invalid pixels and noise pixels in point clouds were removed. In the segmentation step, the soil surface was modeled as a plane in the 3D point cloud and pixels above the plane were treated as outliers and as crop pixels. In the plant extraction step, plant pixels were clustered based on their spatial relationships to one another. In the feature extraction step, a set of reflectance and shape features for describing individual leaves and plant canopies was extracted. In localization refinement, venation features were used to calculate plant center positions. And in the classification step, the extracted features were used to perform crop-versus-weed classification for each extracted plant.

The mechanical weeders were controlled to follow trajectories based on the perceived location of the crop plant trajectories of the rotating tine units. Trajectories were planned which enabled cultivation as close to the crop plants as possible, without coming into contact with or disturbing the plants.

Figure 8 Sample broccoli plant images (16 DAT; 40k lux) at each image processing step including (a) depth and (b) color images; (c) color registration and filtered image; (d) segmented image, with white vegetation pixels; (e) detected plants marked with crosses; and (f) feature-based localization refinement and classification with target crop plants labeled with crosses. Source: adapted from Gai et al. (2019).

Because the interactions between the soil, crop plants, and tine mechanism are so important, a series of experiments were conducted in an indoor soil bin to better understand these interactions. Additionally, the weeding performance of the rotating vertical tine mechanism was studied using small wooden cylinders as simulated weed plants (Johari Jiken, 2016; Kshetri et al., 2019). These cylinders were inserted into the soil, then after the tine mechanism passed by, the location and orientation of each cylinder was observed. Potential weeding rate was analyzed for the rotating tine mechanism at different working soil depths and rotational speeds. Both depth and rotational speed had a significant effect on the potential weeding rate for the mechanism which was found to increase for higher levels of these parameters. Although the width of soil disturbance due to a cylindrical tine is affected by tine diameter and working soil depth, operating parameters such as increased longitudinal and rotational speeds have potential to damage a higher proportion of weed plants.

4 Summary

The work documented in the area of robotic weed management and control demonstrates the potential of automated weeding technology to substantially reduce inputs by treating fields on a much-reduced scale than conventional agriculture, that is, down to a plant or row scale. The following observations can be made about the current state of the art.

- Much work has been directed to the weed/crop plant perception problem. This perception problem has been particularly challenging given the semi-controlled nature of agricultural fields. Variability in lighting, density and species of weed plants, and occlusion of mixtures of plants are some of the main challenges associated with perception systems. Much progress has been reported using combinations of sensors and newer AI approaches.
- A variety of robotic weed control mechanisms have been explored for both chemical and mechanical weed control approaches. In many cases, mechanical weeders show potential to be efficacious under some conditions, but weed control performance studies have been limited.
- Examples of weeders that have machinery behavior technologies included in them are limited. Current robotic weeder examples are limited in being able to achieve goals in the context of variation and uncertainty. For example, there are no reported examples of robotic weeders being able to change tools or adjust tool control due to changing soil conditions.

While these limitations exist, the current rapid development rate of robotics technology will undoubtedly have an impact on the development of robotic weeding technology.

5 Future trends in research

5.1 Future trends in weed perception research

Given the advantages and limitations of the current state of technology for automated or robotic weeding, future research in perception systems will likely focus on either increasing data dimensions or increasing the complexity of algorithms for better weed perception performance. Data dimensions can be increased by fusing sensors from different perspectives or with different sensors that provide complementary information (Barrero and Perdomo, 2018; Gao et al., 2018; Shchez and Marchant, 2000). More complex image processing algorithm can be achieved using deep learning (DL) techniques.

From recent investigations, DL or convolutional neural networks (CNN) is an AI technology that shows promise in agricultural applications. DL uses complex image processing neural networks trained with large amounts of data. Kalimaris and Prenafeta-Boldú (2018) surveyed the literature for research using DL in agriculture and found 40 articles documenting research in this area. All but two of the papers were published during or after 2015. Seven papers focused on weed plant detection (Dyrmann et al., 2016a,b, 2017; Sørensen et al., 2017; McCool et al., 2017; Milioto et al., 2017; Potena et al., 2016).

Compared with traditional pattern recognition methods, higher performance has been reported for CNN methods in crop/weed discrimination and detection. A big advantage of DL is a reduced need to do manual 'feature engineering' to determine the distinguishing features of the plants which allow them to be assigned to the correct class. Because of the complexity of DL models, the technology is able to do feature learning and find distinguishing features in a problem. This capability to learn the plant features which can be used to distinguish crop plants from weed plants could open up the way for substantial advances in the weed perception area. CNN models such as Inception-v3 (Szegedy et al., 2015), GoogLeNet (Szegedy et al., 2014), DenseNet (Huang et al., 2017) and customized models were shown to be effective in crop/weed detection even with uncontrolled illumination and visual defects (Dyrmann et al., 2016a, 2017; McCool et al., 2017; Milioto et al., 2017; Potena et al., 2016).

However, there are some challenges associated with DL. First, DL requires high computational capacity for training and real-time inference. Second, the performance is highly dependent on the quality of the training set. The data set needs to be large enough, and correctly annotated, which usually requires substantial manual labor to collect and annotate images. To improve robustness, the data set must span all conditions such as inconsistent illumination, shadow and occlusion (Kamilaris and Prenafeta-Boldú, 2018). Thus, traditional pattern recognition pipelines are still worth investigation for applications for which it is impractical to have high computational capacity and large data sets. Nevertheless, DL should be a fertile ground for research in the near future.

In the area of weed control mechanism, there is much work to be done. As Merfield (2016) points out, there is a big need for research to investigate the interaction between weeding tools and the soil-weed-crop matrix. More knowledge in this area will enable weeding robots to have a higher degree of intelligence to the point where they can make decisions about the type of tools needed for specific applications or can adjust their strategy based on field conditions at any one point. While Merfield was commenting particularly on a mechanical weed control solution, similar comments and consideration should also be made about chemical weed control as the robotic solutions facilitate new possibilities for reducing application volumes. In all of this, enabling robots and their managers to make decisions in the face of uncertain field environments is wide open for future research.

6 Where to look for further information

More information on robotic weed management and control can be found in journals containing many of the papers cited in this chapter. These journals are often at the interface of engineering and weed science. Examples of such journals include *Transactions of the ASABE*, *Biosystems Engineering*, *Computers and Electronics in Agriculture*, *Precision Agriculture*, and *Weed Research*. The *Journal of Field Robotics* is not exclusively devoted to agricultural robots, but includes agricultural applications and contains many articles about crop plant and weed plant perception. *IEEE* journals, while extremely broad in their technical coverage, also have many technical conference papers on weed detection and classification-related topics. In terms of books, Young and Pierce edited the book *Automation: The Future of Weed Control in Cropping Systems* (2013), which is a good source of information on robotic weeding. This source was written for both the biologist and the engineer and seeks to inform both for solutions needed at this interdisciplinary crossroad. Similarly, Oerke et al. (2010) provides a comprehensive perspective on precision crop protection in the context of heterogeneous crop environments. While it has a broader view than just weed control, this volume does provide much information on weed control systems including those that are used for robotic applications.

7 References

Andújar, D., Ribeiro, Á., Fernández-Quintanilla, C., and Dorado, J. 2011. Accuracy and feasibility of optoelectronic sensors for weed mapping in wide row crops. *Sensors* 11(3), 2304–18. doi:10.3390/s110302304.

Andújar, D., Weis, M., and Gerhards, R. 2012. An ultrasonic system for weed detection in cereal crops. *Sensors* 12(12), 17343–57. doi:10.3390/s121217343.

Andújar, D., Dorado, J., Fernández-Quintanilla, C., Ribeiro, A. 2016. An approach to the use of depth cameras for weed volume estimation. *Sensors* 16(7), 972. doi:10.3390/s16070972.

Åstrand, B., and Baerveldt, A. J. 2002. An agricultural mobile robot with vision-based perception for mechanical weed control. *Autonomous Robots* 13(1), 21–35. doi:10.1023/A:1015674004201.

Barrero, O., and Perdomo, S. A. 2018. RGB and multispectral UAV image fusion for Gramineae weed detection in rice fields. *Precision Agriculture* 19(5), 809–22. doi:10.1007/s11119-017-9558-x.

Bawden, O., Ball, D., Kulk, J., Perez, T., and Russell, R. 2014. A lightweight, modular robotic vehicle for the sustainable intensification of agriculture. In: *Proceedings of the Australasian Conference on Robotics and Automation 2014*, Melbourne, Australia, 2–4 December. Available at: www.araa.asn.au/acra/acra2014/papers/pap147.pdf (accessed on January 31, 2019).

Bawden, O., Kulk, J., Russell, R., McCool, C., English, A., Dayoub, F., Lehnert, C., and Perez, T. 2017. Robot for weed species plant-specific management. *Journal of Field Robotics* 34(6), 1179–99. doi:10.1002/rob.21727.

Bin Ahmad, M. T. 2012. Development of an automated mechanical intra-row weeder for vegetable crops. Iowa State University Graduate Theses and Dissertations, 12278. Available at: lib.dr.iastate.edu/etd/12278.

Blasco, J., Aleixos, N., Roger, J. M., Rabatel, G., and Molto, E. 2002. AE–automation and emerging technologies: robotic weed control using machine vision. *Biosystems Engineering* 83(2), 149–57. doi:10.1006/bioe.2002.0109.

Bogue, R. 2016. Robots poised to revolutionise agriculture. *Industrial Robot: an International Journal* 43(5), 450–6. doi:10.1108/IR-05-2016-0142.

Bond, W., Turner, R. J., and Grundy, A. C. 2003. *A Review of Non-chemical Weed Management*. HDRA, the Organic Organisation, Ryton Organic Gardens, Coventry, UK, p. 81.

Bowman, G. (Ed.). 1997. *Steel in the Field: A Farmer's Guide to Weed Management Tools*. Sustainable Agriculture Network, Beltsville, MD.

Brivot, R., and Marchant, J. A. 1996. Segmentation of plants and weeds using infrared images. *Acta Horticulturae* 406(406), 165–72. doi:10.17660/ActaHortic.1996.406.16.

Chang, J., Clay, S. A., Clay, D. E., and Dalsted, K. 2004. Detecting weed-free and weed-infested areas of a soybean field using near-infrared spectral data. *Weed Science* 52(4), 642–8. doi:10.1614/WS-03-074R1.

Cloutier, D. C., Vanderweide, R. Y., Peruzzi, A., and Leblanc, M. L. 2007. Mechanical weed management. In: Upadhyaya, M. K. and Blackshaw, R. E. (Eds), *Non-Chemical Weed Management*. CAB International, Wallingford, Oxfordshire, England, pp. 111–34.

Dos Santos Ferreira, A., Matte Freitas, D., Gonçalves da Silva, G., Pistori, H., and Theophilo Folhes, M. 2017. Weed detection in soybean crops using ConvNets. *Computers and Electronics in Agriculture* 143, 314–24. doi:10.1016/j.compag.2017.10.027.

Dyrmann, M., Karstoft, H., and Midtiby, H. S. 2016a. Plant species classification using deep convolutional neural network. *Biosystems Engineering* 151, 72–80. doi:10.1016/j.biosystemseng.2016.08.024.

Dyrmann, M., Mortensen, A. K., Midtiby, H. S., and Jørgensen, R. N. 2016b. Pixel-wise classification of weeds and crops in images by using a fully convolutional neural network. In: *International Conference on Agricultural Engineering*, Aarhus, Denmark.

Dyrmann, M., Jørgensen, R. N., and Midtiby, H. S. 2017. RoboWeedSupport - detection of weed locations in leaf occluded cereal crops using a fully convolutional neural network. In: *11th European Conference on Precision Agriculture (ECPA)*, Edinburgh, Scotland.

Dyrmann, M., Christiansen, P., and Midtiby, H. S. 2018. Estimation of plant species by classifying plants and leaves in combination. *Journal of Field Robotics* 35(2), 202–12. doi:10.1002/rob.21734.

ecoRobotix. 2019. Switch to smart weeding. Available at: www.ecorobotix.com (accessed on May 11, 2019).

Fennimore, S. A., and Cutulle, M. 2019. Robotic weeders can improve weed control options for specialty crops. *Pest Management Science*. doi:10.1002/ps.5337.

Foglia, M. M. and Reina, G. 2006. Agricultural robot for radicchio harvesting. *Journal of Field Robotics* 23(6–7), 363–77. doi:10.1002/rob.20131.

Gai, J., Tang, L., and Steward, B. L. 2019. Automated crop plant detection based on the fusion of color and depth images for robotic weed control. *Journal of Field Robotics* (accepted).

Gao, J., Liao, W., Nuyttens, D., Lootens, P., Vangeyte, J., Pižurica, A., He, Y., and Pieters, J. G. 2018. Fusion of pixel and object-based features for weed mapping using unmanned aerial vehicle imagery. *International Journal of Applied Earth Observation and Geoinformation* 67, 43–53. doi:10.1016/j.jag.2017.12.012.

Gerhards, R. and Christensen, S. 2003. Real-time weed detection, decision making and patch spraying in maize, sugarbeet, winter wheat and winter barley. *Weed Research* 43(6), 385–92. doi:10.1046/j.1365-3180.2003.00349.x.

Gobor, Z. 2013. Mechatronic system for mechanical weed control of the intra-row area in row crops. *KI – Künstliche Intelligenz* 27(4), 379–83. doi:10.1007/s13218-013-0265-0.

Griepentrog, H., Nørremark, M., and Nielsen, J. 2006. Autonomous intra-row rotor weeding based on GPS. In: *CIGR World Congress*. Bonn, Germany, p. 7.

Hague, T., Marchant, J. A., and Tillett, N. D. 2000. Ground based sensing systems for autonomous agricultural vehicles. *Computers and Electronics in Agriculture* 25(1–2), 11–28. doi:10.1016/S0168-1699(99)00053-8.

Han, S. F., Steward, B. L., and Tang, L. 2015. Intelligent agricultural machinery and field robots. In: Zhang, Q. (Ed.), *Precision Agriculture for Crop Farming*. CRC Press, Boca Raton, FL, pp. 133–76.

Hassanein, M. and El-Sheimy, N. 2018. An efficient weed detection procedure using low-cost UAV imagery system for precision agriculture applications. In: *International Archives of the Photogrammetry, Remote Sensing and Spatial Information Sciences – ISPRS Archives* (vol. 42), pp. 181–7. doi:10.5194/isprs-archives-XLII-1-181-2018.

Herrera, C. D., Kannala, J., and Heikkilä, J. 2012. Joint depth and color camera calibration with distortion correction. *IEEE Transactions on Pattern Analysis and Machine Intelligence* 34(10), 2058–64. doi:10.1109/TPAMI.2012.125.

Huang, G., Liu, Z., Van Der Maaten, L., and Weinberger, K. Q. 2017. Densely connected convolutional networks. In: *Proceedings – 30th IEEE Conference on Computer Vision and Pattern Recognition (CVPR 2017)*, January 2017, pp. 2261–9. doi:10.1109/CVPR.2017.243.

IEEE. 2019. Robots. *IEEE spectrum*. Available at: robots.ieee.org/learn/ (accessed on February 7, 2019).

Jarvis, J. F. and Grant, E. 2014. Intelligent machine. Available at: www.accessscience.com/content.intelligent-machine/348225 (accessed on November 24, 2014).

Jin, J. and Tang, L. 2009. Corn plant sensing using real-time stereo vision. *Journal of Field Robotics* 26(6–7), 591–608. doi:10.1002/rob.20293.

Johari Jiken, J. B. 2016. Experimental approach to determine the efficacy of a tine mechanism for auto weeding machine. Iowa State University Graduate Theses and Dissertations, 15942. Available at: lib.dr.iastate.edu/etd/15942.

Kamilaris, A. and Prenafeta-Boldú, F. X. 2018. Deep learning in agriculture: a survey. *Computers and Electronics in Agriculture* 147, 70–90. doi:10.1016/j.compag.2018.02.016.

Kise, M., Zhang, Q., and Rovira Más, F. 2005. A stereovision-based crop row detection method for tractor-automated guidance. *Biosystems Engineering* 90(4), 357–67. doi:10.1016/J.BIOSYSTEMSENG.2004.12.008.

Kshetri, S., Steward, B. L., Jiken, J. J., Tang, L., and Tekeste, M. 2019. Investigating effects of interaction of single tine and rotating tine mechanism with soil on weeding performance using simulated weeds. *Transactions of ASABE* (under review).

Kusumam, K., Krajník, T., Pearson, S., Duckett, T., and Cielniak, G. 2017. 3D-vision based detection, localization, and sizing of broccoli heads in the field. *Journal of Field Robotics* 34(8), 1505–18. doi:10.1002/rob.21726.

Kyllo, K. P. 2003. *NASA Funded Research on Agricultural Remote Sensing*. Department of Space Studies, University of North, Dakota.

Lan, Y., Thomson, S. J., Huang, Y., Hoffmann, W. C., and Zhang, H. 2010. Current status and future directions of precision aerial application for site-specific crop management in the USA. *Computers and Electronics in Agriculture* 74(1), 34–8. doi:10.1016/J.COMPAG.2010.07.001.

Langsenkamp, F., Sellmann, F., Kohlbrecher, M., Kielhorn, A., Strothmann, W., Michaels, A., Ruckelshausen, A., and Kielhorn, A. 2014. Tube stamp for mechanical intra-row individual plant weed control. In: *18th World Congress of CIGR (CIGR2014)*, Beijing, China, pp. 16–9.

Lee, W. S., Slaughter, D. C., and Giles, D. K. 1999. Robotic weed control system for tomatoes. *Precision Agriculture* 1(1), 95–113. doi:10.1023/A:1009977903204.

Li, J., and Tang, L. 2018. Crop recognition under weedy conditions based on 3D imaging for robotic weed control. *Journal of Field Robotics* 35(4), 596–611. doi:10.1002/rob.21763.

López-Granados, F. 2011. Weed detection for site-specific weed management: mapping and real-time approaches. *Weed Research* 51(1), 1–11. doi:10.1111/j.1365-3180.2010.00829.x.

López-Granados, F., Jurado-Expósito, M., Peña-Barragán, J. M., and García-Torres, L. 2006. Using remote sensing for identification of late-season grass weed patches in wheat. *Weed Science* 54(2), 346–53. doi:10.1614/WS-05-54.2.346.

McCool, C., Perez, T., and Upcroft, B. 2017. Mixtures of lightweight deep convolutional neural networks: applied to agricultural robotics. *IEEE Robotics and Automation Letters* 2(3), 1344–51. doi:10.1109/LRA.2017.2667039.

Merfield, C. N. 2016. Robotic weeding's false dawn? Ten requirements for fully autonomous mechanical weed management. *Weed Research* 56(5), 340–4. doi:10.1111/wre.12217.

Milioto, A., Lottes, P., and Stachniss, C. 2017. Real-time blob-wise sugar beets vs weeds classification for monitoring fields using convolutional neural networks. In: *Proceedings of the International Conference on Unmanned Aerial Vehicles in Geomatics*, Bonn, Germany.

Moran, M. S., Inoue, Y., and Barnes, E. M. 1997. Opportunities and limitations for image-based remote sensing in precision crop management. *Remote Sensing of Environment* 61(3), 319–46. doi:10.1016/S0034-4257(97)00045-X.

Naïo Technologies. 2019a. Oz weeding robot. Available at: www.naio-technologies.com/en/agricultural-equipment/weeding-robot-oz/ (accessed on January 30, 2019).

Naïo Technologies. 2019b. Autonomous vegetable weeding robot – Dino. Available at: www.naio-technologies.com/en/agricultural-equipment/large-scale-vegetable-weeding-robot/ (accessed on January 30, 2019).

Nørremark, M., Griepentrog, H. W., Nielsen, J., and Søgaard, H. T. 2008. The development and assessment of the accuracy of an autonomous GPS-based system for intra-row mechanical weed control in row crops. *Biosystems Engineering* 101(4), 396–410. doi:10.1016/j.biosystemseng.2008.09.007.

Nørremark, M., Griepentrog, H. W., Nielsen, J., and Søgaard, H. T. 2012. Evaluation of an autonomous GPS-based system for intra-row weed control by assessing the tilled area. *Precision Agriculture* 13(2), 149–62. doi:10.1007/s11119-011-9234-5.

O'Dogherty, M. J., Godwin, R. J., Dedousis, A. P., Brighton, J. L., and Tillett, N. D. 2007. A mathematical model of the kinematics of a rotating disc for inter- and intra-row hoeing. *Biosystems Engineering* 96(2), 169–79. doi:10.1016/j.biosystemseng.2006.10.008.

Oerke, E. C., Gerhards, R., Menz, G., and Sikora, R. A. 2010. *Precision Crop Protection-the Challenge and Use of Heterogeneity* (vol. 5). Springer, Dordrecht, Netherlands.

Peruzzi, A., Martelloni, L., Frasconi, C., Fontanelli, M., Pirchio, M., and Raffaelli, M. 2017. Machines for non-chemical intra-row weed control in narrow and wide-row crops: a review. *Journal of Agricultural Engineering* 48(2), 57–70. doi:10.4081/jae.2017.583.

Peteinatos, G. G., Weis, M., Andújar, D., Rueda Ayala, V., and Gerhards, R. 2014. Potential use of ground-based sensor technologies for weed detection. *Pest Management Science* 70(2), 190–9. doi:10.1002/ps.3677.

Philipp, I. and Rath, T. 2002. Improving plant discrimination in image processing by use of different colour space transformations. *Computers and Electronics in Agriculture* 35(1), 1–15. doi:10.1016/S0168-1699(02)00050-9.

Piron, A., van der Heijden, F., and Destain, M. F. 2011. Weed detection in 3D images. *Precision Agriculture* 12(5), 607–22. doi:10.1007/s11119-010-9205-2.

Potena, C., Nardi, D., and Pretto, A. 2016. Fast and accurate crop and weed identification with summarized train sets for precision agriculture. In: *International Conference on Intelligent Autonomous Systems*, Shanghai, China, pp. 105–21. doi:10.1007/978-3-319-48036-7_9.

Rzevski, G. 2003. On conceptual design of intelligent mechatronic systems. *Mechatronics* 13(10), 1029–44. doi:10.1016/S0957-4158(03)00041-2.

Shchez, A. J. and Marchant, J. A. 2000. Fusing 3D information for crop/weeds classification. In: *Proceedings of the 15th International Conference on Pattern Recognition (ICPR-2000)* (vol. 4). IEEE Computer Society Press, Los Alamitos, CA, pp. 295–8.

Shen, Y., Zhu, H., Liu, H., Chen, Y., and Ozkan, E. 2017. Development of a laser-guided embedded-computer-controlled air-assisted precision sprayer. *Transactions of the ASABE* 60(6), 1827–38. doi:10.13031/trans.12455.

Slaughter, D. C., Giles, D. K., and Downey, D. 2008. Autonomous robotic weed control systems: a review. *Computers and Electronics in Agriculture* 61(1), 63–78. doi:10.1016/j.compag.2007.05.008.

Sørensen, R. A., Rasmussen, J., Nielsen, J., and Jørgensen, R. 2017. Thistle detection using convolutional neural networks. In: *2017 EFITA WCCA Congress*, Montpellier, France, p. 161.

Steward, B. L., Tian, L. F., and Tang, L. 2002. Distance-based control system for machine vision-based selective spraying. *Transactions of the ASAE* 45(5), 1255–62.

Szegedy, C., Liu, W., Jia, Y., Sermanet, P., Reed, S., Anguelov, D., Erhan, D., Vanhoucke, V., and Rabinovich, A. 2014. Going deeper with convolutions. In: *Proceedings of the IEEE Conference on Computer Vision and Pattern Recognition*, pp. 1-9.

Szegedy, C., Vanhoucke, V., Ioffe, S., Shlens, J., and Wojna, Z. 2015. Rethinking the inception architecture for computer vision. In: *Proceedings of the IEEE Conference on Computer Vision and Pattern Recognition*, pp. 2818–26.

Tang, L. and Tian, L. F. 2008. Plant identification in mosaicked crop row images for automatic emerged corn plant spacing measurement. *Transactions of the ASABE* 51(6), 2181-91. doi:10.13031/2013.25381.

Tang, L., Tian, L. F., and Steward, B. L. 2000a. Machine vision-based high-resolution weed mapping and patch-sprayer performance simulation. *SAE Transactions - Journal of Commercial Vehicles* 108(2), 317–26.

Tang, L., Tian, L. F., and Steward, B. L. 2000b. Color image segmentation with genetic algorithm for in-field weed sensing. *Transactions of the ASAE* 43(4), 1019–27. doi:10.13031/2013.2970.

Tang, L., Tian, L. F., and Steward, B. L. 2003. Classification of broadleaf and grass weeds using Gabor wavelets and an artificial neural network. *Transactions of the ASAE* 46(4), 1247–51. doi:10.13031/2013.13944.

Tillett, N. D., Hague, T., Grundy, A. C., and Dedousis, A. P. 2008. Mechanical within-row weed control for transplanted crops using computer vision. *Biosystems Engineering* 99(2), 171–8. doi:10.1016/j.biosystemseng.2007.09.026.

Tippetts, B., Lee, D. J., Lillywhite, K., and Archibald, J. 2016. Review of stereo vision algorithms and their suitability for resource-limited systems. *Journal of Real-Time Image Processing* 11(1), 5–25. doi:10.1007/s11554-012-0313-2.

Torres-Sánchez, J., López-Granados, F., De Castro, A. I., and Peña-Barragán, J. M. 2013. Configuration and specifications of an unmanned aerial vehicle (uav) for early site specific weed management. *PLOS ONE* 8(3), e58210. doi:10.1371/journal.pone.0058210.

Van der Weide, R. Y., Bleeker, P. O., Achten, V. T. J. M., Lotz, L. A. P., Fogelberg, F., and Melander, B. 2008. Innovation in mechanical weed control in crop rows. *Weed Research* 48(3), 215–24. doi:10.1111/j.1365-3180.2008.00629.x.

Weiss, U. and Biber, P. 2011. Plant detection and mapping for agricultural robots using a 3D LIDAR sensor. *Robotics and Autonomous Systems* 59(5), 265–73. doi:10.1016/j.robot.2011.02.011.

Weiss, U., Biber, P., Laible, S., Bohlmann, K., and Zell, A. 2010. Plant species classification using a 3D LIDAR sensor and machine learning. In: *Proceedings of the 9th International Conference on Machine Learning and Applications (ICMLA 2010)*, pp. 339–45. doi:10.1109/ICMLA.2010.57.

Wiedemann, H. T., Ueckert, D. N., and McGinty, W. A. 2002. Spray boom for sensing and selectively spraying small mesquite on highway rights-of-way. *Applied Engineering in Agriculture* 18(6), 661–6.

Wu, S. G., Bao, F. S., Xu, E. Y., Wang, Y.-X., Chang, Y.-F., and Xiang, Q.-L. 2007. A leaf recognition algorithm for plant classification using probabilistic neural network. In: *International Symposium on Signal Processing and Information Technology*, pp. 11–6. doi:10.1109/ISSPIT.2007.4458016.

Young, S. L. and Pierce, F. J. 2013. *Automation: the Future of Weed Control in Cropping Systems*. Springer Science and Business Media, Dordrecht, the Netherlands.

Zaman, Q. U., Esau, T. J., Schumann, A. W., Percival, D. C., Chang, Y. K., Read, S. M., and Farooque, A. A. 2011. Development of prototype automated variable rate sprayer for real-time spot-application of agrochemicals in wild blueberry fields. *Computers and Electronics in Agriculture* 76(2), 175–82. doi:10.1016/j.compag.2011.01.014.

Zhang, C., Staab, E. S., Slaughter, D. C., Giles, D. K., and Downey, D. 2009. Precision automated weed control using hyperspectral vision identification and heated oil. ASABE Paper No. 096365. St. Joseph, MI.

Zhang, Y., Staab, E. S., Slaughter, D. C., Giles, D. K., and Downey, D. 2012. Automated weed control in organic row crops using hyperspectral species identification and thermal micro-dosing. *Crop Protection* 41, 96–105. doi:10.1016/j.cropro.2012.05.007.

Zhu, H., Rosetta, R., Reding, M. E., Zondag, R. H., Ranger, C. M., Fulcher, A., Derksen, R. C., Ozkan, H. E., Krause, C. R., and Canas, L. 2017. Validation of laser-guided variable-rate sprayer for managing insects in ornamental nurseries. *Transactions of the ASABE* 60(2), 337–45. doi:10.13031/trans.12020.

Chapter 9

The use of agricultural robots in orchard management

Qin Zhang and Manoj Karkee, Washington State University, USA; and Amy Tabb, USDA-ARS, USA

1 Introduction

The use of robotic or automated machines in orchard operations is associated primarily with insufficient labour availability and rapidly increasing labour costs in tree fruit production and is critical for improving yield of high-quality fruit with minimal dependence on seasonal human labour. Primarily, mechanized or robotic orchard management operations (after establishment of the trees) include pruning, thinning, spraying and harvesting.

Pruning is an operation to grow fruit trees into a desired shape for improving sunlight penetration into the canopies, supporting more effective (e.g. more human- and machine-friendly) orchard operations, and regulating location and amount of fruiting sites that can lead to improved yield and quality of fruit. Pruning is physically cutting and removing parts of a tree, such as cutting branches back and/or removing small limbs entirely following some guidelines. This operation requires workers to have a specific level of knowledge and skill to perform the job well. Pruning is performed both in the winter (also called 'dormant pruning') and in the summer. Currently, research and development of robotic pruning is focussed primarily on dormant

http://dx.doi.org/10.19103/AS.2019.0056.14

pruning, therefore, we will focus on dormant pruning for the remainder of this chapter.

Fruit trees' bloom often includes many more flowers than are required for producing desired yield and quality of a commercial crop. Crop thinning is a practice to manage/control crop load by removing a portion of the bloom and/or young fruit on the tree either selectively or non-selectively to improve the size and quality of the remaining fruit. Thinning is commonly performed during and shortly after the bloom period to reduce the number of set fruit using either chemical thinners or physical means. However, thinning can occur later in the season to remove poor quality fruit or those growing too close together to ensure good quality and size of the remaining ones using only physical means. Currently research and development of robotic thinning technologies is more focussed on blossom thinning, and therefore, we will focus on this aspect of thinning in this chapter.

Pest control is an essential operation in orchard management to protect the yield and quality of fruit from various types of pests. One solution to achieve effective pest control is to spray recommended pesticides at the correct time and at correct rate on the target area. Due to the confined environment in tree fruit orchards, spraying pesticides is inherently hazardous for the operators driving a sprayer. Robotic sprayers are probably the best solution to ensure operator safety and efficiency in pest control.

Fruit harvesting for fresh market consumption requires a large seasonal semi-skilled workforce which has created a critical labour-related risk of not having enough workers to perform this time-sensitive task. In addition, harvesting labour is already the most significant variable cost in tree fruit production. For example, labour represented 46% of variable production costs on a full production year of 'Gala' apples (Gallardo et al., 2009). While vast efforts have been directed to developing robotic solutions for fresh market fruit harvesting, the technology has achieved limited success, primarily due to inadequate *harvesting speed, efficiency* and *harvesting-induced fruit damage*. Research and development by both the public and private sectors continue to attempt to fill the gap of having a satisfactory solution for harvesting tree fruit crops using robotic machines.

Robotic management will only make operational and economic sense when all or most of the field operations are completed by means of corresponding robotic machines (or multipurpose robotic machines) giving producers minimal dependence on seasonal human labour. As a complete harvesting process of tree fruit crops includes picking the fruit from the trees and loading it into bins, then transporting the filled bins out of orchards, robotic harvesting should include robotic transportation of the bins to complete the process.

The following sections of this chapter will provide an overview of robotic technologies for major tree fruit production tasks.

2 Robotic pruning

As briefly discussed in Section 1, pruning involves cutting away of tree branches, and is performed for a diversity of horticultural and economic reasons, including to balance the vegetative and reproductive growth of the plant, constrain the plant size, manipulate the canopy resulting in good fruit quality, size and yield, and set the optimal crop load for the next season (Schupp et al., 2017). Since fruit trees are perennial, pruning is a cumulative process, with the results of one year's pruning influencing the size and production of the plant for years to come.

Within fruit species, there are differences with regard to where reproductive flower buds, and consequently, fruit, will appear. As a result, pruning protocols differ by species. For instance, apple trees will flower and fruit on the last season's growth (called one-year-old wood). However, peach will only flower and fruit on growth that is two years old at the time of harvest. These differences mean that any autonomous system for pruning must take into account species-dependent differences so that the next season's crop is not inadvertently pruned off.

Trellis systems and tree training systems are also a major factor driving how pruning is accomplished in practice. We will not go into the detail of individual systems here, but briefly mention that we will focus on systems that aim to automate selective pruning, where individual branches are precisely removed from the central trunk. Hedging, or mass removal via cutting bars, will not be covered, though it is sometimes an operation performed in orchards or vineyards.

Robotic pruning as interpreted in the literature thus far consists of perception of the shape of the tree, implementation of the pruning protocol, robotic navigation and cutting of the branch. Established computer vision techniques for estimating object shape with high accuracy typically rely on objects with easily identifiable features, such as corners or edges, a consistent or predictable image acquisition environment or object smoothness. Dormant pruning occurs during the period when the trees are without leaves, the object is thin and the data is acquired outside. These characteristics violate many of the assumptions of established techniques. For these reasons, new methods have been developed for perceiving tree canopies. Finally, the spacing of trees and permanent trellis structures restrict visual access to the tree canopy, and images can usually be acquired only from one side of the tree at a time.

While not autonomous, a motion tracker that works via the relative displacement of an electromagnetic field was used by Sinoquet et al. (1997) to model walnut trees and by Vougioukas et al. (2016) to model pear trees for automated harvest. The sensor can be used to record three-dimensional (3D) positions in space at branch junctions, which can be used to sparsely represent tree shape.

Time-of-flight (ToF) of light-based sensors have been most frequently used for sensing tree shape. Karkee and Adhikari (2015) described the use of a ToF camera to capture apple tree shape. The sensor has a small field of view, so a pan-and-tilt stage is used to acquire multiple images, which are combined in a data pre-processing step. Image processing and skeletonization steps are performed, resulting in the tree structure of trunk and primary branches. This system is also used in Karkee et al. (2014) to implement a pruning protocol based on branch spacing and branch length.

The Microsoft Kinect sensor is a commodity ToF sensor used in a range of vision tasks, and was employed by Elfiky et al. (2015) and Akbar et al. (2016b) for reconstructing apple tree shape as well. Approaches range from using the KinectFusion API (Izadi et al., 2011) that comes with the Microsoft Kinect 2 for surface reconstruction, and then skeletonization and feature extraction steps as in Elfiky et al. (2015), to an approach that estimates the skeleton and branch diameter features from the multiple depth images, which are then merged together. In Akbar et al. (2016a), one Kinect depth image was used to estimate branch diameters. A dataset containing Kinect data from apple trees was described in Akbar et al. (2016b).

Medeiros et al. (2017) used a laser detection and ranging sensor, or light detection and ranging (lidar), for estimating tree shape for a pruning application. In that work, a two-dimensional lidar unit was mounted on a rotary and linear stage to acquire large depth maps of the scene. Tree regions were modelled as cylinders, which were then used to extract various geometric features of the canopies including diameters of the primary branches.

A shape from inconsistent silhouette approach was used by Tabb (2013) and Tabb and Medeiros (2017) to reconstruct the shape of fruit trees. Colour cameras were used to acquire images of apple trees in an outdoor environment. Because automatic segmentation of the scene between tree and non-tree regions is required, a background unit was used to block the appearance of other rows of trees. While in Tabb (2013) multiple cameras were used in laboratory conditions, in Tabb and Medeiros (2017) a robot was used for data acquisition and is shown in Fig. 1. Features computed include branch segment diameter, length, angle with respect to parent and branch segment position.

Many research articles surveyed above estimated the branch diameter feature, some also computed the branch angle feature (Karkee et al., 2014; Karkee and Adhikari, 2015; Tabb and Medeiros, 2017). A subject treated in Schupp et al. (2017) concerns the development and testing of a quantifiable protocol for robotic pruning. The protocol, called the limb-to-trunk ratio (LTR), consisted of summing the cross-sectional area of all primary branches, and dividing this sum by the cross-sectional area of the trunk at a certain distance from the graft union. To achieve a target LTR, an autonomous system would select the branch with the largest cross-sectional area to remove, remove it, recalculate LTR and repeat, until the target is achieved.

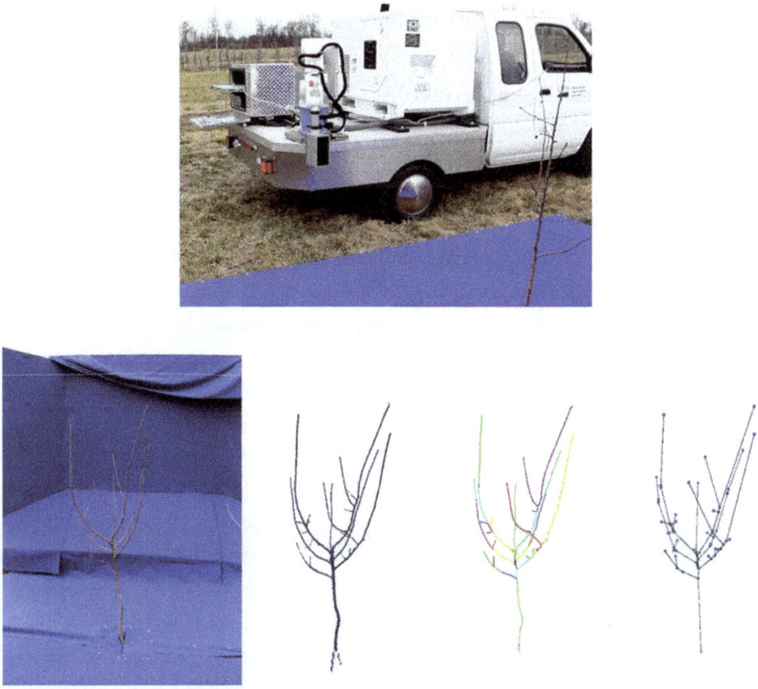

Figure 1 Top row: Robotic vision system for estimating shape features of trees. Bottom row: image of a tree, reconstruction generated by a system, curve skeleton and graph representation, all used to compute shape features. Source: Adapted from Tabb and Medeiros (2017).

While the perception systems necessary for robotic pruning have been in development under various groups around the world, a major bottleneck to real-time implementation is computational time. For example, Medeiros et al. (2017) reported that data acquisition time was an hour for two trees. In Tabb and Medeiros' system (2017), the data acquisition time was 1.5 min per tree, and on average the run time was 8.5 min for small trees using a high-performance computer. None of these research activities considered navigation of a robot to cutting points. Further research and development is necessary for autonomous pruning to be a future reality.

3 Robotic thinning

Fruit trees produce many more flowers than are required for a commercial crop with desired yield, fruit size and fruit quality. Thinning of flowers is performed to regulate the number of fruit that will set, and as a result, influences fruit size and quality. Thinning is also required to offset the tendency of some varieties of fruit crops towards alternate year bearing. The current practice is to thin the crop with chemicals, or manually. When chemical thinning, growers, extension specialists and crop consultants use flowering counts to schedule thinning sprays as a component of crop load management. In apples, two models used in predicting thinning sprays require flowering as input (Yoder et al., 2013).

The small size of the target, the flowers, relative to the large size of the canopy poses some challenges for implementing robotics and automation solutions to this orchard task. Robotic and automated thinning has so far focussed on flower thinning, including introducing elements of autonomy to mass removal systems such as string thinners with the development of robotic thinning arms and end effectors. There also has been work on detecting flowers in colour or multispectral images for two purposes: precise flower estimates as inputs to thinning models and 3D estimation of canopy and flowers for robotic thinning.

String thinners were developed in the 2000s to mass remove flowers in high-density fruit production systems (Baugher et al., 2010). String thinners are composed of a rotating spindle with plastic strings; the system is attached to a tractor which navigates through the rows in orchards. As the rotating spindle and strings pass by trees, flowers are removed. The angle of the string thinner, as well as the angular speed (revolutions per minute), can be adjusted to control the thinning efficacy. To develop sensing systems for automating some aspects of the operation, Aasted et al. (2011) used lidar to sense the canopy location for positioning the thinner appropriately in tree canopies trained in perpendicular V configurations. Gebbers et al. (2013) used a stereo camera to assess flower density on trees to support automated adjustments of thinner location and its spinning rate in real-time for optimal thinning in apple trees.

Computer vision approaches for detecting flowers in two-dimensional colour images of fruit trees are concerned with either applications of chemical thinning, or as a module of robotic thinning. Some of these approaches will be discussed again in Section 4 in the context of robotic spraying. For such applications, images were acquired using hand-held or ground-based vehicles. One of the first efforts in this category, Aggelopoulou et al. (2011), aimed to adjust chemical thinning rates at the block level for a specific fruit yield in apples. Colour images of trees while flowering were collected, as well as yield information, and correlation established between these two entities. The motivation in Hočevar et al. (2014) was orientated towards tree-specific management in apples, using tractor-mounted colour cameras. They performed thresholding and morphological operations to estimate the number of flower clusters in acquired images. As part of a larger study, Underwood et al. (2016) detected almond flowers from colour images using thresholding. Dias et al. (2018a,b) detected apple and peach flowers from colour images using convolutional neural networks (CNNs) and illustrated in

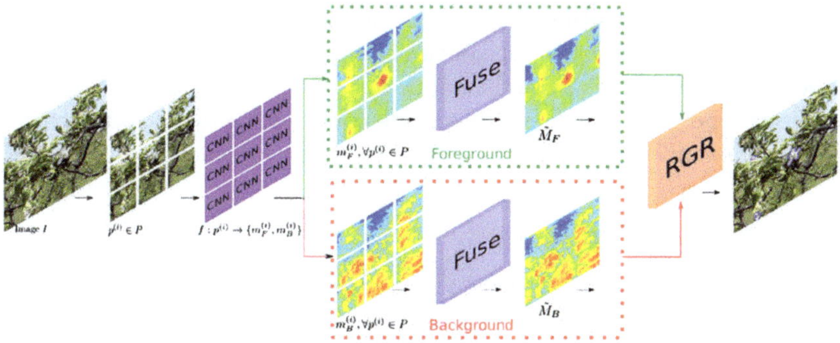

Figure 2 Schematic of a flower classification vision system from Dias et al. (2018b). Input images are divided and a CNN classifies pixels as flower class (foreground) or not (background).

Fig. 2. Tabb and Medeiros (2018) and Dias et al. (2018c) are two public datasets with colour images of fruit tree flowers, the latter with ground truth data, for experimentation by the community.

Some work made use of multispectral image sensors to detect flowers. For example, Horton et al. (2017) used multispectral information in the colour and near infrared range captured using an unmanned aerial system (UAS) to detect flowers using contrast enhancement and thresholding methods in peach orchards. In peaches, flowers are visible before leaves, which makes an aerial survey approach feasible. Wouters et al. (2015) explored multispectral imaging as well for detecting many stages of pear flowers from a mobile, ground-based unit, with wavebands from the visible and near infrared range. Imaging occurred at night with artificial illumination, and canonical correlation analysis was used to create models to detect flower parts.

Robotic systems for flower removal will require knowledge of the canopy structure as well as the location of flowers. There have been efforts to investigate the 3D reconstruction of canopies and flowers, as a component of a robotic thinning system, using the perpendicular V training system in peaches. Emery et al. (2010) developed a method to detect simplified peach canopies via structured light in a laboratory setting, and Nielsen et al. (2012) used a specialized stereo reconstruction technique, combined with flash illumination and night-time operation, to reconstruct peach canopy locations and map flower locations in more realistic outdoor situations.

Other efforts in this area were concerned with the development of the robotic components for flower thinning. Yang (2012) in a master's thesis developed and evaluated a one-fourth-scale table-top robotic arm for the manoeuvrability necessary for thinning. Lyons et al. (2015) used a similar arm as Yang (2012) and demonstrated a brush end effector for peach thinning that kept the branch damage at a minimal level.

Even when chemical thinning is performed, green fruit thinning is necessary and done manually in many fruit species after the threat of frost is past, and in some cases with a mechanical thinner (Miller et al., 2011). However, there are no efforts reported in the literature to date concerning robotic approaches to the green fruit thinning problem, neither green fruit detection nor its removal.

4 Robotic spraying

As discussed in Section 1, spraying agrochemicals on orchard crops is one of the most important field operations to protect crops from various types of pests and diseases and to provide desired nutrients and other inputs such as plant growth regulators. Precise application of chemicals (i.e. right amount of spraying at the right location and at the right time, sometimes referred to as 3R) is essential to make sure inputs are applied effectively to achieve desired efficacy. Chemical application is an operation with a risk for adversely affecting human and environmental health, so health and safety regulations, such as limit on chemical residue levels on produce and the limit of chemicals movement away into neighbouring fields, atmosphere and/or water bodies must be considered. Therefore, precision spraying has been and will continue to be a crucial technology in tree fruit production. With rapidly advancing computational power and sensing technologies, and revolutionizing artificial intelligence techniques such as CNNs or deep learning, robotic systems show promise for spraying agrochemicals on orchard crops with a higher level of precision than previously achieved.

Orchard spraying could use both ground and aerial robotics (e.g. targeted site-specific chemical application with UASs) in achieving the desired level of precision. In addition to applying agrochemicals, precision spraying has also been investigated for artificial (also called mechanical) pollination of orchard crops. A robotic spraying system, in general, consists of a target sensing system, and a spraying control system to accurately and precisely deliver the right amount of materials to targeted areas in the canopy. In the following several subsections, recent advancements in various components and aspects of robotic spraying of orchard crops will be discussed.

For a targeted chemical application system, which attempts to match chemical application to canopy characteristics at different sections of the canopies without regard to individual canopy objects or parts, non-imaging sensors such as ultrasonic, infrared and lidar sensors have been used. These sensors can characterize canopies in terms of canopy depth, size, shape and foliage density at a relatively coarse resolution (depending on the number of sensors or scanning resolution used). For example, Jeon et al. (2011) used infrared sensors to detect canopy depth and other canopy parameters such as canopy volume. As illustrated in Fig. 3, the sprayer could therefore adjust the

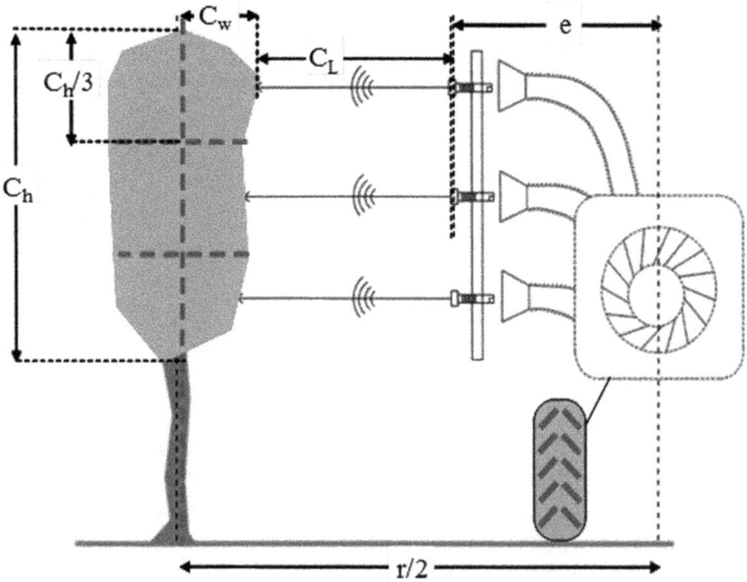

Figure 3 Ultrasonic sensing of various parameters for variable application of canopy inputs; r = inter-row spacing in orchards, C_h = canopy height, C_w = canopy depth, C_L = distance to canopy surface from sensor and e = distance to centre of the row from sensor. Source: Adapted from Jeon et al. (2011).

amount of chemicals applied to crop canopies in terms of the sensed canopy parameters to achieve a satisfactory result. Liu and Zhu (2016) tested a laser scanner for measuring the shape and size of various types of tree canopies and achieved a coefficient of variance of up to 6.8% in measuring the tree height and other dimensions.

If the precision needed is higher or inputs are desired to be targeted to specific canopy objects (e.g. flower, fruit, leaves, branches, trunks) rather than broadcasting to a specific canopy region, a machine vision system with the capability to detect and locate individual canopy objects would be necessary. For example, robotic pollination of cherry flowers requires accurate detection and localization of individual flower stigmas. Colour cameras, multi- and hyper-spectral cameras and thermal cameras are some of the commonly used sensors for detecting canopy objects/parts whereas stereo vision cameras and laser scanners are the commonly used sensors for estimating 3D locations of objects. For example, Aggelopoulou et al. (2011) processed colour images to estimate the number of flowers in apple trees and found it could be correlated with fruit yield with a correlation coefficient (*r*) of 0.78. A machine vision system was developed by Braun et al. (2010), constructing composite images from a sequence of colour images and using a Bayesian classification to segment out crop foliage to determine spatial distribution of foliage in grapevines. In addition, contours and edges of flower parts (e.g. petals) have been used in detecting flowers in tree canopies (Hong and Choi, 2012). Multispectral imaging has also been investigated for detecting various parts of the canopies.

These machine vision systems utilizing geometric, colour and spectral features of crop canopies suffer from the challenges of variable lighting, uncertainty of field environments, inherent biological variability of canopies, occlusion of objects and variation in object shape, size, colour and other parameters. To overcome some of these challenges in processing orchard images, deep learning techniques have been investigated widely in the last decade. Different variations of CNNs such as Faster R-CNN (Ren et al., 2017) have been used leading to improved accuracy and robustness in detecting desired canopy objects and/or regions (Ahn et al., 2018).

As also discussed in Section 3, 3D laser ranging and stereo vision systems have been used widely for localizing flower, fruit and other parts of the canopies in orchard environments (Emery et al., 2010; Aasted et al., 2011). Berenstein et al. (2010) used statistical measures, learning and shape matching to distinguish and localize foliage and grape clusters in grape canopies and achieved an accuracy of 90% in distinguishing grape clusters from foliage. Dey et al. (2012) used structures-from-motion to create 3D structure of grapevine canopies, which were then used to detect and locate various parts of grapevines including leaves, branches and fruit. As mentioned in Section 3, Nielsen et al. (2012) developed a sensor system to map the 3D structure of blooming peach trees, which was used to distinguish flowers from tree limbs using stereo vision and colour images. They achieved an accuracy of <1 cm in locating flowers in the trees. Similarly, Gebbers et al. (2013) described a stereo vision system that used colour contrast between blossoms, leaves and branches to detect and locate these canopy objects.

In many of these studies, two or more sensors were used for object detection and localization, which requires precise calibration and registration to project location information estimated by a 3D sensor onto objects detected using a 2D image. In recent years, commodity RGB-D (Red, Green, Blue-Depth) sensors, such as the Microsoft Kinect and Intel RealSense, have been increasingly popular in a variety of settings because of their low cost and the public availability of software development kits (SDKs). These sensors have the depth and colour hardware components pre-registered to one another. Some groups have started using these types of sensors in orchard settings. Xiao et al. (2017) used a commodity RGB-D sensor to detect foliage using colour information and to estimate leaf area density and average distance to canopy surfaces using depth information. Dong and Isler (2018) used such a sensor and mapping algorithm to create 3D maps of orchard blocks where depth information from two sides of the row is merged; traits autonomously measured include canopy volume, tree height and trunk diameter. Zhang et al. (2017) combined a commodity RGB-D sensor with CNNs to localize branches in apple trees.

Supported by an ultrasonic-, infrared-, laser- or lidar-based canopy sensing system, it is possible to implement effective, targeted spraying by turning the nozzles on and off in terms of detected presence and absence of canopies in front of the nozzles. Gil et al. (2007) and Escolà et al. (2013) reported a few examples of using such automated technologies for spraying agrochemicals on the targeted zone at a desired rate on the canopy of various types of orchard crops. Zhang et al. (2018a) provided a comprehensive review on those technologies, and the number or the type of both the sensors and the nozzles installed on a sprayer can be optimized based on the crop types and canopy structures or architectures.

A precision spraying system developed by Xiao et al. (2017) could vary pesticide application rate based on detected 'leaf area density' and 'distance-to-canopy' using an RGB-D camera. This sprayer was tested in peach trees, apricot trees and grapevines, which showed improved efficiency and reduced waste in spraying pesticides.

The ground spray systems discussed so far lack the capability to spray input materials to specific targets such as fruit or individual flowers. There have been some efforts at developing robotic solutions to spray targeted canopy objects or specific parts. One system targeted herbicide application to grapevine suckers, which grow at the base of grapevine trunks (Fig. 4; Kang et al., 2012). The system included a camera system to detect trunks and suckers of grapevines and a set of nozzles to spray chemicals to only targeted

Figure 4 Left: A precision spraying system developed and tested by Washington State University for targeted application of pesticides and other pest control materials to grapevines. Right: Comparison of the amount of chemical used with a conventional broadcasting method and targeted, precision application to grapevine trunks.

areas of the trunks to control sucker plants and insects, such as cutworm. The system was tested in wine grapes in Washington State showing a chemical saving of up to 92% compared to a band broadcasting of chemicals with a regular sprayer. Berenstein et al. (2010) developed a machine vision-based automated system for precision spraying of hormones to fruit and pesticide to foliage without impacting non-target areas. A similar system was investigated by Oberti et al. (2016) for controlling powdery mildew in grape canopies. Tests in a greenhouse environment achieved an accuracy of 85–100% in detecting the diseased areas resulting in up to 35% saving in chemical use. Further research and development is warranted to make these systems robust and widely applicable to targeted flowers, fruit or canopy parts where the problem is detected. Research and development efforts need to focus on both sensing systems that can identify and locate the presence of specific pests or issues in crop canopies and a robotic system that can precisely apply the desired amount of chemicals to target objects or canopy parts.

Other than the application of agrochemicals, artificial pollen application (also called mechanical pollination) is another important application of spraying technology in producing fruit, berry, nut and vegetable crops. There are primarily two approaches investigated for artificial pollination: aerial platforms and ground platforms. Aerial platforms include various designs of small UASs which often apply a bombing technology to spray pollens down onto the crop canopies from above. An innovative newer approach of aerial pollination is the use of a fleet of nano-aerial robots to mimic bee behaviour as artificial pollinators. Berman et al. (2011) has developed and tested robotic bees that showed potential for pollinating crops when a swarm of those vehicles are used in orchards. Similar development of aerial robots has been reported by Abutalipov et al. (2016).

Artificial pollination using ground robots/vehicles is very much like agrochemical applications using electrostatic sprayers. The machines are often designed to spray pollen suspended in a liquid or dust medium to target canopy areas. Different types of sprayers have been tested to optimize the type and size of nozzles, operating pressure, flow rate, carrier medium and distance to bloom so that the desired density of pollen could be sprayed to targeted canopy areas. For example, Whiting (2016) has tested an electrostatic sprayer for artificial pollination and found that supplemental application of pollen (suspended in a sucrose and boron solution) increased the pollen density on flower stigma by three-fold compared to natural pollination with bees (Fig. 5). Though there have been several systems evaluated for broadcasting pollen to target crop canopies including some commercialization efforts (e.g. a mechanical blower system commercialized by PollenPlus™, New Zealand), research and development of robotic systems (with manipulators and end effectors) for selective pollination of a desired type and number of flowers

Figure 5 An electrostatic sprayer retrofitted by Dr. Mathew Whiting and his team at Washington State University to apply pollens; being evaluated in a cherry orchard in Prosser, WA. Picture by TJ Mullinax, Good Fruit Grower.

have been limited. One example system was developed and tested by Barnett et al. (2017) in kiwifruit. The vision system included a camera and a CNN-based image processing system for flower detection. The robotic machine included an autonomous platform and a spray manipulator with various degrees of freedom. The system was able to detect more than 70% of the flowers and over 80% of the flowers were pollinated with the robotic system.

5 Robotic harvesting

Harvesting is a labour-intensive and time-sensitive operation in orchard crops. Scientists and engineers have been investigating robotic systems for harvesting various types of crops including apples (e.g. Rabatel et al., 1995), citrus (e.g. Harrel et al., 1989), kiwifruit (e.g. Bazley, 2017) and cherries (e.g. Tanigaki et al., 2008). Unfortunately, so far, no commercial success has been achieved. In recent years, however, there have been heightened interest and investment in research and developments from both public and private organizations. With the latest developments in sensing, computing and artificial intelligence technologies, robotic harvesting of fruit appears now to be a viable technology.

A robotic harvesting system for orchard crops consists generally of a vision system to detect and locate target fruit (and obstacles when needed), a manipulation system to reach the target fruit, an end effector to detach fruit from branches and collect them onto a conveyance system and a conveyance system to bring harvested fruit onto a container/bin (some of the components are depicted in Fig. 6). In this section, some of the latest efforts in various aspects of robotic fruit harvesting are discussed. In addition, an alternative

Figure 6 A robotic harvesting system developed at Washington State University by a team led by Manoj Karkee and Qin Zhang. Source: Adapted from Silwal et al. (2017).

method of automated fruit harvesting, a targeted shake-and-catch system, will be introduced. More detailed discussion on past efforts in fruit harvesting, conveyance and container filling technologies can also be found in review articles and book chapters published recently including Bac et al. (2014), Gongal et al. (2015), Zhang et al. (2016) and Karkee et al. (2017).

The first step in robotic harvesting is to detect the fruit and estimate their 3D location in tree canopies so that an end effector could reach the target fruit and detach it from the tree. A wide range of studies can be found in detecting fruit and obstacles using specific features including colour, shape, size, edge and texture and using various thresholding and classification techniques such as Bayesian classifier and neural networks (e.g. Silwal et al., 2014; Tabb et al., 2006). However, these specific feature-based techniques have achieved only limited success due to issues including occlusion, fruit clustering, crop and canopy variability, and unstructured, uncertain and variable lighting conditions. To address the challenge of fruit clustering, Changhui et al. (2017) used a convex hull technique to identify individual citrus fruit and their centres (assuming circular fruit) in images with overlapped fruit clusters. Wang et al. (2017) proposed an image enhancement technique involving the wavelet transform and Retinex principle (McCann, 2016) to minimize the issues with variable lighting conditions.

In recent years, deep learning has been used successfully to address various challenges in fruit detection (e.g. Zhang et al., 2017; Chen et al., 2017). Stein et al. (2016) used deep learning networks for mango detection and localization whereas Halstead et al. (2018) used an end-to-end deep learning network to detect the number and quality of sweet peppers. Chen et al. (2017) used a fully CNN (F-CNN) to detect and count oranges and apples with a mean intersection of union of 0.813 and 0.838 respectively for oranges and apples. Häni et al. (2018) also used CNNs to localize apple fruit from multiple tree varieties.

Fusion of colour and 3D information has also been investigated recently for multi-class canopy image classification. Tao and Zhou (2017) used support vector machines and a genetic algorithm to classify apple tree canopy images into fruit, branches and leaves. Similarly, Zhang et al. (2018b) used colour and depth information collected by a Microsoft Kinect sensor to train a Faster R-CNN, which showed better accuracy in classifying branches, trunks and background compared to the same without depth information.

Approaching a fruit is a critical step in robotic harvest, which primarily involves defining an optimal path to move the picking end effector to the target fruit to complete the fruit detachment from the tree. Sequence or prioritization of fruit picking could be defined using the Travelling Salesman Problem (TSP), which can be solved using various algorithms such as nearest neighbour (Kizilates and Nuriyeva, 2013) that can minimize the total distance travelled, time taken, energy used or some weighted combination of parameters that define the performance of a robotic system (Silwal et al., 2017). Once the sequence is known, visual surveying could be used to approach the target fruit (Ringdahl et al., 2018). With visual surveying, approaching a fruit would involve detecting the fruit and its position (often in 2D image space) frequently using an end effector-based imaging system and changing the manipulator joint positions such that the target fruit remains at the desired image coordinates all the time.

An alternative method to visual surveying is using a global camera system, by which a camera is mounted at a fixed position to take images at the beginning of a number of harvest cycles. Fruit position is then estimated at the beginning of the harvesting cycle for all the fruit available in the given field of view or the workspace. Once the initial or current position of the end effector and final position (fruit location) are available, an inverse kinematics is used to estimate new positions for all the joints of the robotic manipulator that would move the end effector to the desired final position and orientation. One challenge with this technique is to detect and locate fruit accurately at the beginning and to have an accurate calibration between camera coordinates and manipulator coordinates so that the end effector could reach the fruit precisely.

Fruit detachment from the tree is the core in robotic harvest of tree fruit. It commonly uses a fruit picking end effector to exert desired force and movement onto a fruit to detach it from the tree. There are different types of end effector techniques being used in fruit picking. One technique is to detach fruit using mechanical end effectors with a human-like hand and fingers. A soft palm is required to prevent the fruit from being bruised. Different designs have been used in a mechanical hand including different number of fingers and actuator to operate the fingers (Davidson et al., 2016). One of the ways by which the fingers are actuated are electric motors running each joint in a hand individually, which requires a number of actuators per hand making it relatively slower, and more complex and expensive. Another way would be to use a tandem design (under actuated hand) where one central motor would drive all the fingers using a cable such that the hand can be close to the target fruit confirming the fruit size and shape as the desired amount of force/tension is exerted in the cable (Davidson et al., 2016).

A mechanical hand may also use soft robotic materials and pneumatic actuation (Shintake et al., 2018). Fingers in soft robotic hands are designed with a hollow space inside the fingers with a wrinkled surface on the outer side. As the hollow space is filled with air, the fingers close on the target fruit as the wrinkled surface expands outward due to pressurized air. These kinds of fingers and the hand can operate much faster than the motor-operated, mechanical hands. In addition, soft hands provide a level of cushioning to the fruit being detached. However, generally, these fingers tend to be thicker than traditional counterparts, making it challenging to harvest closely spaced fruits or fruits in clusters.

Another type of end effector being used is the scissors type which cuts the stem to detach the fruits. For example, Bac et al. (2017) designed a four-fingered hand following the 'Fin Ray principle' for grabbing the fruit and then mounted a pair of scissors above the hand to cut a stem. However, detecting and locating stems for cutting with scissors is a highly challenging machine vision task. To address this issue, devices such as cup-shaped, conformal scissors that close around a fruit to cut the stem irrespective of its position (Rosenberg, 1975) may be used. This technique is more suitable for fruit varieties with long stems. Once the fruit is detached and delivered using a robotic or mass harvesting system onto a conveyance system, they have to be conveyed, decelerated at the end of the conveyance and collected onto a container. There have been various successful efforts in this area, details of which can be found in Zhang et al. (2018c) and Karkee et al. (2017).

6 Robotic fruit transportation

To implement robotic harvesting with a minimal seasonal workforce, it is desirable to use a robotic carrier to transport empty or full fruit containers/bins to and from the working zone. A functional robot carrier requires having basic functionalities of (1) autonomous navigation, (2) intelligent in-orchard management and (3) bin handling. Fruit bins used in the Pacific Northwest (PNW) region of the United States are typically designed to hold about

400 kg fruit during harvesting, transportation and storage. Therefore, two basic requirements of a robotic bin carrier are its capability of handling a load up to 500 kg and effective manoeuvrability in an orchard environment (Ye et al., 2017). Current robotic technology is capable of meeting this need by providing a robotic self-propelled fruit bin carrier system for transporting bins autonomously (Hamner et al., 2011).

The navigation functions of a robot bin carrier consist of (1) guiding the bin carrier with an empty bin into a tree alleyway where harvesting is performed; (2) moving to deliver and place the empty bin in the alleyway of an appropriate location in a harvesting zone; and (3) carrying a full bin out of the alleyway and delivering it to a repository station. A capable navigation system will require having both a GPS system for navigating the carrier between the repository and target alleyways, and a distance scanning system (either lidar or ultrasonic sensors) for detecting the boundaries and obstacles within the alleyway where GPS signals are often compromised by trees (Ye et al., 2017).

An intelligent management system, including an automated robot control system, was used by Zhang et al. (2015) to autonomously plan and guide multiple robotic bin carriers in handling bins in the orchard to support efficient harvesting operations. This intelligent management system coordinates multiple robots working faultlessly and efficiently through making two decisions of (1) which full bin should be picked up and returned to the repository; and (2) where it should carry an empty bin. To plan for efficient coordination in multi-robot management, Kalra et al. (2005) developed and validated an algorithm based on a market-based framework, which applied an auctions approach. This approach was found to be efficient for sequential management and implementation: once the intelligent management system makes a decision, it will proceed to complete its designated task and only makes another decision when it has finished its current task (Zhang et al., 2015).

The manoeuvrability of a robotic bin carrier concerns bin handling within confined fruit tree alleyways. Typical bin handling operations require the robot to (1) load an empty bin at a bin repository station; (2) lift the empty bin and go over a detected filled bin on its way to place it at an appropriate location; (3) position itself properly to readily pick up the filled bin on the way back; (4) pick the filled bin reliably and quickly; and (5) carry the picked bin securely out of the alleyway back to the repository (Ye et al., 2018). Implementing all the basic functions effectively and reliably in commercial orchard environments could be very challenging, mainly due to the confined working space in fruit tree alleyways, and the interference of randomly grown and deformable tree canopies.

Researchers found that use of a four-wheel-independent-steering system on wheeled field robots could effectively solve this problem (Bak and Jakobsen, 2004; Nagasaka et al., 2004). Such a steering system allows a wheeled robot

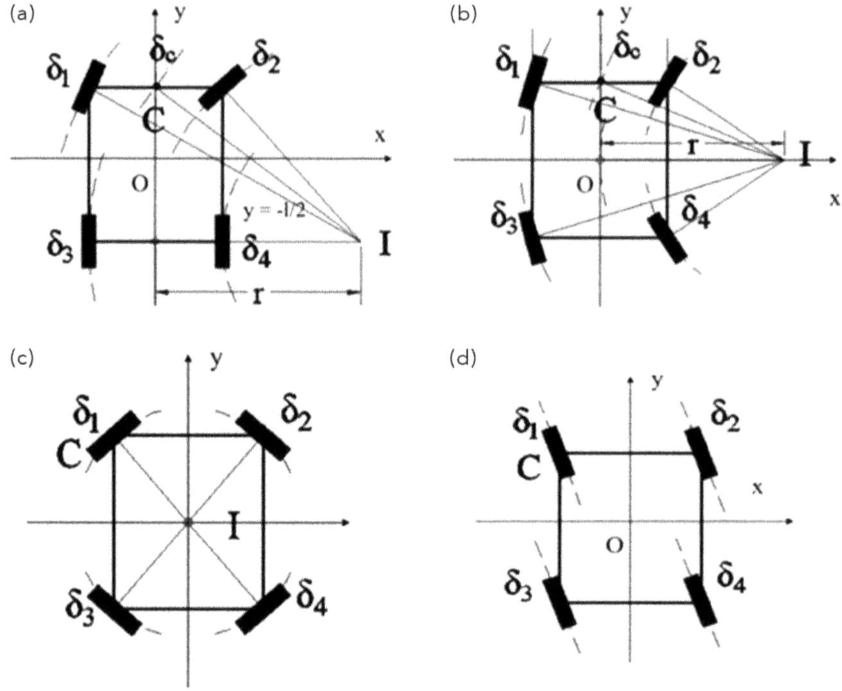

Figure 7 Four steering modes of four-wheel-independent-steering for wheeled robot: (a) Ackermann steering, (b) active-front-and-rear-steering, (c) spinning and (d) crab steering.

to be manoeuvred in four steering modes of two-wheel-steering (front or back Ackermann steering), four-wheel active-front-and-rear steering, spinning (the robot rotated about its geometric centre) and crab steering (all the wheels have the same steering angles) as illustrated in Fig. 7. This steering mechanism, when controlled properly, could achieve the maximum manoeuvrability of a wheeled robot in confined working spaces (Oksanen and Backman, 2013; Ye et al., 2018).

7 Future trends and conclusion

After more than 50 years of persistent research and development of automated and robotic orchard machines, there seems to be an accelerated success in the past decade. Such an acceleration may be attributed to the substantial advancement in agricultural robotics enabled by powerful and robust computing, sensing and automation technologies created for other industries. Various companies around the world have shown interest in developing and commercializing robotic systems for crop production in recent years. Venture capital and other funding sources such as USDA-Small Business Innovation

Research programmes have also provided funding to encourage companies developing robotic solutions for production agriculture, including fruit and vegetable harvesting, weed control and fruit tree pruning. A couple of examples include Agrobot S.L. (Huelva, Spain), which has successfully commercialized a strawberry harvesting robot (Agrobot.com), and Robotics Plus, a New Zealand-based high-tech company, which has introduced a robotic kiwifruit harvesting machine to the market (Bazley, 2017).

A major reason for the slow progress and adoption of robotic technologies to tree fruit production could be the remaining, specific technical challenges for agricultural robots. Among those challenges, one of the most difficult is the need for the robot to interact with deformable plants in varying operational conditions. Agricultural robots often need to handle targets that might move away from its detected position when the actuator starts to act on the target object. The plant deformation could be induced either by the natural stimuli or induced by robot-plant interaction, which could result in the end effector missing the target or failing to form a required reaction to complete the designated tasks.

Another challenge to be solved is operational safety and efficiency. One basic requirement for any operation using a robot or other unmanned autonomous machinery is the safety of both the people working on sites with the robot and the robot itself. Production agriculture operations are generally performed in natural, varying and unstructured environments bringing additional challenges to achieve a sensitive, dynamic and robust job site situation awareness. Robots' incomplete situation awareness may cause large safety hazards. Such incomplete situation awareness may also be attributable to current robotic systems' low efficiency to perform the designated tasks.

Economic concern always plays a big role in adopting new technologies for end users in agriculture. Farmers often wish to keep the total cost of adopting a new technology at a level comparable to their current operations. For example, the average cost for growers to pay to seasonal apple pickers in the state of Washington, USA was up to US$28 for a full bin (typical size of $1.2 \times 1.2 \times 0.6$ m^3 with around 420 kg of fruit at full load) in 2017 (Bacon, 2017). While this target price for adopting robotic harvesting today could be very challenging, with the continuously increasing harvest labour cost (e.g. a rapid increase from US$13 per bin in 2001 to US$28 in 2017), the break-even time may come more quickly than anticipated. Nevertheless, minimizing total grower costs, including the equipment purchases, maintenance, insurance, risks of crop loss caused by equipment reasons and all the relevant operation costs are always important concerns for robotic operations in production agriculture.

To relieve farmers of the dependency on human labour in production, multiple robots are needed to perform different field tasks, during specific limited times, resulting in a very low equipment utilization efficiency. As with any other commercial product, agricultural robots will be manufactured and marketed in a standard initial setting suitable for as broad a market as possible to ensure the economics of manufacturing are sustainable when these robots are commercialized. The cost per unit time usage would be very high if they want to own and maintain the machines. These economic scenarios set a difficult bar for many small farms to overcome in adopting robotic machines.

Another challenge is the adequate skill and knowledge required to implement robotic field operations effectively and efficiently. Cultural practices, field environments, crop varieties planted and farming preferences or habits can vary widely for farmers from different parts of the world, or even between neighbouring farms. This may lead to the need for customized initialization to make operation-specific adjustments, and more importantly to create adequate robotic operation plans, before a robot product could be optimized for a particular end user. With a large percentage of ageing farmers worldwide who lack knowledge of computing and robot technologies, a significant change in farmer and worker skillset will be needed.

Either technical or operational solutions could remove these challenges; however, more studies are needed to make such potential solutions practically usable. For example, if we could create a trustworthy situation awareness and adaptive mechanism with self-learning capability as a technical solution, it could be possible to remove the need for robotic machine initialization to meet user-specific operation conditions as well as make robotic machines able to react to detected situation changes similar to a human worker. Another possible technical approach is the adoption of a design philosophy of tractor-implement systems to build agricultural robots on a common platform providing power, sensing, task management and control capabilities to support a fleet of interchangeable task-specific end effectors for conducting different field operations such as harvesting, pruning, training, thinning and spraying.

While most farmers normally possess their own machinery to perform field operations, different adoption models could be worthy of investigation for robotized farming. For example, while possessing a fleet of robots could still be technically and economically feasible for some large farms, will a community co-op model, namely different farmers owning different robots and sharing use of those robots to reduce capital investment, be feasible? Or will commercialized robotic field operation services, sometimes termed contract farming, be another solution? Nevertheless, either option may create a new industry, namely agricultural robot support networks, and bring in professionals of agricultural robot support to rural areas.

Other than the progress in robotic technology and their application, the importance of developing some types of machine-friendly architectures of tree canopies can never be overemphasized. Such machine-friendly architectures are obtained through training and pruning to convert randomly grown thick canopies to more uniform and accessible canopies. Over the past several decades, numerous different types of training and pruning systems have evolved around the world leading to many different types of more accessible tree architectures (Fideghelli et al., 2003). While today's robotic technology could find ways to penetrate a randomly grown thick canopy of a fruit tree to perform some specific orchard operation, a more uniform and accessible canopy would simplify the mechanism of a robotic machine for performing those operations more effectively. It would not only reduce the price of the robotic machine at purchase, but more importantly could help to improve the effectiveness, efficiency and robustness while potentially enhancing the safety of the machine being used in commercial orchards.

In summary, robotic technologies for tree fruit production have never been so close to being practically applicable. While there are challenges remaining to be solved before robots could be commercially adopted in orchard operations, it is expected that robots would make their way into some of the field operations in tree fruit production in the near future. Advancement in agricultural robotic technology may also bring in some important societal and economic impacts to rural, agricultural areas. For example, it might bring a demographic change in rural populations as new farming styles create a new industry to support robotic farming, which will bring in well-paid, year-around jobs both for supporting the farming and servicing of machines to replace close-to-minimally paid, seasonal field labourers. Such a change in the employment force will bring in more well-educated, young professionals to work in rural areas, which in turn will bring in more societal services and professionals, such as health care facilities, schools, retail and restaurant services leading to potential boom in rural area economies.

8 Where to look for further information

Currently, the researchers working in the field of orchard robotics come from a variety of disciplines; consequently, these researchers belong to a varied set of professional societies. The readers are referred to the following professional societies and technical committees, their journals and conferences for publications on the topic of agricultural robots in orchard management.

- American Society of Agricultural and Biological Engineers (https://www.asabe.org/), relevant journal *Transactions of the ASABE* and annual conferences.
- European Society for Agricultural Engineers (https://www.eurageng.eu/) and annual conferences.
- IEEE Robotics and Automation Society (http://www.ieee-ras.org/), relevant conferences are IROS and ICRA (proceedings are refereed and published), most relevant journal is *Robotics and Automation Letters*.

- IEEE Robotics and Automation Society's Technical Committee on Agricultural Robotics and Automation (http://ieeeagra.com/), webinars, some of which concern orchard topics, are available.
- International Federation of Automatic Control (IFAC, https://www.ifac-control.org/) and its congress, conferences and workshops, such as Bio-Robotics and AgriControl Conferences.
- IFAC journals, as well as IFAC-PapersOnLine (https://www.ifac-control.org/publications/ifac-papersonline.net).
- Other journals are *Journal of Field Robotics* (https://onlinelibrary.wiley.com/journal/15564967), *Computers and Electronics in Agriculture* (https://www.journals.elsevier.com/computers-and-electronics-in-agriculture) and *Biosystems Engineering* (https://www.journals.elsevier.com/biosystems-engineering).

Centres of expertise on this topic include:

- Washington State University Center for Precision & Automated Agricultural Systems (CPAAS) – http://cpaas.wsu.edu/.
- Australian Centre for Field Robotics (https://sydney.edu.au/engineering/our-research/robotics-and-intelligent-systems/australian-centre-for-field-robotics.html).

9 References

Aasted, M. M., Dise, R. J., Baugher, T. A., Schupp, J. R., Heinemann, P. H. and Singh, S. (2011). Autonomous mechanical thinning using scanning LIDAR. *2011 ASABE Annual International Meeting*, 7004(11), Paper No: 1111792. American Society of Agricultural and Biological Engineers, St. Joseph, MI. https://doi.org/10.13031/2013.38170

Abutalipov, R. N., Bolgov, Y. V. and Senov, H. M. (2016). Flowering plants pollination robotic system for greenhouses by means of nano copter (drone aircraft). In *2016 IEEE Conference on Quality Management, Transport and Information Security, Information Technologies, IT and MQ and IS 2016* (pp. 7–9). https://doi.org/10.1109/ITMQIS.2016.7751907

Aggelopoulou, A. D., Bochtis, D., Fountas, S., Swain, K. C., Gemtos, T. A. and Nanos, G. D. (2011). Yield prediction in apple orchards based on image processing. *Precision Agriculture*, 12(3), 448–56. https://doi.org/10.1007/s11119-010-9187-0

Ahn, H. S., Dayoub, F., Popovic, M., MacDonald, B., Siegwart, R. and Sa, I. (2018). An overview of perception methods for horticultural robots: From pollination to harvest. Retrieved from http://arxiv.org/abs/1807.03124

Akbar, S. A., Elfiky, N. M. and Kak, A. (2016a). A novel framework for modeling dormant apple trees using single depth image for robotic pruning application. In *Proceedings - IEEE International Conference on Robotics and Automation* (Vol. 2016 – June, pp. 5136–42). https://doi.org/10.1109/ICRA.2016.7487718

Akbar, S. A., Chattopadhyay, S., Elfiky, N. M. and Kak, A. (2016b). A novel benchmark RGBD dataset for dormant apple trees and its application to automatic pruning. In

IEEE Computer Society Conference on Computer Vision and Pattern Recognition Workshops (pp. 347-54). https://doi.org/10.1109/CVPRW.2016.50

Bac, C. W., Van Henten, E. J., Hemming, J. and Edan, Y. (2014). Harvesting robots for high-value crops: State-of-the-art review and challenges ahead. *Journal of Field Robotics*, 31, 888-911. https://doi.org/10.1002/rob.21525

Bac, C. W., Hemming, J., van Tuijl, B. A. J., Barth, R., Wais, E. and van Henten, E. J. (2017). Performance evaluation of a harvesting robot for sweet pepper. *Journal of Field Robotics*, 34, 1123–39. https://doi.org/10.1002/rob.21709

Bacon, D. (2017). The apple pickers of the Yakima Valley. *The Atlantic*. Retrieved 22 June 2017, from https://www.theatlantic.com/business/archive/2017/06/apple-pickers-yakima-photo-essay/530967/ (accessed on 7 September 2018).

Bak, T. and Jakobsen, H. (2004). Agricultural robotic platform with four wheel steering for weed detection. *Biosystems Engineering*, 87, 125–36. https://doi.org/10.1016/j.biosystemseng.2003.10.009

Barnett, J., Seabright, M., Williams, H., Nejati, M., Scarfe, A., Bell, J. and Duke, M. (2017). Robotic Pollination-Targeting kiwifruit flowers for commercial application. In *PA17 International Tri-Conference for Precision Agriculture*.

Baugher, T. A., Schupp, J., Ellis, K., Remcheck, J., Winzeler, E., Duncan, R., Johnson, S., Lewis, K., Reighard, G., Henderson, G., Norton, M., Dhaddey, A. and Henderson, G. (2010). String blossom thinner designed for variable tree forms increases crop load management efficiency in trials in four United States peach-growing regions. *HortTechnology*, 20(2), 409–14.

Bazley, T. (2017). Robotic hands gently snatch kiwifruit future. Retrieved 6 June 2018, from http://www.roboticsplus.co.nz/robotic-hands-snatch-kiwifruit-future

Berenstein, R., Shahar, O. Ben, Shapiro, A. and Edan, Y. (2010). Grape clusters and foliage detection algorithms for autonomous selective vineyard sprayer. *Intelligent Service Robotics*, 3(4), 233–43. https://doi.org/10.1007/s11370-010-0078-z

Berman, S., Kumar, V. and Nagpal, R. (2011). Design of control policies for spatially inhomogeneous robot swarms with application to commercial pollination. In 2011 IEEE International Conference on Robotics and Automation. doi:10.1109/ICRA.2011.5980440

Braun, T., Koch, H., Strub, O., Zolynski, G. and Berns, K. (2010). Improving pesticide spray application in vineyards by automated analysis of the foliage distribution pattern in the leaf wall. In *1st Commercial Vehicle Technology Symposium*.

Changhui, Y., Youcheng, H., Lin, H., Sa, L. and Yanping, L. (2017). Overlapped fruit recognition for citrus harvesting robot in natural scenes. In *Robotics and Automation Engineering (ICRAE), 2017 2nd International Conference on* (pp. 398–402). IEEE.

Chen, S. W., Shivakumar, S. S., Dcunha, S., Das, J., Okon, E., Qu, C., Taylor, C. J. and Kumar, V. (2017). Counting apples and oranges with deep learning: A data-driven approach. *IEEE Robotics and Automation Letters*, 2(2), 781-8. https://doi.org/10.1109/LRA.2017.2651944

Davidson, J. R., Silwal, A., Hohimer, C. J., Karkee, M., Mo, C. and Zhang, Q. (2016). Proof-of-concept of a robotic apple harvester. In *IEEE International Conference on Intelligent Robots and Systems*. https://doi.org/10.1109/IROS.2016.7759119

Dey, D., Mummert, L. and Sukthankar, R. (2012). Classification of plant structures from uncalibrated image sequences. In *Applications of Computer Vision (WACV), 2012 IEEE Workshop on* (pp. 329–36). IEEE.

Dias, P. A., Tabb, A. and Medeiros, H. (2018a). Apple flower detection using deep convolutional networks. *Computers in Industry*, 99, 17–28. https://doi.org/10.1016/j.compind.2018.03.010

Dias, P. A., Tabb, A. and Medeiros, H. (2018b). Multispecies fruit flower detection using a refined semantic segmentation network. *IEEE Robotics and Automation Letters*, 3(4), 3003–10. doi:10.1109/LRA.2018.2849498

Dias, P. A., Tabb, A. and Medeiros, H. (2018c). Data from: Multispecies fruit flower detection using a refined semantic segmentation network. *Ag Data Commons*. http://dx.doi.org/10.15482/USDA.ADC/1423466

Dong, W. and Isler, V. (2018). Tree morphology for phenotyping from semantics-based mapping in orchard environments. arXiv: 1804.05905 [cs.RO].

Elfiky, N. M., Akbar, S. A., Sun, J., Park, J. and Kak, A. (2015). Automation of dormant pruning in specialty crop production: An adaptive framework for automatic reconstruction and modelling of apple trees. In *IEEE Computer Society Conference on Computer Vision and Pattern Recognition Workshops*. https://doi.org/10.1109/CVPRW.2015.7301298

Emery, K. G., Faubion, D. M., Walsh, C. S. and Tao, Y. (2010). Development of 3-D range imaging system to scan peach branches for selective robotic blossom thinning. In *American Society of Agricultural and Biological Engineers Annual International Meeting 2010*, ASABE 2010.

Escolà, A., Rosell-Polo, J. R., Planas, S., Gil, E., Pomar, J., Camp, F., Llorens, J. and Solanelles, F. (2013). Variable rate sprayer. Part 1 – Orchard prototype: Design, implementation and validation. *Computers and Electronics in Agriculture*, 95, 122–35. https://doi.org/10.1016/j.compag.2013.02.004

Fideghelli, C., Sartori, A. and Grassi, F. (2003). Fruit tree size and architecture. *Acta Horticulturae*, 622, 279–93. https://doi.org/10.17660/ActaHortic.2003.622.26

Gallardo, R. K., Taylor, M. and Hinman, H. (2009). Cost estimates of establishing and producing gala apples in Washington. Washington State University Extension Factsheet FS005E, January 2010.

Gebbers, R., Pflanz, M., Betz, A., Hille, B., Mattner, J., Rachow-Autrum, T., Özyurtlu, M., Schischmanow, A., Scheele, M., Schrenk, J. and Schrenk, L. (2013). OptiThin – Implementation of precision horticulture by tree-specific mechanical thinning. *33. GIL-Jahrestagung in Potsdam 2013 – Massendatenmanagement in Der Agrar- Und Ernährungswirtschaft*.

Gil, E., Escolà, A., Rosell, J. R., Planas, S. and Val, L. (2007). Variable rate application of plant protection products in vineyard using ultrasonic sensors. *Crop Protection*, 26, 1287–97. https://doi.org/10.1016/j.cropro.2006.11.003

Gongal, A., Amatya, S., Karkee, M., Zhang, Q. and Lewis, K. (2015). Sensors and systems for fruit detection and localization: A review. *Computers and Electronics in Agriculture*, 116, 8–19. https://doi.org/10.1016/j.compag.2015.05.021

Halstead, M., McCool, C., Denman, S., Perez, T. and Fookes, C. (2018). Fruit quantity and quality estimation using a robotic vision system. Retrieved from http://arxiv.org/abs/1801.05560

Hamner, B., Bergerman, M. and Singh, S. (2011). Autonomous orchard vehicles for specialty crops production. In 2011 ASABE Annual International Meeting, American Society of Agricultural and Biological Engineers.

Häni, N., Roy, P. and Isler, V. (2018). Apple counting using convolutional neural networks. In *IEEE/RSJ International Conference on Intelligent Robots and Systems*.

Harrell, R. C., Slaughter, D. C. and Adsit, P. D. (1989). A fruit-tracking system for robotic harvesting. *Machine Vision and Applications*, 2, 69–80.

Hočevar, M., Širok, B., Godeša, T. and Stopar, M. (2014). Flowering estimation in apple orchards by image analysis. *Precision Agriculture*, 15(4). https://doi.org/10.1007/s11119-013-9341-6

Hong, S. W. and Choi, L. (2012). Automatic recognition of flowers through color and edge based contour detection. In *2012 3rd International Conference on Image Processing Theory, Tools and Applications, IPTA 2012*. https://doi.org/10.1109/IPTA.2012.6469535

Horton, R., Cano, E., Bulanon, D. and Fallahi, E. (2017). Peach flower monitoring using aerial multispectral imaging. *Journal of Imaging*, 3(1), 2. https://doi.org/10.3390/jimaging3010002

Izadi, S., Kim, D., Hilliges, O., Molyneaux, D., Newcombe, R., Kohli, P., Shotton, J., Hodges, S., Freeman, D., Davison, A. and Fitzgibbon, A. (2011). KinectFusion: Real-time 3D reconstruction and interaction using a moving depth camera. In *Proceedings of UIST*, pp. 559–68.

Jeon, H. Y., Zhu, H., Derksen, R., Ozkan, E. and Krause, C. (2011). Evaluation of ultrasonic sensor for variable-rate spray applications. *Computers and Electronics in Agriculture*, 75(1), 213–21. https://doi.org/10.1016/j.compag.2010.11.007

Kalra, N., Ferguson, D. and Stentz, A. (2005). Hoplites: A market-based framework for planned tight coordination in multirobot teams. In *Proceedings - IEEE International Conference on Robotics and Automation*. https://doi.org/10.1109/ROBOT.2005.1570274

Kang, F., Wang, H., Pierce, F. J., Zhang, Q. and Wang, S. (2012). Sucker detection of grapevines for targeted spray using optical sensors. *Transactions of the ASABE*, 55(5), 2007–14.

Karkee, M. and Adhikari, B. (2015). A method for three-dimensional reconstruction of apple trees for automated pruning. *Transactions of the ASABE*, 58(3), 565–74.

Karkee, M., Adhikari, B., Amatya, S. and Zhang, Q. (2014). Identification of pruning branches in tall spindle apple trees for automated pruning. *Computers and Electronics in Agriculture*, 103, 127–35. https://doi.org/10.1016/j.compag.2014.02.013

Karkee, M., Davidson, R. J. and Silwal, A. (2017). Mechanical harvest and in-field handling of tree fruit crops. In: Zhang, Q. (Ed.), *Automation in Tree Fruit Production: Principles and Practice*. CABI, Wallingford, UK, pp. 179–233.

Kizilateş, G. and Nuriyeva, F. (2013). On the nearest neighbor algorithms for the traveling salesman problem. In: *Advances in Computational Science, Engineering and Information Technology*. Springer International Publishing, pp. 111–18.

Liu, H. and Zhu, H. (2016). Evaluation of a laser scanning sensor in detection of complex-shaped targets for variable-rate sprayer development. *Transactions of the ASABE*, 59(5), 1181–92.

Lyons, D. J., Heinemann, P. H., Schupp, J. R., Baugher, T. A. and Liu, J. (2015). Development of a selective automated blossom thinning system for peaches. *Transactions of the ASABE*, 58(6), 1447–57.

McCann, J. (2016). Retinex theory. In: Luo, M. R. (Ed.), *Encyclopedia of Color Science and Technology*. Springer, New York, NY, pp. 1118–25.

Medeiros, H., Kim, D., Sun, J., Seshadri, H., Akbar, S. A., Elfiky, N. M. and Park, J. (2017). Modeling dormant fruit trees for agricultural automation. *Journal of Field Robotics*, 34(7), 1203–24. https://doi.org/10.1002/rob.21679

Miller, S. S., Baugher, T. A., Schupp, J. R. and Wolford, S. D. 2011. Performance of mechanical thinners for bloom or green fruit thinning in peaches. *HortScience*, 46(1), 43–51.

Nagasaka, Y., Zhang, Q., Grift, T. E., Kanetani, Y., Umeda, N. and Kokuryu, T. (2004). An autonomous field watching-dog robot for information collection in agricultural fields. *ASAE Annual International Meeting 2004*.

Nielsen, M., Slaughter, D. C. and Gliever, C. (2012). Vision-based 3D peach tree reconstruction for automated blossom thinning. *IEEE Transactions on Industrial Informatics*, 8(1), 188–96. https://doi.org/10.1109/TII.2011.2166780

Oberti, R., Marchi, M., Tirelli, P., Calcante, A., Iriti, M., Tona, E., Hočevar, M., Baur, J., Pfaff, J., Schütz, C. and Ulbrich, H. (2016). Selective spraying of grapevines for disease control using a modular agricultural robot. *Biosystems Engineering*, 146, 203–15. https://doi.org/10.1016/j.biosystemseng.2015.12.004

Oksanen, T. and Backman, J. (2013). Guidance system for agricultural tractor with four wheel steering. In *IFAC Proceedings Volumes (IFAC-PapersOnline)*. https://doi.org/10.3182/20130327-3-JP-3017.00030

Rabatel, G., Bourely, A., Sevila, F. and Juste, F. (1995). Robotic harvesting of citrus: State-of-art and development of the French Spanish EUREKA project.

Ren, S., He, K., Girshick, R. and Sun, J. (2017). Faster R-CNN: Towards real-time object detection with region proposal networks. *IEEE Transactions on Pattern Analysis and Machine Intelligence*, 39(6), 1137–49. https://doi.org/10.1109/TPAMI.2016.2577031

Ringdahl, O., Kurtser, P. and Edan, Y. (2018). Evaluation of approach strategies for harvesting robots: Case study of sweet pepper harvesting. *Journal of Intelligent & Robotic Systems*, 1–16.

Rosenberg, B. (1975). U.S. Patent No. 3,902,306. U.S. Patent and Trademark Office, Washington DC.

Schupp, J. R., Winzeler, H. E., Kon, T. M., Marini, R. P., Baugher, T. A., Kime, L. F. and Schupp, M. A. (2017). A method for quantifying whole-tree pruning severity in mature tall spindle apple plantings. *HortScience*, 52(9), 1233–40.

Shintake, J., Cacucciolo, V., Floreano, D. and Shea, H. (2018). Soft robotic grippers. *Advanced Materials*, 30(29), 1707035.

Silwal, A., Gongal, A. and Karkee, M. (2014). Identification of red apples in field environment with over the row machine vision system. *Agricultural Engineering International: CIGR Journal*, 16(4).

Silwal, A., Davidson, J. R., Karkee, M., Mo, C., Zhang, Q. and Lewis, K. (2017). Design, integration, and field evaluation of a robotic apple harvester. *Journal of Field Robotics*, 34(6), 1140–59. https://doi.org/10.1002/rob.21715

Sinoquet, H., Rivet, P. and Godin, C. (1997). Assessment of the three-dimensional architecture of walnut trees using digitising. *Silva Fennica*, 31(3), 265–73. https://doi.org/10.14214/sf.a8525

Stein, M., Bargoti, S. and Underwood, J. (2016). Image based mango fruit detection, localisation and yield estimation using multiple view geometry. *Sensors*, 16(11). https://doi.org/10.3390/s16111915

Tabb, A. (2013). Shape from silhouette probability maps: Reconstruction of thin objects in the presence of silhouette extraction and calibration error. In *Proceedings of the IEEE Computer Society Conference on Computer Vision and Pattern Recognition*. https://doi.org/10.1109/CVPR.2013.28

Tabb, A. and Medeiros, H. (2017). A robotic vision system to measure tree traits. In *IEEE/RSJ International Conference on Intelligent Robots and Systems*. https://doi.org/10.1109/IROS.2017.8206497

Tabb, A. and Medeiros, H. (2018). Video data of flowers, fruitlets, and fruit in apple trees during the 2017 growing season at USDA-ARS-AFRS. *Ag Data Commons*. http://dx.doi.org/10.15482/USDA.ADC/1416008

Tabb, A., Peterson, D. and Park, J. (2006). Segmentation of apple fruit from video via background modeling. ASABE paper No. 063060. ASABE, St. Joseph, MI.

Tanigaki, K., Fujiura, T., Akase, A. and Imagawa, J. (2008). Cherry-harvesting robot. *Computers and Electronics in Agriculture*, 63(1), 65–72.

Tao, Y. and Zhou, J. (2017). Automatic apple recognition based on the fusion of color and 3D feature for robotic fruit picking. *Computers and Electronics in Agriculture*, 142, 388–96. https://doi.org/10.1016/j.compag.2017.09.019

Underwood, J. P., Hung, C., Whelan, B. and Sukkarieh, S. (2016). Mapping almond orchard canopy volume, flowers, fruit and yield using lidar and vision sensors. *Computers and Electronics in Agriculture*, 130, 83–96. https://doi.org/10.1016/j.compag.2016.09.014

Vougioukas, S. G., Arikapudi, R. and Munic, J. (2016). A study of fruit reachability in orchard trees by linear-only motion. *IFAC-PapersOnLine*. https://doi.org/10.1016/j.ifacol.2016.10.051

Wang, C., Tang, Y., Zou, X., SiTu, W. and Feng, W. (2017). A robust fruit image segmentation algorithm against varying illumination for vision system of fruit harvesting robot. *Optik-International Journal for Light and Electron Optics*, 131, 626–31.

Whiting, M. (2016). Mechanical pollination for yield security. Project report to Washington Tree Fruit Research Commission. http://jenny.tfrec.wsu.edu/wtfrc/PDFfinalReports/2016FinalReports/WhitingPollinationFinal.pdf.

Wouters, N., De Ketelaere, B., Deckers, T., De Baerdemaeker, J. and Saeys, W. (2015). Multispectral detection of floral buds for automated thinning of pear. *Computers and Electronics in Agriculture*, 113, 93–103. https://doi.org/10.1016/j.compag.2015.01.015

Xiao, K., Ma, Y. and Gao, G. (2017). An intelligent precision orchard pesticide spray technique based on the depth-of-field extraction algorithm. *Computers and Electronics in Agriculture*, 133, 30–6. https://doi.org/10.1016/j.compag.2016.12.002

Yang, F. O. (2012). Development of a table-top robot model for thinning of fruit. Master's thesis, University of Illinois Urbana-Champaign, Champaign, IL.

Ye, Y., Wang, Z., Jones, D., He, L., Taylor, M., Hollinger, G. and Zhang, Q. (2017). Bin-dog: A robotic platform for bin management in orchards. *Robotics*, 6(2), 12. https://doi.org/10.3390/robotics6020012

Ye, Y., He, L., Wang, Z., Jones, D., Hollinger, G. A., Taylor, M. E. and Zhang, Q. (2018). Orchard manoeuvring strategy for a robotic bin-handling machine. *Biosystems Engineering*, 169, 85–103. https://doi.org/10.1016/j.biosystemseng.2017.12.005

Yoder, K. S., Peck, G. M., Combs, L. D., Byers, R. E. and Smith, A. H. (2013). Using a pollen tube growth model to improve apple bloom thinning for organic production. *Acta Horticulturae*, 1001, 207–14. https://doi.org/10.17660/ActaHortic.2013.1001.23

Zhang, Y., Ye, Y., Wang, Z., Taylor, M. E., Hollinger, G. A. and Zhang, Q. (2015). Intelligent in-orchard bin-managing system for tree fruit production. In *Proceedings of the Robotics in Agriculture Workshop at the 2015 IEEE International Conference on Robotics and Automation, Seattle, WA, USA* (Vol. 30).

Zhang, Z., Heinemann, P. H., Liu, J., Baugher, T. A. and Schupp, J. R. (2016). The development of mechanical apple harvesting technology: A review. *Transactions of the ASABE*, 59(5), 1165–80.

Zhang, J., He, L., Karkee, M., Zhang, Q., Zhang, X. and Gao, Z. (2017). Branch detection with apple trees trained in fruiting wall architecture using stereo vision and regions-convolutional neural network (R-CNN). In *2017 ASABE Annual International Meeting* (p. 1). American Society of Agricultural and Biological Engineers.

Zhang, Z., Wang, X., Lai, Q. and Zhang, Z. (2018a). Review of variable-rate sprayer applications based on real-time sensor technologies. In *Automation in Agriculture-Securing Food Supplies for Future Generations*. InTech.

Zhang, J., Gao, G., Karkee, M., Zhang, X., He, L. and Zhang, Q. (2018b). Branch detection for apple trees trained in fruiting wall architecture using depth features and regions-convolutional neural network (R-CNN). *Computers and Electronics in Agriculture* (Under review).

Zhang, Z., Pothula, A. K. and Lu, R. (2018c). A review of bin filling technologies for apple harvest and postharvest handling. *Applied Engineering in Agriculture*, 34(4), 687–703.

Chapter 10

Advances in automated in-field grading of harvested crops

Jose Blasco, María Gyomar González González, Patricia Chueca and Sergio Cubero, Instituto Valenciano de Investigaciones Agrarias (IVIA), Spain; and Nuria Aleixos, Universitat Politècnica de València, Spain

1 Introduction

Mechanized harvesting devices for grains, beans, cotton or some fruits have been available since the early 1960s. Today, for industrial purpose, mechanical harvesting based on canopy and trunk shakers are widely used for some fruits and vegetables. However, most fruit and vegetables produced for the fresh market have to be harvested manually. Harvesting in the orchard represents between 30% and 60% of the total production costs (Ruiz-Altisent et al., 2004). Manual harvesting of tree crops using ladders and picking buckets or bags is still prevalent in some countries (Zhang et al., 2017a). In this operation, workers gently detach the fruits and put them into picking bags or in boxes on the ground. When a box is full, it is transported manually to an area where tractors or trucks are waiting to transport them to post-harvest facilities. This picking pattern is inefficient and involves dead times. In recent years, however, cultivation methods are changing. Low-density unstructured orchards are being transformed into high-density structured orchard systems, which facilitate orchard mechanization (Xu, 2016). As a result, these traditional harvesting methods are gradually being replaced by advanced machines (Bochtis et al.,

http://dx.doi.org/10.19103/AS.2019.0056.16

2014), robotics (Bechar and Vigneault, 2017) or harvest-assist platforms (Lu et al., 2017).

Robotic systems are being extensively used in agriculture (Bac et al., 2014; Blasco et al., 2002; Conesa-Muñoz et al., 2016; Gonzalez-de-Santos et al., 2017) but they present a series of drawbacks that slow down their introduction as automatic harvesters in agriculture. One of the main problems is the difficulty to detect the fruit in a tree, either because the fruit colour is similar to the leaves or because the fruit is partially or totally occluded by them. An alternative to robotic harvesting is the use of mobile platforms to assist in harvesting tasks. Although they cannot complete the harvesting, they can help in manual harvesting and increase labour efficiency by up to 60% (Zhang et al., 2017b).

Since the early proposals by Jutras and Coppock (1958), little progress has been made in the use and possibilities of the technology to assist in the harvesting of different fruits. Obviously, they can optimize the transportation operations of the harvested fruit in addition to increasing the speed of the manual operation. However, in recent years, important advances have been made in geolocation and navigation technology, together with the progress made in the creation of new sensors for agriculture and a gradual reduction in prices. Currently, these sensors are getting smaller and cheaper, allowing them to be incorporated into agricultural vehicles, including harvest-assist platforms. Thus, these machines, in addition to facilitating the harvesting operations, can capture crop data and associate them with the exact point of collection in a precise manner. One of the technologies that has evolved the most is computer vision. Generally, in agricultural robotics this technology is commonly used for autonomous navigation (Shalal et al., 2013), crop inspection tasks (Mazzetto et al., 2010), the prediction of harvest or the presence of pests and diseases (Zarco-Tejada et al., 2018) or weed control (Fernández-Quintanilla et al., 2018). Now, the advances in computer vision technology allow the incorporation of high-performance inspection systems capable of pre-grading the fruit in the field. One of the main advantages of these platforms is that, simultaneously, they could also carry out other tasks in addition to fruit harvesting, such as product inspection and pre-sorting.

The range of vision devices that can be built into agricultural vehicles is quite wide, depending on the purpose of the inspection (Zhao et al., 2016). Most of them include colour cameras, thermal cameras, stereovision systems, a laser range finder, time-of-flight (TOF) or light-based 3D cameras (Gongal et al., 2015). More recently, the use of hyperspectral systems has been increasing (Lorente et al., 2012). However, the open field presents complex conditions that cannot be controlled. One of the main problems identified for computer vision systems when they have to acquire images in outdoor conditions is the changes in lighting owing to clouds or the different positions of the sun, a complex and unstructured background, the presence of dust and dirt, limited energy or the advance of the vehicle over uneven terrain (Bechar, 2010; Tremblay et al., 2011). As a result, the precise choice of these devices is therefore essential to ensure their proper operation in the field (Kubinger et al., 2017), and algorithms capable of adapting to different natural conditions are also needed (Cubero et al., 2016).

A key advantage of the harvest-assist platforms is that they normally incorporate conveyor belts that transport the fruit from the point where they are manually dropped into the machine to bin fillers (Zhang et al., 2017a, 2018). The electronic sorters found in many packing houses have computer vision systems that inspect each piece of fruit at a very high speed to determine its quality. In a similar way, vision systems can be installed on these conveyor belts in order to analyse the fruit while it is being harvested so as to determine its quality and certain physical properties such as size or colour, and, if possible, separate it into different categories.

In vision systems for in-field sorting, the part of the conveyor belts where the cameras are installed can be covered, and therefore the scene captured by the cameras can be illuminated artificially. Although a relatively controlled environment can be developed, some of the limiting factors still remain, such as the presence of dust, limited space and the movement of the machine. Moreover, the fruit must be inspected at a high speed and the images must be processed in real time so that the fruit can be separated properly. Therefore, powerful processing units together with optimized image processing algorithms are required (Thanpattranon et al., 2016). Another important factor is the need to capture different views of the fruit as it passes under the camera in order to inspect most of the fruit surface. This can be achieved either by using several cameras with different angles or by rotating the fruit as it passes through the vision system, which therefore makes the mechanics more complex (Cubero et al., 2016).

The agricultural machines where these sorting systems are installed present other limitations that affect productivity, such as restrictions due to the small amount of space available to install the vision systems or the limited energy provided by the machine. Sensors can be small, but the need for an inspection chamber to isolate the scene from natural light or to protect the electronics from dust can increase the height of the machine if the design is not optimized properly. Moreover, due to the restrictions of working on a machine, there is a restraint regarding the possible design of the conveyor belts, which limits the number of fruits that can be transported simultaneously and thereby affects the speed of inspection, and it is also necessary to ensure that they do not cause damage to the fruits by impact or friction. On the other hand, the energy consumed by all the systems (electrovalves, conveyor belts, lighting, computers, cameras, electronics etc.)

should not exceed the maximum power that can be supplied by the motor of the machine. If this is not taken into account, the battery will be discharged too quickly. Another problem to keep in mind is that the outside temperature can be high in some seasons of the year. Thus, the heat accumulated inside the inspection chamber, together with that produced by radiation of the equipment, can damage the cameras and the electronics, so it is important to design adequate ventilation systems, which at the same time prevent dust from entering the systems.

2 Advantages of in-field sorting

While quality is one of the factors that directly affect the marketable value of the product and that intrinsically influence the producer's profits (Qiao et al., 2005), consumers are increasingly aware of and demand guidelines for good practices, sustainability and traceability in the agricultural products they acquire. This makes it necessary to keep track of the origin and the quality of the fruit from the first stages of crop production (Cubero et al., 2016). Mixing healthy fruits with others that do not reach the minimum quality to be marketed makes the whole batch more susceptible to invasion by contamination, diseases or pests during post-harvest storage (Lu et al., 2017), which could result in an important loss of production for growers. Moreover, defective fruit is transported to the post-harvest facility, pre-sized, cleaned and washed, waxed and treated for disease control (El-Ramady et al., 2015) and finally rejected or transported to the food processing industry to convert it into juices, jams or other products. During this process, the machines that move the fruit during post-harvest handling are exposed to contamination. In addition to the health risk for healthy fruits, this implies unnecessary spending on fruit that will later be removed from the produce to be marketed and hence a reduction in the grower's profits and competitiveness.

These systems are of special interest when the fruit is harvested using mechanical methods, such as canopy shakers, because these devices are not selective and gather all the fruit regardless of its condition (Torregrosa et al., 2014). Thus, in-field pre-sorting allows inferior quality fruit to be separated from marketable fruit in the field after harvesting. This helps growers to achieve cost savings in post-harvest sorting, grading and storage by, for instance, early detection of fruit that must not be marketed due to infection or contamination or by identifying those fruits that can be shipped directly to the food processing industry (Mizushima and Lu, 2011).

An additional benefit is the detection and separation of fruit that requires certain post-harvest treatments. This is especially important, for instance, at the beginning of the citrus season, when the fruits have reached the optimum stage of internal maturity but still maintain a green colour that prevents them from being commercialized. As a result, depending on their colour at harvest, they will need different de-greening treatments. Moreover, the data obtained from the fruits linked to the collection point makes it possible to create maps associated with the plantation and with the terrain that provide information about the yield, the quality and the crop status.

3 Harvest-assist platforms

The use of machines to assist harvest started in the 1950s for citrus, as a means to increase efficient picking in the groves and reduce labour costs (Whitney et al., 2012). Machines incorporating moveable platforms at different levels on which the pickers stand were proposed to replace the traditional ladders used to pick the highest fruits, which allowed continuous removal of fruits (Jutras and Coppock, 1958). These machines based on self-propelled units achieved an increase in performance of around 40% when compared to conventional methods (Coppock and Jutras, 1960). However, Seamount and Opitz (1973) compared a multi-man platform with ladder picking in citrus crops and they did not observe any increase in productivity with the platform. Similar results were obtained by Davey and Lahiff (1978), who compared two types of multi-man platforms with hand/ladder picking of 'Valencia' oranges. A possible reason for this conclusion could be the difficulty to manoeuvre around the large tree canopies that were common in commercial orchards at that time (Allshouse, 1970). Therefore, to increase the productivity of these machines an essential factor is to provide the crop with a more uniform shape, but also to design machines that better matched the size of the trees and the planting frameworks. In fact, the harvest-assist platform is a technology that originally responded to a mid-1960s apple marketing crisis by planting high-density systems in espaliers trained into flat two-dimensional forms (Hutchinson, 1977; Brooke Peterson, 1989). More recent studies have demonstrated that the use of these machines in crops with this structure improves the harvest efficiency by between 35% and 70% (Baugher et al., 2009; Elkins et al., 2011). On the other hand, it should be kept in mind that harvest-assist equipment also improves harvesting conditions for pickers (Ferreira et al., 2018).

The technology has advanced since then. The designs of the machines have been improved in terms of occupational health and safety, ergonomics, gentle handling of the fruit, efficiency or speed, and currently there are harvest platforms on the market designed for different crops. Moreover, in recent years, computer vision systems have been optimized so as to allow them to be included in such systems. As a result, the inspection of the fruit in-field while it is being harvested offers an added value to the traditional methods of fruit harvesting.

3.1 In-field sorting on harvest-assist platforms

In-field inspection is a process that can be carried out for selective harvesting or to perform a pre-sorting of the just-harvested fruit in order to classify it into different categories. In the first case, four crop environments can be distinguished with different degrees of complexity for harvesting robots: indoor, greenhouse, orchard and open field. The open field is one of the most challenging places for picking robots to operate in, since computer vision systems are affected by constantly changing natural conditions (Leu et al., 2017). This affects both the autonomous navigation systems and the correct identification of the piece to be picked. In the case of selective harvesting, where the robot must analyse images to determine whether the product is ready to be picked, even more difficult and complex algorithms are needed. Leu et al. (2017) developed a robotic system for selective harvesting of green asparagus that included a computer vision system to recognize the size of asparagus stalks in order to distinguish and gather only asparagus ready for harvesting. A self-propelled machine for broccoli harvesting was created by Blok et al. (2016) in 2016. The size, texture and colour of the broccoli head were inspected by computer vision and only those units that had reached commercial standards were harvested using a pneumatic manipulator. In both cases, the influence of the natural light was reduced by locating the detector system under the machine in the case of the asparagus, and using a black curtain and artificial light in the case of broccoli. This emphasizes the need to effectively control the illumination of the scene in order to improve the efficiency of such outdoor imaging systems. These conditions can be achieved more easily by the computer vision systems mounted on these machines, this being one of the main advantages of inspecting the fruit once it has been gathered.

The two main fruit crops for which on-board pre-sorting systems have been developed are citrus fruits and apples. Kohno et al. (2011) developed a computer vision system mounted on a mobile harvest-assist platform for citrus fruits so as to obtain crop information. The device consisted of a differential global positioning system to produce maps relating the quality of each fruit to the tree, conveyor belts to transport the fruit, a colour camera, a NIR spectrometer and a computer as a processing unit, all powered by an electric motor and manually driven by an operator. The system was also able to measure some external and internal qualities of the fruit, such as its diameter and sugar content. This sorting system stored the information and generated maps and reports of the harvested fruit but did not have any grading system to sort the fruit in situ.

A different concept that adds the capability of sorting the fruit into three different categories was presented by Cubero et al. (2010). While the system developed by Kohno et al. (2011) could only process the fruit one by one, taking up to 12 s to inspect a single fruit without the possibility of further sorting, this system processed eight fruits per second, inspecting the fruit by size and colour, and sorting it into three categories according to combinations of these properties. To save energy and reduce the complexity they used light-emitting diodes (LEDs) and a smart camera with processing capabilities and digital I/O devices. As a result, no additional computer or electronics were needed to sort the fruit (Vidal et al., 2013). Later, a solar panel was added to improve efficiency and the algorithms were optimized (Cubero et al., 2014). More recently, the design has been improved to make it more compact, versatile and easier to use in the field but reducing the sorting categories from three to two. This system is explained in detail in the next section.

For apples, Mizushima and Lu (2011) presented a system for grading the fruit in the field that could sort the fruit according to size. This sorting system was based on a colour camera and LED lighting and was capable of inspecting the fruit at a rate of four units per second. A mobile prototype machine was developed that was able to sort apples by colour and size (Lu et al., 2016, 2017). The fruit dropped by the workers was singulated before reaching the sorting system. The machine vision was composed of a digital colour camera capable of inspecting 15 images per second in a scene illuminated by LEDs. Once inspected, the fruit was distributed in different bin fillers according to the quality parameters estimated by the vision system. To reduce the damage in the fruit due to the fall into the bin filler, Zhang et al. (2017a) developed a sensorized bin filler to monitor and measure the status of the apples falling into the bin. Bruise damage by impact or friction is a critical factor in the development of a harvest and in-field pre-sorting prototype, and reducing the damage caused to the fruit during inspection and transport is a priority. Pothula et al. (2018) studied how to improve the bin filling in this machine by using an impact recording device in order to reduce damage to the apples, especially when the bin filler is empty.

4 Case study: in-field pre-sorting of citrus

An example of an autonomous mobile platform to assist in the harvesting of citrus fruits that can inspect and sort the fruit in the field has been developed by the Agricultural Engineering Centre of the Valencian Institute of Agricultural Research (IVIA). During harvesting, the machine is driven past the crops at a very low speed by remote control. At the front of the machine there is a hopper where the pickers who harvest the fruit on foot deposit the fruit from their bags or boxes. In addition, two raised platforms at different heights (1.25 m and 1.90 m) located on the sides of the machine allow workers to harvest fruits located in the higher parts of the tree. To facilitate the work, the platforms can be extended laterally by means of a hydraulic system to approach the tree. The harvesters located on these platforms deposit the fruit on conveyor belts that take all this fruit to the hopper of the machine. All the fruit in the hopper is then raised by a conveyor belt to the sorting area located at the top of the machine, where individualizers align the fruit on four lanes of rotating roller conveyors. The conveyor that raises the fruit has been carefully designed so that it spreads the fruit across the entire width of the belt and does not raise an excessive amount of fruit that can saturate the individualizers. Later, the roller conveyors move the fruit at high speed through the field of vision of two artificial vision cameras (each covering two lanes). The rollers make the fruit rotate while the camera takes the images, so that most of the surface of the fruit can be inspected. After inspection, the fruit can be separated into two different categories depending on different quality parameters that can be programmed by the user.

Two prototypes have already been developed. The first prototype (Fig. 1) allowed the fruit to be separated into three different categories, but this increased the size, cost and complexity of the machine. Moreover, the benefit to be gained between separating the fruit into two or three categories was not high for the grower. In addition, the fruit was transported by only two lanes and inspected using only one camera. The need for high speed sometimes caused an excessive accumulation of fruit at the entrance to the vision system. Therefore, it was decided to redesign the machine to transport the fruit by four lanes, inspect it with two cameras and separate it into just two categories.

Figure 1 First prototype of the harvest-assist platform for citrus developed at IVIA.

Therefore, the second prototype, which is currently undergoing testing, contains two zones where the fruit can be deposited into bins. One is located in the central area of the machine and is intended to collect the fruits for the second category or that needs de-greening, and the other is located in the rear area of the machine, where the high-quality fruit is gently deposited in the bin using a pinwheel. In this machine, instead of keeping the bin initially in an elevated position and lowering it as it fills to avoid strong impacts, the fruit is deposited gently in the pinwheel using a conveyor belt with fingers. The pinwheel is placed at the beginning in a low position close to the floor of the bin and it is elevated as it is filled (Fig. 2). Empty bins are transported on a trailer. When a bin is filled, it is deposited on the ground and another empty bin enters the machine from the trailer. These actions are carried out automatically by means of commands from the remote control.

The vision system consists of an inspection chamber that houses two colour industrial cameras and a lighting system based on strips of LEDs. The first prototype was equipped with only one camera with a built-in microprocessor and digital input and outputs to control the sorting system. In this way, since all the image acquisition and processing was carried out in the camera, there was no need for an additional computer. This scheme is simpler and avoids adding

Figure 2 Bin filling area of fruit in a better category.

greater complexity to the whole system. However, the camera is more expensive and the processing capacity is limited, which prevented, for example, defects in the fruits from being identified despite optimizing the algorithms to make them extremely fast (Vidal et al., 2013).

To light the scene, there are a total of four strips of 500 mm per lane, two above and two on the sides, each of them with 50 LEDs with a power rating of 0.3 W and a colour temperature of 6000 K. Due to the limited energy of the machine, the LEDs were powered by means of a stroboscopic mode synchronized with the image acquisition, which allowed a saving of about 93% of the energy consumed by the LEDs. Polarizing filters were used in front of the camera lens and the lamps to achieve cross-polarization and prevent the appearance of bright spots that could hide defects or alter the natural colour of the fruit.

In the traditional design of computer vision systems for fruit inspection, the camera is located vertically above the conveyor belt, which requires a high inspection chamber to cover a scene that includes several fruits. This design increases the height of machines that are already high. The easy solution is to use a wide-angle lens but then other problems such as distortion of the image arise and this makes it difficult to perform accurate measurements of the fruit. Another solution is to reduce the height of the inspection chamber by changing the geometry. Instead of the typical vertical arrangement, the cameras are placed parallel to the movement of the fruit and a mirror is used to keep the distance between the camera and the scene, as depicted in Fig. 3. Thus, the total height of the inspection chamber is reduced to only 330 mm from the conveyor belt to the top of the cover (Cubero et al., 2010).

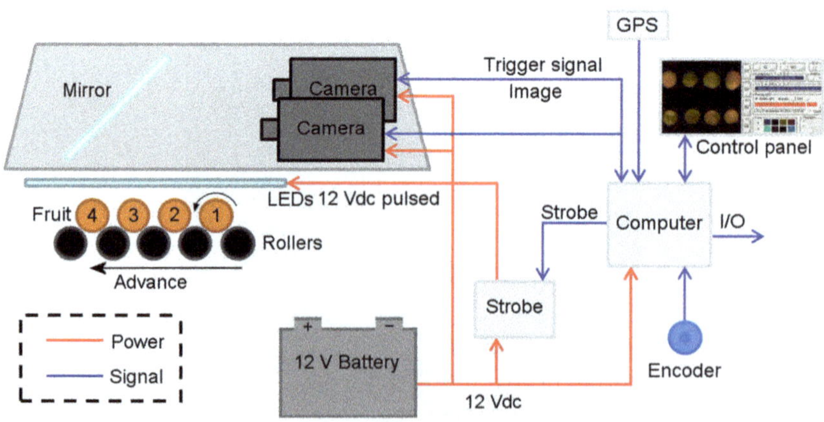

Figure 3 The computer vision system. The numbers inside the fruits represent the four positions in which the system acquires the images of each fruit.

Each camera captures a scene composed of eight fruits (four in each lane). The images are captured each time a fruit advances the equivalent distance between two rollers, that is, the positions of the fruits are always known to be the same. To achieve this, the trigger of the cameras and the illumination is synchronized to the conveyor by using a shaft encoder. In Fig. 3, fruit 1 is first captured in this position. In the next image, all the fruits have advanced so fruit 1 will be located in position 2 showing another part of the surface and so on until it leaves the scene. As the fruits rotate while advancing, four images are captured showing different faces of the fruit. Although the rotation movement is controlled, it cannot be guaranteed that 100% of the surface of the fruit can be seen, since the rotation is influenced by the size of the fruit or the transport movements, but it is precise enough to achieve the objectives of a pre-sorting in the field.

To increase the processing speed, instead of looking for each fruit individually, as the image is captured in a synchronized way, it is divided into eight regions where each fruit is expected to be found. The segmentation of the image is performed with a supervised technique that needs previous training using test images. This training is carried out manually, by selecting regions of interest and associating the colour of the pixels inside them to the different classes into which the image is to be divided: these classes are the background, orange skin (light and dark), green skin (light and dark) and defect (light and dark), but the detection of defects is currently still under development. From the selection and manual assignment of the classes, a Bayesian linear discriminant analysis is performed to obtain functions that allow classification of any possible colour within one of the predefined classes (Blasco et al., 2009). Once this step has been completed, all the possible colour values of the image are classified and this classification is stored in a table. During the in-line inspection of the harvested fruit, the segmentation algorithm only has to obtain the colour of a pixel and check the class to which it belongs in the table, this being an extremely fast process.

After segmentation, the image is divided into a number of regions of interest (classes). Following this step, an analysis is performed within each of the eight zones of the image to find the fruits (sum of all the classes except the background). For each fruit, the index of the citrus colour (ICC) (Jiménez-Cuesta et al., 1981) is estimated but only using the regions of the skin (defects are not taken into account for greater accuracy). As the ICC is calculated using the Hunter Lab colour space, as shown in Eq. (1), the average RGB of each side of the fruit is calculated and the resulting value converted to *L, a, b* values in order to estimate the index (Vidal et al., 2013). Finally, parameters to estimate the sizes of the fruit are also calculated (largest and smallest diameter, average diameter and area). The colour is calculated as the average of the four images captured for each fruit but the size is estimated with the largest of the four values (Cubero et al., 2014). The ongoing development will soon include the number and size of the defects, and the total defective area calculated from the sum of all images.

$$ICC = \frac{(1000 \times a)}{(L \times b)} \tag{1}$$

The classification of the fruit into one of the possible categories and the parameters and thresholds to decide whether the fruit falls into one category or another can be selected by the grower through a control panel. In addition, the platform incorporates a GNSS receiver (global navigation satellite system) that allows all the data obtained for each fruit to be associated with the location in which it was collected, taking into account the estimation of the time elapsed since the fruit is harvested to the moment it is inspected. This will allow them to be used for statistical purposes and to create quality maps of the fruit that provide the grower with a valuable tool to detect or sense some problems such as water stress, vegetative status and potential presence of plagues or diseases that allow decisions to be made about the handling of the crop.

In practice, at the beginning of the season, the classification system allows the fruit to be separated by two levels of colour so that when it arrives at the packinghouse it can be separated for different de-greening treatments (or no treatment) of an appropriate duration, depending on the colour at harvest. As the season advances, it becomes more important to be able to perform in-field detection and separation of fruits not reaching the minimum quality required for the fresh fruit market. This fruit can then be trashed or delivered to the food processing industry directly from the field. Figure 4 shows the second prototype under development.

Figure 4 Second prototype of the harvest-assist platform for citrus developed at IVIA. The computer vision system is shown at the top without the cover.

5 Summary

Harvest-assist platforms have been used since the 1950s to optimize the efficiency of harvesting and also to improve the pickers' conditions. Currently, the most advanced machines can incorporate efficient computer vision systems installed on the conveyor belts that transport the fruit from the picking points to the bin fillers. These systems allow in-field inspection and pre-grading of the harvested fruit. Detecting and separating a particular fruit in the field offers a series of advantages. On the one hand, separating fruits by size, colour or presence of external defects, depending on their need for storage or further treatments on arrival at the packinghouse, makes it possible to optimize subsequent operations. On the other hand, separating low-quality fruits in the field prevents them from being mixed, transported or treated together with higher quality fruits, thus saving the costs of handling these fruits.

However, inspection systems in the field also present a series of problems and limitations in terms of space, energy and computational power, which requires important efforts to be made in the design of the systems and equipment. In addition, image analysis algorithms must be optimal to achieve the speeds required to inspect and sort all the harvested fruit. Finally, the separation systems should not damage the fruit, since separation by quality makes more sense when the fruit is destined to the fresh market and any defect or damage caused to the fruit should therefore be avoided.

6 Future trends in research

Agriculture tends towards a higher degree of mechanization and automation. In the particular case of technology applied to automatic picking and selection in the field, the advancement of technology, communications and the processing of data to create decision-making tools all play in favour of this automation.

Within the scope of in-field sorting systems, the incorporation of a new up-to-date technology involving particular vision systems will play an important role. The technology used for in-field sorting takes advantage of the advances in the computer vision systems developed to be incorporated into the electronic sorters used in packinghouses. The use of thermal and multispectral cameras that can obtain measures of the fruit at certain wavelengths will allow fruits to be sorted according not only to their external appearance but also to their internal properties. The advances in information and communication technology will make it possible to equip these machines with the ability to connect and interact with the packinghouse during the harvest operations. Hence, when the fruit arrives at the packinghouse, its condition and external and internal qualities will already be known and therefore the subsequent tasks will be streamlined. This can result in a more efficient management of warehouses leading to the creation of optimal market strategies to increase profits while still satisfying the demands of consumers and markets. Ideally, these machines could even have the capacity to perform a pre-harvest selection, so that fruits are only picked on demand. Only those fruits that are needed according to their ripeness state or market needs at that time will be gathered.

The capabilities of these machines to separate fruits into categories are limited. Therefore, beyond classification into quality categories, which allows considerable savings by preventing lower quality fruit from going to the packinghouse together with healthy fruit, it is important to analyse all the information that the inspection systems of these machines can offer on the whole harvest. At present, the information extracted from the inspection of the fruit, both during and after harvest, is lost once the fruit is classified into commercial categories. However, within the framework of Industry 4.0, the information on the quality and composition of each fruit combined with the exact point of collection, together with the information extracted from the traceability of each fruit in post-harvest from the moment it enters the warehouse until it reaches the market, must allow the generation of knowledge that makes it possible to create new tools for crop management. In addition, this management must be more sustainable, more competitive and more oriented to what the consumer demands.

Moreover, this on-board technology to inspect harvested fruit can be combined with other sensors capable of inspecting different crop or soil features. At present, there are applications that obtain different types of maps related to the vegetative state of the crop, crop prediction, soil quality, presence of pests or diseases, and so on. The information obtained from the field inspection of the fruit provides these maps with an added value, by including the real quality of the harvest obtained at the exact picking point. In this way, it would be possible to compare the real influence of these parameters on the final result of the harvest and make more scientific decisions about crop management.

7 Where to look for further information

Most of the available information is focused on robotics in agriculture, on harvest-assist platforms and on post-harvest fruits grading systems, separately. The pre-grading of fruits and vegetables in the field during the harvest is an emerging combination of these worlds and little information is still available. Some centres of expertise include:

- The Agricultural Engineering Centre of the Instituto Valenciano de Investigaciones Agrarias (IVIA) in Spain (http://www.ivia.gva.es/es/centro-de-agroingenieria), headed by Professor Dr. Jose Blasco.

- The USDA-ARS Sugarbeet and Bean Research Unit in Michigan (USA) (https://www.ars.usda.gov/midwest-area/east-lansing-mi/sugarbeet-and-bean-research), headed by Dr. Renfu Lu.
- Graduate School of Agriculture, Kyoto University (Japan) (http://www.aptech.kais.kyoto-u.ac.jp/e/summary/sum6.html), headed by Professor Naoshi Kondo.

Most related scientific works are included in the references section of this chapter. However, among them the below works are probably the most recommendable for further reading.

- Lu, R., Zhang, Z. and Pothula, A. K. (2017). Innovative technology for apple harvest and in-field sorting. *Fruit Quarterly*, 25(2), 11–14. Available at: http://nyshs.org/wp-content/uploads/2017/12/Lu-Pages-11-14-from-NYFQ-Book-Summer-2017-2.pdf
- Cubero, S., Aleixos, N., Albert, F., Torregrosa, A., Ortiz, C., García-Navarrete, O. and Blasco, J. (2014). Optimised computer vision system for automatic pre-grading of citrus fruit in the field using a mobile platform. *Precision Agriculture*, 15(1), 80–94.
- Kohno, Kondo, N., Iida, M., Kurita, M., Shiigi, T., Ogawa, Y., Kaichi, T. and Okamoto, S. (2011). Development of a mobile grading machine for citrus fruit. *Engineering in Agriculture, Environment and Food*, 4(1), 7–11.

8 References

Allshouse, G. W. (1970). A picking platform to aid in fruit harvesting. Progress Report. PA Agric. Exp. Stat., 301, bibl. 8.

Bac, C. W., van Henten, E. J., Hemming, J. and Edan, Y. (2014). Harvesting robots for high-value crops: State-of-the-art review and challenges ahead. *Journal of Field Robotics*, 31(6), 888–911.

Baugher, T., Schupp, J., Lesser, K., Harsh, R. M., Seavert, C., Lewis, K. and Auvil, T. (2009). Mobile platforms increase orchard management efficiency and profitability. *Acta Horticulturae*, 824, 361–4.

Bechar, A. (2010). Robotics in horticultural field production. *Stewart Postharvest Review*, 6(3), 1–11.

Bechar, A. and Vigneault, C. (2017). Agricultural robots for field operations. Part 2: Operations and systems. *Biosystems Engineering*, 153, 110–28.

Blasco, J., Aleixos, N., Roger, J. M., Rabatel, G. and Moltó, E. (2002). Robotic weed control using machine vision. *Biosystems Engineering*, 83(2), 149–57.

Blasco, J., Cubero, S., Gómez-Sanchis, J., Mira, P. and Moltó, E. (2009). Development of a machine for the automatic sorting of pomegranate (*Punica granatum*) arils based on computer vision. *Journal of Food Engineering*, 90(1), 27–34.

Blok, P. M., Barth, R. and van den Berg, W. (2016). Machine vision for a selective broccoli harvesting robot. *IFAC-PapersOnLine*, 49(16), 66–71.

Bochtis, D. D., Sørensen, C. G. C. and Busato, P. (2014). Advances in agricultural machinery management: A review. *Biosystems Engineering*, 126, 69–81.

Brooke Peterson, A. (1989). *Intensive Orcharding*. Good Fruit Grower, Yakima, Washington DC.

Conesa-Muñoz, J., Bengochea-Guevara, J. M., Andujar, D. and Ribeiro, A. (2016). Route planning for agricultural tasks: A general approach for fleets of autonomous vehicles in site-specific herbicide applications. *Computers and Electronics in Agriculture*, 127, 204–20.

Coppock, G. E. and Jutras, P. J. (1960). Mechanizing citrus fruit harvesting. *Transaction of the ASAE*, 3(2), 130–2.

Cubero, S., Moltó, E., Gutiérrez, A., Aleixos, N., García-Navarrete, O. L., Juste, F. and Blasco, J. (2010). Real-time inspection of fruit on a mobile harvesting platform in field conditions using computer vision. *Progress in Agricultural Engineering Science*, 6, 1–16. doi:10.1556/Progress.6.2010.1

Cubero, S., Aleixos, N., Albert, F., Torregrosa, A., Ortiz, C., García-Navarrete, O. and Blasco, J. (2014). Optimised computer vision system for automatic pre-grading of citrus fruit in the field using a mobile platform. *Precision Agriculture*, 15(1), 80–94.

Cubero, S., Lee, W. S., Aleixos, N., Albert, F. and Blasco, J. (2016). Automated systems based on machine vision for inspecting citrus fruits from the field to postharvest – a review. *Food and Bioprocess Technology*, 9(10), 1623–39.

Davey, P. C. and Lahiff, P. (1978). Mechanical aids for citrus harvesting. *Proceedings of the International Society of Citriculture*, 2, 105–6.

Elkins, R. B., Meyers, J. M., Duraj, V., Miles, J. A., Tejeda, D. J., Mitcham, E. J., Biasi, W. V., Asín, L. and Abreu, J. (2011). Comparison of platform versus ladders for harvest in Northern California pear orchard. *Acta Horticulturae*, 909, 241–9.

El-Ramady, H. R., Domokos-Szabolcsy, É., Abdalla, N. A., Taha, H. S. and Fári, M. (2015). Postharvest management of fruits and vegetables storage sustainable agriculture reviews. In: E. Lichtfouse (Ed.), *Sustainable Agriculture Reviews*, vol. 15. New York, NY: Springer, pp. 65–152.

Fernández-Quintanilla, C., Peña, J. M., Andújar, D., Dorado, J., Ribeiro, A. and López-Granados, F. (2018). Is the current state of the art of weed monitoring suitable for site-specific weed management in arable crops? *Weed Research*, 58(4), 259–72.

Ferreira, M. D., Sanchez, A. C. Braunbeck, O. A. and Santos, E. A. (2018). Harvesting fruits using a mobile platform: A case study applied to citrus. *Engenharia Agrícola, Jaboticabal*, 38, 293–9

Gongal, A., Amatya, S., Karkee, M., Zhang, Q. and Lewis, K. (2015). Sensors and systems for fruit detection and localization: A review. *Computers and Electronics in Agriculture*, 116, 8–19.

Gonzalez-de-Santos, P., Ribeiro, A., Fernandez-Quintanilla, C., Lopez-Granados, F., Brandstoetter, M., Tomic, S., Pedrazzi, S., Peruzzi, A., Pajares, G., Kaplanis, G., Perez-Ruiz, M., Valero, C., del Cerro, J., Vieri, M., Rabatel, G. and Debilde, B. (2017). Fleets of robots for environmentally-safe pest control in agriculture. *Precision Agriculture*, 18(4), 574–614.

Hutchinson, A. (1977). Establishment costs and management of high density orchards. *Proceedings of the Annual Meeting Massachusetts Fruit Growers Associations Inc.*, 83, 55–68.

Jiménez-Cuesta, M. J., Cuquerella, J. and Martínez-Jávega, J. M. (1981). Determination of a color index for citrus fruit degreening. *Proceedings of the International Society of Citriculture, Tokyo (Japan)*, 2, 750–3.

Jutras, P. J. and Coppock, G. E. (1958). Mechanization of citrus fruit picking. *Florida State Horticultural Society*, 71, 201–4.

Kohno, Y., Kondo, N., Iida, M., Kurita, M., Shiigi, T., Ogawa, Y., Kaichi, T. and Okamoto, S. (2011). Development of a mobile grading machine for citrus fruit. Engineering in Agriculture, *Environment and Food*, 4(1), 7–11.

Kubinger, W., Peschak, B., Woeber, W. and Sulz, C. (2017). Machine vision systems for robotic systems in agricultural robotics and agricultural engineering. *Elektrotechnik und informationstechnik*, 134(6), 316–22.

Leu, A., Razavi, M., Langstädtler, L., Ristić-Durrant, D., Raffel, H., Schenck, C. and Kuhfuss, B. (2017). Robotic green asparagus selective harvesting. *IEEE/ASME Transactions on Mechatronics*, 22(6), 2401–10.

Lorente, D., Aleixos, N., Gómez-Sanchis, J., Cubero, S., García-Navarrete, O. L. and Blasco, J., (2012). Recent advances and applications of hyperspectral imaging for fruit and vegetable quality assessment. *Food Bioprocess Technology*, 5, 1121–42.

Lu, R., Pothula, A. K., Vandyke, M., Mizushima, A. and Zhang, Z. (2016). System for sorting fruit. U.S. Patent 15/256,931.

Lu, R., Zhang, Z. and Pothula, A. K. (2017). Innovative technology for apple harvest and in-field sorting. *Fruit Quarterly*, 25(2), 11–14.

Mazzetto, F., Calcante, A., Mena, A. and Vercesi, A. (2010). Integration of optical and analogue sensors for monitoring canopy health and vigour in precision viticulture. *Precision Agriculture*, 11(6), 636–49.

Mizushima, A. and Lu, R. (2011). Development of a cost-effective machine vision system for in-field sorting and grading of apples: Fruit orientation and size estimation. ASABE Paper No. 1110723. St. Joseph, MI: ASABE.

Nieuwenhuizen, A., Hofstee, J. W., and Van Henten, E. J. (2010). Adaptive detection of volunteer potato plants in sugar beet fields. *Precision Agriculture*. 11. 433–447. doi: 10.1007/s11119-009-9138-9

Pothula, A. K., Zhang, Z. and Lu, R. (2018). Design features and bruise evaluation of an apple harvest and in-field presorting machine. *Transactions of the ASABE*, 61, 1135–44.

Qiao, J., Sasao, A., Shibusawa, S., Kondo, N. and Morimoto, E. 2005. Mapping yield and quality using the mobile fruit grading robot. *Biosystems Engineering*, 90(2), 135–42.

Ruiz-Altisent, M., Ortiz-Cañavate, J. and Valero, C. (2004). Fruit and vegetables harvesting systems. In: R. Dris and S. M. Jain (Eds), *Production Practices and Quality Assessment of Food Crops, Vol. 1: Preharvest Practice*. Dordrecht, The Netherlands: Kluiwert Academic Publishers, pp. 261–85.

Seamount, D. T. and Opitz, K. W. (1973). The Sunkist 2C platform harvester. *Citrograph*, 59, 55–8.

Shalal, N., Low, T., McCarthy, C. and Hancock, N. (2013). A review of autonomous navigation systems in agricultural environments. In: *SEAg 2013: Innovative Agricultural Technologies for a Sustainable Future*, 22-25 September 2013, Barton, Western Australia.

Thanpattranon, P., Ahamed, T. and Takigawa, T. (2016). Navigation of autonomous tractor for orchards and plantations using a laser range finder: Automatic control of trailer position with tractor. *Biosystems Engineering*, 147, 90–103.

Torregrosa, A., Albert, F., Aleixos, N., Ortiz, C. and Blasco, J. (2014). Analysis of the detachment of citrus fruits by vibration using artificial vision. *Biosystems Engineering*, 119, 1–12.

Tremblay, N., Fallon, E. and Ziadi, N. (2011). Sensing of crop nitrogen status: Opportunities, tools, limitations, and supporting information requirements. *HortTechnology*, 21(3), 274–81.

Vidal, A., Talens, P., Prats-Montalbán, J. M., Cubero, S., Albert, F. and Blasco, J. (2013). In-line estimation of the standard colour index of citrus fruits using a computer vision system developed for a mobile platform. *Food and Bioprocess Technology*, 6(12), 3412–19.

Whitney, J. D., Hyman, B. R. and Roka, F. M. (2012). The past, present and future of citrus mechanical harvesting. *Acta Horticulturae*, 965, 183–8.

Xu, K. 2016. Making apple trees friendlier for mechanized harvesting. *New York Fruit Quarterly*, 24(3): 29-32.

Zarco-Tejada, P. J., Camino, C., Beck, P. S. A., Calderon, R., Hornero, A., Hernández-Clemente, R., Kattenborn, T., Montes-Borrego, M., Susca, L., Morelli, M., Gonzalez-Dugo, V., North, P. R. J., Landa, B. B., Boscia, D., Saponari, M. and Navas-Cortes, J. A. (2018). Pre-visual symptoms of *Xylella fastidiosa* infection revealed in spectral plant-trait alterations, *Nature Plants*, 4, 432–9

Zhang, Z., Pothula, A. K. and Lu, R. (2017a). Development and preliminary evaluation of a new bin filler for apple harvesting and in-field sorting machine. *Transactions of the ASABE*, 60(6), 1839–49.

Zhang, Z., Pothula, A. K. and Lu, R. (2017b). Economic evaluation of apple harvest and in-field sorting technology. *Transactions of the ASABE*, 60(5), 1537–50.

Zhang, Z., Pothula, A. K. and Lu, R. (2018). A review of bin filling technologies for apple harvest and postharvest handling. *Applied Engineering in Agriculture*, 34(4), 687–703.

Zhao, Y., Gong, L., Huang, Y. and Liu, C. (2016). A review of key techniques of vision-based control for harvesting robot. *Computers and Electronics in Agriculture*, 127, 311–23.

Chapter 11

Advances in using robots in forestry operations

Ola Lindroos and Omar Mendoza-Trejo, Swedish University of Agricultural Sciences (SLU), Sweden; Pedro La Hera, Swedish University of Agricultural Sciences (SLU) and The Cluster of Forest Technology, Sweden; and Daniel Ortiz Morales, Cranab, Sweden

1 Introduction

Forestry is a term that spans a wide range of activities, from the agricultural-like cultivation of rapidly growing trees to the harvesting of pristine forests. Hence, the operations related to forestry are similarly diverse. When focusing on managed forests, the fundamental processes are similar to the cultivation of any product: to establish, manage and harvest a particular crop. However, the time span for the rotation cycle and the size of the harvested crop differs

http://dx.doi.org/10.19103/AS.2019.0056.18
Published by Burleigh Dodds Science Publishing Limited, 2019.

substantially for forestry compared with other cultivated products. Moreover, how forestry operations are carried out varies greatly around the world. Current practices are adapted to complex, locally variable conditions, such as geophysical characteristics (terrain), management regimes, tree properties, climate, ownership structure, industrial infrastructures, labour availability and capacity, and societal rules for acceptable practices.

Since the twentieth century, timber harvesting has progressed from being entirely manual and animal-powered to being fully mechanized and to some extent automated (Silversides, 1997). Although many aspects of forestry operations are highly mechanized, there are still some that are rarely mechanized (e.g. planting), and many geographical regions where mechanized options are not applied or viable (e.g. tree felling in steep terrain and of large trees).

An interest in automated forestry operations developed soon after the first applications of mechanization. Examples can be seen in the symposium on 'Forest Harvesting Mechanization and Automation' held in 1974 (Silversides, 1974) by a division of the International Union of Forest Research Organizations (IUFRO) and a Swedish workshop on 'Automation and Remote Controlling of Forest Machinery' held in 1983 (Uusijärvi, 1985). More than a decade later, ideas for fully automated, but supervised, logging systems were described (Hallonborg, 1997). More recent publications have summarized state-of-the-art developments and considered possible future steps (e.g. Bayne and Parker, 2012; Billingsley et al., 2008; Hellström et al., 2009; Lindroos et al., 2017; Milne et al., 2013; Parker et al., 2016). There have been plenty of innovative projects over the years attempting to automate the technology used in forestry operations. Some early examples are machines aimed at autonomously carrying out tree felling, delimbing and piling of stems (Golob, 1981), and robots aimed at autonomously performing weeding, pre-commercial thinning and thinning of young coniferous stands (Kourtz, 1996). However, few innovations involving automation technology have successfully reached the market.

Forestry falls within the category of outdoor applications, which is still a challenging area for the robotics community. Forest is a highly unstructured environment, and very few developments seen in the field of robotics are directly applicable to this situation. Most developments in current robotics research take place in simpler outdoor applications than those existing in forestry, for example driving cars, rescue missions, exploration of hazardous environments and surveillance. The main focus in these applications is the ability to provide precise data, such as global and local positioning, as most of them are for use around cities, with easy access to the internet or some other type of communication. The technology to achieve this within forestry operations does not yet exist. Unlike other industries, forestry is still a young research field in the application of robotics, because the data needed to develop advanced autonomous or semi-autonomous applications is still absent. For instance, there is lack of computer vision ability, and global and local positioning capability; additionally, current forestry machines have little or no sensing equipment to understand their own work. Hence, most research effort is aimed at acquiring this type of data, such as research focusing on developing sensors and carriers for forestry machinery (e.g. Lideskog et al., 2015). Advances in methods of collecting, handling and using data are, then, the stepping stones to smart machines and eventually automation. Although there is still a long journey to go before some level of fully automated machines is reached, there are plenty of interesting advancements within the field of forestry operations.

This chapter presents the current state of robotics in advanced, but yet conventional, forestry operations. The focus is on commercially available products, with some selected examples of ongoing research as an indication of what can be expected in the near future in terms of teleoperated, semi-autonomous and fully autonomous forestry machines. Less attention is given to the many creative and futuristic concepts that constantly arise across various media, because of the less likely chance that they will someday actually become part of forestry operations. However, the trends and possible directions for future technologies are discussed in Section 13.

When addressing automation it is important to mention that the factors that need to be taken into account to achieve automation coincide with efforts to improve the current, manual, forestry operations, such as occupational health and safety issues and improving work efficiency (by analysing and improving work methods and providing decision support systems (DSS)). Thus, development efforts will directly benefit operators in many ways, even before, in the far future, resulting in fully autonomous forestry operations.

To be able to understand the automation process for forestry operations, it is necessary to be familiar with the conditions under which the processes take place. Thus, Section 2 describes the particular challenges to automation of forestry operations. We then present different sections based on the different fields of research required to solve those challenges. The chapter focuses primarily on timber harvesting systems, although automation of planting and silvicultural machinery is addressed to some extent in Section 13. Automating forestry operations is a global activity, but this chapter focuses particularly on the Nordic cut-to-length (CTL) harvesting system, which uses harvesters (Fig. 1) and forwarders (Fig. 2). That is partly because the authors are based in Sweden, but also because the Nordic CTL system is the most technologically advanced in the world.

Figure 1 A Nordic CTL harvester, Komatsu 901XC, in an unstructured, sloping environment demonstrating a self-levelling cabin function. Photo courtesy of Komatsu Forest AB.

Figure 2 A Nordic CTL forwarder, Ponsse ElephantKing, in an example of harsh work conditions encountered during forestry operations: darkness, snow and an unstructured environment. Photo courtesy of Ponsse Oyj.

2 Challenges to using robots in forestry operations

Within the field of forestry operations research, the themes of Forestry 4.0 and Internet of Things are hot topics. In this respect, research is keeping up with contemporary technological development. This might seem at odds with the fact that there are still parts of the forestry operations that are not even mechanized, and that there are limited possibilities for automatically monitoring production within some mechanized operations. In fact, within a global context, most forestry operations rely on robust low-tech equipment, which is far from being connected to the internet. This global variation in levels of technology can also be found in agriculture, but the reasons for this are somewhat different.

The major challenge to using autonomous machines in forestry operations is the work environment, which is off-road, covers rough terrain and is located in remote areas. Outdoor work over rough terrain makes automation extremely difficult, compared with automation in controlled environments. The appearance of the forest environment can also vary greatly, as a result of light conditions, precipitation, wind and so on. Thus, a given location might look very different depending on whether, for instance, the surrounding trees are fully leaved, still and well lit, or whether the leaves have been shed and the branches are moving yet barely visible because of a snowstorm. Moreover, the time span of forest management means visits to any particular forest area are rare, so investment in infrastructure is normally kept to a minimum. When any work is carried out, it involves challenging decisions related to production as well as management of environmental impacts and social values.

The actual work carried out by forestry operations has not only many similarities with agricultural operations but also many essential differences. When focusing on managed forests, the re-establishment of a forest stand involves many familiar cultivation processes, such as soil preparation (scarification), planting or sowing, suppression of unwanted plants (e.g. weeding and pre-commercial thinning) and the management of pests and diseases.

Although a very important part of forestry operations, the early phases of the silvicultural cycle (e.g. the rotation cycle) are given little attention in this chapter, because the processes are either too similar to agricultural practices (e.g. mechanical site preparation) to warrant a separate focus or they have not yet been mechanized within forestry (e.g. planting, weeding and pre-commercial thinning). There are exceptions, but they normally occur in plantations with relatively short rotation periods and management regimes that make them similar to agriculture.

The latter part of the silvicultural cycle includes the harvesting of various sizes of tree. With large-sized trees, there are obviously specific demands for the machines used to handle them. Hence, these operations differ substantially from harvesting operations in agriculture. The main focus of this chapter is therefore operations related to tree harvesting.

Tree harvesting can be divided into five distinct work elements: (1) accessing/reaching the tree, (2) felling the tree, (3) debranching the tree, (4) cross-cutting the stem/tree (bucking) and (5) transporting the stem/log/tree to a roadside landing. All five elements have to be carried out to enable the delivery of roundwood logs for further processing, but in what order and where they are carried out can differ greatly between operations. Irrespective of which work element(s) a given machine is designed to carry out autonomously, one of the first technological challenges is being capable of advanced localization and decision-making. For instance, the machine should be capable of knowing where it is located, and the status and location of its parts. It should also be aware of the surrounding environment, and how the work objects (trees/stems/logs) are placed within it, their qualitative features and so forth. Consequently, it should possess the computing ability to decide how to carry out the work. In other words, an intelligent machine has to possess basic human operator abilities through sensing and computing. And, as mentioned, this needs to be possible under the most challenging conditions possible: an unstructured and highly variable outdoor environment.

Forestry machines need to be robust to endure the demanding conditions in which they operate. The automation challenges that need to be overcome are in many cases, similar to those encountered in the development of space and military applications (unstructured, unknown terrain to be navigated under extreme conditions), but with a much smaller research budget. As only a few thousand dedicated forestry machines are produced around the globe annually, the resources available for development are limited. And, naturally, the machines must be cost efficient, which has resulted in two distinct types of product: cheap low-tech machines and expensive high-tech machines. Nordic CTL machines are an example of the latter, for which the philosophy is that an expert operator with a higher salary needs a highly productive machine that can output at a low cost per unit. Thus, the more advanced machines are rather expensive, costing several hundred thousand euros, with small margins for increasing machine costs. Thus, the cost of any additional function has to be covered by increased machine or operator productivity. However, with autonomous machines the cost of the operator will be excluded, which can allow for higher machine costs at the same productivity, or lower productivity at the same machine cost if the machine can work for longer days.

This short overview just skims the surface of the conditions under which forestry operations are conducted, with the aim of introducing the main challenges encountered in the automation of forestry operations. It also serves as an introduction to the structure of this chapter, in order to discuss the advances in automation of forestry operations. Readers interested in more information on the challenges of forestry operations and the drivers for technological advancement are directed to, for example Häggström and Lindroos (2016) and Lindroos et al. (2017).

3 Knowing the state of the machine

Most forestry machines, irrespective of whether they are more or less advanced, share some common features. A diesel engine generates the power to operate hydraulic cranes and tools, and often the locomotive transmission also includes a hydraulic step.

Many machines are still controlled by very old, but robust, techniques (e.g. an electro-mechanical open loop hydraulic control), with the help of distributed controller area network (CAN)-based control systems. In the more advanced machines there has been, however, rapid development of hardware and software to improve the human–machine interaction. Current developments are addressing the need to monitor daily operations automatically in terms of, for instance, production levels and fuel consumption (Manner et al., 2016a,b; Pierzchała et al., 2018b). To this end, machines are equipped with traditional sensor hardware, similar to many other machines in different industries. Nordic CTL machines can provide detailed information on their performance, including fuel consumption, production and idling time (e.g. Eriksson and Lindroos, 2014; Manner et al., 2016a,b). Such data is saved automatically and can be transferred wirelessly, if such facilities are available at the work site. Thus, automatic follow-up of conventional operations is advancing rapidly, although many challenges remain. Ownership of the data produced, for instance, is under discussion for the Nordics.

At the other end of the spectrum, current developments involving the actual use of automation technology are largely focused on the use of intelligent hydraulic valves: a technology that has only recently appeared on the market. This type of valve is equipped with digital electronics and dedicated sensing hardware, with the potential for the software to be improved for dynamic motion control of the machine (Mathworks, 2016; Danfoss, 2015). Such sensors enable analysis of the operators' conventional work as well as the possibility of improving the working methods and automating the movements of the crane. This will be addressed further, mainly in Section 8.

Automatically levelling cabins is a standard function of many forestry machines (Fig. 1), which can be seen as an early application of a sensor feedback control system. Forestry machines have also been equipped with sensors in research projects to identify, for instance, the position of the crane (Hyyti et al, 2018; Lindroos et al., 2015), wheel slippage (Ringdahl et al., 2012b; Suvinen and Saarilahti, 2006), soil damage (Melander and Ritala, 2018) and machine slope (Visser and Berkett, 2015). With such information, it is possible to adapt the driving to compensate for the encountered conditions. For example, some machine manufacturers are starting to use sensor feedback control algorithms to improve the working performance of machines.

4 Knowing where the machine is located

The ability to locate the machine within the forest is improving, and satellite navigation systems are standard equipment for conventional Nordic CTL machines. The location of the machine's movement is automatically logged and saved, together with information on the production at each location (e.g. volume of various assortments). Global mapping is normally not a problem in current state-of-the-art operations. However, autonomous operations will also require high-precision local mapping. For this, the standard global navigation satellite system (GNSS) equipment of current forestry machines suffers from the challenge of locating the machines correctly under the forest canopy. Thus, high localization accuracy currently requires additional effort, and is normally only carried out at an experimental level (Melander and Ritala, 2018; Lindroos et al., 2015).

Such experiments include efforts to develop simultaneous location and mapping (SLAM) algorithms suitable for the highly variable forest environment. There have been several experimental platforms in forest environments investigating the possibility of equipping forestry machines with such a technique (e.g. Pierzchała et al., 2018a). The use of unmanned aerial vehicles (UAVs) as scouts to compile local maps for forestry machines has also been suggested (Talbot et al., 2017). However, so far there has been no progress with commercially available products.

5 Knowing the location of surrounding objects

Currently, no conventional forestry machine is able to sense its surroundings. However, there are plenty of ongoing research into detecting trees (Fig. 3) (e.g. Brunner and Gizachew, 2014; Hauglin et al., 2017; Oveland et al., 2018; Wells, 2018), obstacles (Lideskog, 2018; Puliti et al., 2018), people (Ostovar et al., 2016) and what effects the machines have on the terrain (Marra et al., 2018; Melander and Ritala, 2018; Salmivaara et al., 2018). There is a huge variation in the methods used, with typical techniques including computer vision via two-dimensional (2D) and three-dimensional (3D) terrestrial laser scanners (Talbot et al., 2017). Thanks to recent advancements in the areas of computer vision, technological development is expected to progress rapidly within forestry as well. The expectation is that ultimately the operator will be informed about the location and amount of trees that are surrounding the machine as well as receiving suggestions about which trees should be harvested. Similar

Figure 3 A photometry-based example of ongoing research on tree identification. Image courtesy of Lucas Wells.

technology will be fundamental for the automation of crane work. Although advancing fast, the research within this field is still at an early stage, and identifying objects in the forest is very challenging. Commercial products are not likely to be available within the coming decade.

6 Knowing how to plan the work

Because of the rapid progress in remote sensing as well as sensor technology, there has been progress with computational planning. Contemporary research comprises a plethora of DSS for various aspects of forestry work. The development of DSS will in the first instance support operators, but, more importantly in the context of this chapter, the algorithms of DSS can also be used for autonomous decision-making. Thus, the successful development of DSS can be seen as a prerequisite for autonomous forestry operations. Here, we will focus on the decisions needed to choose which trees to harvest, how to plan the movements during the harvest and where to take the harvest.

The challenges of tree selection are well-known for human operators because of the variability in tree features and the conditions in which they grow, and there are only simple heuristics for addressing this task (Vestlund and Hellström, 2006; Vestlund et al., 2005, 2006; Contreras and Chung, 2013).

Thus, given that there are difficulties in even sensing trees reliably and rapidly in research settings, DSS for tree selection is not likely to enter conventional forestry operations within the next decade.

When it comes to DSS for which paths to take during forestry work, there has been more progress, and this topic has received much attention during the last decade because of increasing concerns about soil damage. Soil damage is likely to be reduced if the operator can be guided into making better choices of where to drive, and how often. Aerial light-detecting and ranging (LiDAR) data has enabled high-precision digital elevation maps, which in turn has enabled depth-to-water (DTW) maps (e.g. Ågren et al., 2014). To provide even more information on soil-bearing capacity, machine learning has been used to combine aerial data with field-measured data (Pohjankukka et al., 2016). Although successful to a certain level of prediction accuracy, the conclusion is that the accuracy would be improved substantially with real-time data (e.g. from sensors mounted on the machine). Digital DTW maps are currently used in Nordic CTL machines to provide the operator with information about the expected soil conditions of an area; however, the operator still needs to plan the driving manually.

Algorithms for computational planning of where machines should be driven during harvesting are under development for Nordic CTL machines (Mohtashami et al., 2012; Flisberg et al., 2007). However, it is a complex problem when trying to produce an optimum route for the full operation of, for example, forwarding the harvested logs in a stand while simultaneously considering both environmental and economic factors (Hosseini et al., 2019). Software that finds the best (especially the driest) main path from the roadside landing to given points in a stand is already available, and there is ongoing research to produce DSS that takes the full network into consideration.

DSS for locating the roadside landing, to which the harvest can be taken, is comparatively well developed. Such location problems are often integrated with an analysis of the road network design, where the challenge is the trade-off between road construction costs and off-road extraction costs (e.g. Contreras and Chung, 2007; Grigolato et al., 2017; Søvde, 2015). Thus, there is relatively good support for knowing where the machine should start and where to extract the harvested products. In practice, however, the machines are not usually equipped with such DSS because the location of roadside landings is not normally an operational problem, being solved at earlier planning phases.

7 Moving around in the forest

Once there is a planned path, the next challenge is to be able to follow it. In forest environments paths are rarely straight or flat and obstacles are common. In addition, the vehicle itself is likely to have problems moving about because of the large quantity of logging residue (branches, treetops etc.) on the ground surface and potentially high variability in traction and soil-bearing capacity. Furthermore, the possibility of adapting the environment so that it complies with automation requirements is strongly limited in forestry because of ecological and social concerns. Thus, forestry operations include more complex in-field decisions than typical agricultural operations. The challenges can be divided into the machine's ability to move in rough terrain, and to follow a specified path.

7.1 Locomotion in rough terrain

Forestry operations can be executed by either ground-based machines or non-ground-based machines. Conventional ground-based machines are wheeled or tracked.

For some time there has been an abundance of impressive robotic concepts that can move in outdoor environments, such as the walking robots from Boston Dynamics (e.g. Raibert et al., 2008) and the ANYmal (Hutter et al., 2017). These are important stepping stones towards the automation of forestry operations, but their focus is often general in character. However, there are a few walking concepts that are dedicated to forestry operations, such as the PlusTech Ltd (now John Deere Ltd) harvester of the 1990s (Billingsley et al., 2008) and the more recent Portalharvester (Fig. 4) (Anon., 2013; Erler, 2013). The

Figure 4 The Portalharvester, a walking concept machine with two tripod legs and a sliding cab. Photo courtesy of Christian Knobloch.

benefits of walking machines, compared with wheeled and tracked machines, include improved negotiation of some obstacles and terrains, although such machines are limited in terms of complexity, fuel consumption and so on (Billingsley et al., 2008). From a soil damage perspective, the benefit is that only soil compression points are created and not continuous tracks. Thus, avoidance of tracks prevents the risk of blocking off roots and water from certain areas by walls of compacted soil.

Conventional Nordic CTL machines are wheeled, often with bogies and various systems to improve operator comfort when driving over sloping and uneven terrain. Self-levelling cabins have long been available (Fig. 1) and the latest commercial enhancement is active suspension (Ponsse, 2017), such that dynamic motion control is employed to make the hydraulic suspension system even out over obstacles.

Aerial forestry operations take place around the globe, but because of higher costs they are mainly used when ground-based operations are not an option. There are several conventional systems available, such as cable yarding (Lindroos and Cavalli, 2016) and heli-logging (Bigsby and Ling, 2013). Balloons were an option until the 1970s (Peters, 1973), while recent advances in UAVs mean they can be used in forestry for various monitoring purposes (Torresan et al., 2017). Given the large loads that have to be carried when harvesting or extracting trees, current UAV technology is unlikely to be used for such purposes, at least in the near future.

Tree-based locomotion is a solution that falls between ground and aerial locomotion. The brachiating (tree-to-tree moving) robot developed in New Zealand was inspired by the way monkeys move (Parker et al., 2016) (Fig. 5). As with aerial systems, it avoids soil damage and is not affected by how rough or steep the terrain is. However, as with UAVs, the work that can be carried out by climbing machines is probably limited in relation to harvesting purposes. The development of a tree-to-tree moving machine capable of tree felling might be feasible; however, the weight of logs that could be carried while climbing is probably limited, although some innovative solutions have been proposed by having several machines that can co-operate (see Section 13).

7.2 Following a planned path

An autonomous vehicle must know where it is at any time, and how it should manoeuvre to follow the chosen path. Systems for detecting obstacles, human beings, animals and other machines are also needed. Moreover, changed or unexpected conditions necessitate the ability to react and adapt to the input, in terms of re-planning the path as well as manoeuvring to follow the new path.

Although there is a plethora of research on how to plan suitable and efficient paths in unstructured terrain (e.g. Norouzi et al., 2017) and on how

Figure 5 The brachiating robot is designed to move from tree to tree. Photo courtesy of Richard Parker.

to follow given (human-made) paths in forest terrain by using various kind of sensors (e.g. Leidenfrost et al., 2013; Giusti et al., 2016), little has been implemented in actual forestry settings. In fact, most research has been focused on following actual paths (roads) in the forest, in which there is a distinct difference between the path and the surrounding forest. During forestry operations, however, planned paths are seldom visible. Thus, it is not surprising that there is currently no conventional forestry machines that are able to follow planned paths autonomously (apart from some cable-guided systems, see Section 10). However, it has been shown that a conventional forwarder equipped with some additional hardware and software can autonomously repeat a previously driven path with centimetre precision (Ringdahl et al., 2011). In fact, it would probably be difficult for professional human operators to have better path-tracking accuracy than the autonomous machine in that study.

Path following normally requires a combination of techniques to assess the vehicle's current position, velocity, steering angles and goal position. A major problem is that the vehicle's wheels can slip and slide considerably on the unstructured terrain. Another source of error, related to the articulated joint design, is that the motions of the front and rear parts of the vehicle relative to the ground are uncertain. When the steering angle changes the two parts move differently depending on factors such as weight distribution and ground conditions. The errors in position and heading estimates accumulate into total errors that increase with time.

8 Reaching and handling the trees

Most ground-based forestry machines are equipped with a manipulator in the form of a hydraulic crane by which the trees are reached and handled. Traditionally, the cranes are operated by controlling each link separately, which is an advanced task for an operator because there are many possible combinations of link movements that can result in the same overall movement and final position of the tool at the end of the crane. However, there has been rapid advancement in how cranes are controlled. Over recent years, several entry-level products have appeared on the market, such as the following:

- *Cranes equipped with motion sensors*, providing entry-level products that use improved motion control software (Cranab, 2015).
- *Hydraulic valves equipped with digital electronics*, providing entry-level products that use improved software for dynamic motion control of the machine (Mathworks, 2016; Danfoss, 2015).
- *Basic boom-tip control*, where the operator receives computer support in order to carry out expertly co-ordinated end-effector movements with less effort (John Deere, 2013).
- *Reduced crane vibrations*, making the operation of the crane more comfortable (John Deere, 2013; La Hera and Ortiz Morales, 2015).

Among the examples listed above, the concept of boom-tip control has long been anticipated. In 2013, John Deere became the first forestry machine manufacturer to offer their solution: smooth and intelligent boom control (SBC & IBC) systems, first for forwarders and then for harvesters. At the same time, Cranab released their Cranab Intelligent System (CIS), comprising cranes with integral sensors. Simultaneously, different producers of hydraulic valves have released products involving sensors and computers, resulting in a technology known as 'intelligent valve'. This combination of sensors in cranes and intelligent hydraulics provides sufficient technology for more machine manufacturers to develop their own automated crane functions. All these examples are entry-level solutions, opening the door to automation. Various concepts for automated crane functions have been trialled and/or implemented in test beds (Ortiz Morales, 2015; Hansson and Servin, 2010). For example, in relation to intelligent boom control, there is a large number of different boom-tip control algorithms that can be implemented with a machine, because these algorithms respond to selectable optimization options such as minimum kinetic energy control, minimum potential energy control, failure recovery, strength optimization and fuel consumption (La Hera, 2011; Westerberg, 2014; Ortiz Morales, 2015).

9 Converting trees into products

9.1 Decision support for processing

The development of value optimization algorithms for the processing of felled trees into logs (delimbing and bucking) was initiated during the 1960s (e.g. Smith and Harrel, 1961; Pnevmaticos and Mann, 1972). In the late 1980s, the algorithms became commercially available as computerized DSS (Anon., 1988), and since then algorithms have been part of the standard equipment in Nordic harvesters. The input is continuous measurement of diameters along the length of the stem, by use of the feeding rollers and/or delimbing knifes. The measured diameters are used to calculate a stem's taper and to predict the shape of the remaining part of the stem. Thus, optimization decisions for the assumed stem can be taken without first measuring the whole stem's diameter and then going back and bucking the logs. The optimization process is carried out continuously and updated based on the flow of new diameter recordings, while the stem is fed through the harvester head at a speed of circa 5 m/s.

This DSS can already be used for autonomous delimbing and bucking under rather simplistic conditions. The system could, for instance, be set to process trees automatically based on predefined log length(s). However, in most situations the value of bucking optimization depends on the tree species and quality, which are currently recorded by the operator. Thus, until sensors can replace the operator's input for that information, the current processing system will remain a DSS. Although there is no apparent advancement with such sensors, developments related to tree recognition are likely to be transferable and of benefit to processing activities.

9.2 Improved characterization of product properties

Data collected by conventional harvesters can already provide some industry-relevant information regarding tree features (sizes, volumes, assortments etc.), but at an aggregated level because individual logs are not tracked. Moreover, technological developments have made it possible for harvesters to gather data while bucking trees into logs, including data on wood properties that had not previously been considered, such as stiffness (Murphy, 2014).

Although this avenue of development might not lead directly to advances in automation, the ongoing development of new wood products and processes such as bio-refining means that it may become increasingly important to identify, select and sort trees with desirable chemical properties. Indeed, trees of the same species can vary in their chemical composition (e.g. Arshadi et al.,

2013). Data on the chemical composition of wood has hitherto rarely been used during industrial processing, but the ongoing development in tree recognition is likely to enable detection of detailed tree and wood properties before harvest. However, it is not obvious that such features should be identified in the forest. With the increasing number of desired characteristics from a given tree and from given stands, it is less likely that sorting can be carried out efficiently in the forest. Thus, the future might see machine systems focused on harvesting and delivering stems (maybe even with branches) to centralized wood yards, where the products are then identified, captured and delivered to customers. If so, it will be easier to automate the work of processing trees, because that work will be carried out in a controlled environment. However, the rest of the workflow will be just as, or even more, problematic, if larger units (whole stems with branches) are to be handled.

10 Extracting logs or trees to roadside landings

Extraction work follows a cyclic procedure, in which the machine travels out unloaded, accumulates the load, travels back loaded and unloads at a roadside landing. Cable yarders, which are used for extracting timber on steep terrain, have for some decades been available commercially in semi-autonomous versions. The carriage autonomously travels the cable out and in, but is loaded and unloaded with operator input. The input of the former cable yarder operator has now been overtaken by two other operators involved in the cable yarding. During the accumulation and unloading phases, the choker-setter and the processor, respectively, have complete control of the yarder by means of a wireless radio remote controller.

In addition to increased worker productivity (with one less operator), the advantages of automation include improved management of acceleration and deceleration and speed management along the extraction path. Moreover, the computer can stop the movement of the carriage in 1/100th of a second if the tension-monitoring system detects that a turn is stuck. Conversely an operator is likely to take up to 2 s, and in that time the machine can become severely shock-loaded. Examples of remotely controlled tower yarders with autonomously travelling cabins include the Konrad KMS 12Uxii, the Valentini V1500 and the Griefenberg TG 1100.

A similar concept, but ground-based, is the Konrad 'Pully', which is a semi-autonomous/remotely operated forwarder or skidder that runs along a wire rope that connects the steep-slope harvester and the landing/roadside area (Konrad, 2017). The Pully requires operator input to be loaded in the forest and to be unloaded at roadside.

The existing products all rely on a cable for guidance along the path to travel autonomously. Self-navigating extracting forest machines are still only at an early experimental stage (see the examples in Section 7.2).

11 Remote-controlled operations

Being able to control the machine remotely can be seen as an important step in making a machine autonomous, in the sense that it enables new designs and technological solutions to be considered when operator safety and comfort do not have to be taken into account. Moreover, it is likely to facilitate a transition to controlling (or rather supervising) several machines from one centralized control hub. However, research into teleoperated forestry machines (Milne et al., 2013; Westerberg and Shiriaev, 2013) and unmanned self-navigating vehicles (Ringdahl et al., 2011; Hellström et al., 2009; Vestlund and Hellström, 2006) has highlighted the challenges in making sensors perceive and understand the structure of the 'natural' forest landscape. Nevertheless, a number of forestry machines already have remote-controlled operating systems in place, for productivity and safety reasons. These remote-controlled forestry machines range from those that rely on the operator's own direct vision, to those that rely on camera vision.

A number of European skidder manufacturers provide a remote-control feature as a complement to cab control. During operations, this is typically used to reposition the skidder while the operator is out of the cab, for example when pulling out the winch rope while setting chokers.

In New Zealand, a teleoperated winch-assisted John Deere 909 feller-buncher has been trialled. The machine has been retrofitted to enable full teleoperation from a purpose-built control booth trailer (Fig. 6). To provide the operator with a sense of the terrain slope, the system includes an artificial horizon line and a 'heads-up display' overlaid on one of the screens. It has been tested successfully during harvesting operations (Parker et al., 2016). With the operator taken out of the cab, the system can be pushed onto steeper slopes, but the remote-control system itself does not improve stability and tractability on steep slopes.

Larger specialist companies are also involved in retrofitting machines with remote controls or teleoperation. For instance, the Applied Research Associates' Modular Robotic Appliqué Kit (M-RAK) has been used to teleoperate a Caterpillar 521B feller-buncher at the Fort Bragg army site to clear timber on firing ranges without putting the workers in danger. The M-RAK provides video of both the operation and the surrounding environment, and enables control at a range of about 2.4 km.

In experimental settings, the range for control is not a problem. Forwarder cranes have been set up and controlled worldwide. However, to enable successful teleoperation of conventional operations problems of information presentation and visibility need to be solved. For instance, the viewing angle and abstraction level have been shown to affect operator performance (Westerberg and Shiriaev, 2013).

Figure 6 The control booth for a teleoperated winch-assisted John Deere 909. Photo courtesy of Keith Raymond.

Another example of a remote-controlled forest machine was the Swedish 'Beast system', in which an unmanned harvester was controlled from a forwarder located in direct proximity to the harvester. The remote control worked well, but to have a harvester and forwarder co-operating for direct loading of logs has proved to be unsuccessful (e.g. Lindroos, 2012; Ringdahl et al., 2012a).

A recent development in commercialized remote control is the HiVision system for self-loading timber trucks (Hiab, 2018). An operator is still present in the truck but, by using augmented reality goggles, the crane can be operated from the truck cabin instead of from a crane cabin.

12 Conclusion

Ongoing research and development in the automation of forestry operations is contributing to the sustainable supply of fossil-free raw material for circular bio-economies of the future. Smart machines in forestry operations will facilitate the much-needed growth of feedstock and sustainable harvest, with better consideration of economic, ecological and ergonomic needs. The advances already made in particular address needs related to labour shortage, increased safety requirements and increased productivity targets. The current level of automation in forestry operations is very low, as even the level of mechanization is low in many parts of the world. There are logical reasons for the slow rates of development, but there has nevertheless been substantial progress, and there is no shortage of innovation. Indeed there is a great range of innovation focus, which is to be expected because forestry practices are a complex mixture of adaptations to complicated, locally variable conditions.

There is, however, a large step from innovation to a commercial product for forestry operations. The automation challenges that need to be overcome are in many cases similar to those encountered in the development of space and military applications (unstructured, unknown terrains to be mastered under extreme conditions), but with a much smaller research budget. Moreover, to be commercially viable, the innovations need to be profitable compared to conventional forestry operations.

The challenges to developing the first stages of automated forestry operations are related to the availability of data, so there is currently a strong focus on the integration of sensor technology and development of control systems, to manage machine movements efficiently. The difficulty lies in making sensors perceive and understand the structure of the 'natural' forest landscape, to enable both remote-controlled and autonomous operations. Moreover, development in the automatic detection of qualitative features of trees and logs is needed, to enable automated decisions on where and how to move in the forest, which tree to harvest and what products can be processed from a given tree.

Current research is starting to consider the hardware requirements and initial software needed for automation. However, transitioning towards this technology will not be easy, because developing software and redesigning all the hydraulics and embedded electronics in conventional forestry machines will be challenging, particularly when trying to make a profit during the process.

Therefore, as we enter the world of operator-assisted operations, this could still take at least a decade to complete. During this time, however, improvements can be expected in control performance, particularly precision boom movements using motion sensors and operator-assistance software.

However, the operator will still have an essential role in the correct use of these tools, and many difficult movements will still need to be carried out manually.

13 Future trends

The use of autonomous machinery for forestry operations is still considered to be several decades away. Naturally, there will have to be a gradual transition to systems with a decreased level of operator input. It is also likely that some forestry operations will be automated earlier than others because of the nature of the work. Naturally, the advances will benefit in general advances in robotics on how to make machines plan and execute actions autonomously. However, in this section we address some forestry-related challenges that is likely to not be covered in general robotics.

There will probably be rapid development of machines that can gather forestry information (e.g. UAVs). However, it will likely take some time before there are unmanned and autonomous machines that can carry out some actual physical forestry work. Activities that require the fewest decisions will be easiest to carry out autonomously. For instance, automating the process of scarification has good potential, because it is 'only' about navigating and creating suitable regeneration spots. Moreover, it is an activity that can be ergonomically challenging for the operator, so there is motivation for removing the operator from the machine. However, the work does require a machine that is able to traverse the ground and adapt the scarification to soil conditions.

Some operations have not yet been successfully mechanized, like planting on natural soils. Technically, mechanical planting can be carried out, but the logistics of efficient handling of fragile plants remains a challenge to complete mechanization (e.g. Ersson et al., 2011, 2014). Once mechanized, hitherto theoretical advances on automated planting might be implemented (Lideskog, 2018).

When it comes to pre-commercial and commercial thinning, the challenges relate to the choices of which trees to harvest and which to leave, as well as what products to make. As with the harvesting of mature trees, tree sizes can be a challenge, not least because failing to handle a large tree successfully is likely to be harmful to the machine and anything in its proximity.

It will therefore probably be easiest to automate operations in which all trees within a stand need to be felled, but without any product recovery. This should preferably be done without having to traverse the ground, and the trees should be small. Hence, using UAVs to help weed out small trees in areas where no trees should be growing may be the first autonomous forestry operation. This process 'only' requires the capacity to navigate a defined area, identify the trees and use a tool to cut the trees. For small trees, small tools and UAV carriers would suffice. A typical application would be weeding under power lines and along rights of way. An operator could supervise a swarm of autonomous machines, and move with them along the area to be weeded. Naturally the required robotic advances would be substantial, especially in the area of computer vision, but the forestry-related challenges would be considerably small.

When it comes to harvesting activities, it is likely that automated extraction work will be expanded to vehicles without cable guidance. Such 'shuttles' would autonomously transport logs or stems to the roadside. However, the work of loading and unloading will probably take time to automate, because of the challenges in automating the gripping work as well as identifying different tree assortments.

To be economically viable, any innovations in automation need to cut costs in some way, or increase income. As the hardware is likely to be more expensive initially, operator productivity has to be increased, or running costs such as fuel consumption has to be decreased. The first step is likely to be semi-automation, in which some processes can be carried out quickly, efficiently and autonomously, while an operator will continue to carry out the work that is more difficult to automate, such as the gripping of logs, and locating logs on the ground correctly. Eventually, the operator will co-ordinate a number of work units, either by controlling several cranes on a machine or controlling several machines remotely. The challenge will then be to reach a level of automation, that means the operators have a supervisorial role rather than carrying out the work themselves.

14 Acknowledgements

This text was partly financed by Sweden's Innovation Agency VINNOVA (project Auto^2), the Swedish Energy Agency (project HAFSBiT, 48003-1), the Swedish Foundation for Strategic Environmental Research MISTRA (program Mistra Digital Forest) and the Kempe Foundations (project JCK-1713).

15 Where to look for further information

15.1 Journals and conferences

The ongoing development in forestry operation robotics is taking place within two different branches of research: robotics and forest operations. The latter is an applied field for advances in forestry, and the scientific work is often published in journals such as the *International Journal of Forest Engineering*, *Croatian Journal of Forest Engineering* and more general forestry journals. Applied, but not necessarily forestry related, advances can also be found in journals such as the *Journal of Field Robotics*. Naturally, basic advances in robotics can be found in journals such as the Institute of Electrical and Electronics Engineers' (IEEE) *IEEE Transactions on Robotics*, *IEEE Transactions on Mechatronics* and *IEEE Transactions on Automatic Control*, and at conferences such as the IEEE

International Conference on Robotics and Automation (ICRA) and the IEEE International Conference on Intelligent Robots and Systems (IROS). However, as it is assumed that readers will be able to find general, basic robotics-related information relatively easily, we are providing further information on the more limited, and probably less familiar, field of forestry operations.

Forest operation research networks are formed predominantly through organizations and conferences in which applied advances in forestry operations robotics are presented.

The IUFRO's Division 3 – Forest Operations Engineering and Management unites people with an interest in advancing forestry operations. Among other networking activities, Division 3 arranges sessions within the IUFRO World Congress, which is held at 5-year intervals and is one of the largest global forest events, attended by more than 2000 participants. The next congress will be held in 2019 in Brazil. More information about IUFRO and Division 3 can be found at https://www.iufro.org/.

There are a number of international Anglophone conferences dedicated to advances in forestry operations, for example:

FORMEC, which is held annually in Europe, www.formec.org.

COFE, which is held annually in North America, http://cofe.org/.

The Precision Forestry Symposium is dedicated to new technology for data collection, automation, robotics and information and communications technology (ICT). It is held every 2 or 3 years in the southern hemisphere; the most recent, fourth, symposium, was held in 2017, https://www.sun.ac.za/english/faculty/agri/forestry/upcoming-conferences/precision-forestry-symposium.

The Forest Engineering Conference rotates across the globe, and is held every 3 or 4 years. The seventh conference will be conducted in 2021 in Italy, and the most recent one was held in 2018 in New Zealand, http://www.foresteng.canterbury.ac.nz/FEC2018.shtml.

For those interested in regular updates on advances in forestry operations, the free newsletter *Logging On* is available at https://loggingon.net/.

15.2 Futuristic concepts

The internet abounds with exotic and thought-provoking robotic concepts, although there are not many focusing on forestry operations. A small selection of concepts is presented here; they are far from being implemented, but might serve as inspiration for potential development.

The now almost 20-year-old animation of STINA, the pre-commercial thinning robot, demonstrates some of the challenges in forestry. It is not actually mentioned in the video, but Sten Gellerstedt was one of the researchers for the concept: https://www.youtube.com/watch?v=t8Qb5t3NIyc.

Published by Burleigh Dodds Science Publishing Limited, 2019.

Scion's brachiating robots provide an interesting take on possible methods of tree-to-tree harvesting: https://www.youtube.com/watch?v=mzzE70ymuTU&t=22s.

For work that is difficult to mechanize, the operators themselves could be strengthened to carry out the work in a less strenuous and safer way, as exemplified by Aegis, an exoskeleton for logging workers: www.jasonplee.com/#/aegis/.

The SLOPE project (www.slopeproject.eu) spans a wide range of possible uses of sensors and digitization, although not specifically for automation. However, its video provides a good exposé of some of the technologies and applications tested, especially in the fields of planning and product quality identification: https://youtube/Ml8nQzf17Mc.

The Virtual Forest project (http://www.virtueller-wald.de/en/the-virtual-forest/) draws on the potential of gathering detailed data to model and visualize forests and forestry operations to provide decision support: https://www.dw.com/en/the-virtual-forest/av-17904981.

Interested readers would also like to keep an eye on the two Swedish projects Auto[2] and Mistra Digital Forest, both with the aim to advance the automation in forestry operations and with strong engagement from forest industry, machine manufacturers and academic institutions.

16 References

Ågren, A. M., Lidberg, W., Strömgren, M., Ogilvie, J. and Arp, P. A. 2014. Evaluating digital terrain indices for soil wetness mapping – A Swedish case study. *Hydrology and Earth System Sciences* 18(9), 3623–34. doi:10.5194/hess-18-3623-2014.

Anon. 1988. 50 years of Forest Technology R&D in Sweden. Redogörelse nr. 6. Kista, Sweden: Forskningsstiftelsen Skogsarbeten (In Swedish).

Anon. 2013. Vollmechanisiertes Fällen und Aufarbeiten mit Portalharvester; Rücken mit Flachlandseilkran (Fully mechanized harvesting with the portalharvester, transporting with flat-land cable crane). *AFZ-DerWald* 18, 50–9.

Arshadi, M., Backlund, I., Geladi, P. and Bergsten, U. 2013. Comparison of fatty and resin acid composition in boreal lodgepole pine and Scots pine for biorefinery applications. *Industrial Crops and Products* 49, 535–41. doi:10.1016/j.indcrop.2013.05.038.

Bayne, K. M. and Parker, R. J. 2012. The introduction of robotics for New Zealand forestry operations: Forest sector employee perceptions and implications. *Technology in Society* 34(2), 138–48. doi:10.1016/j.techsoc.2012.02.004.

Bigsby, H. and Ling, P. 2013. Long-term productivity of helicopter logging in Sarawak. *International Journal of Forest Engineering* 24(1), 24–30. doi:10.1080/19132220.2013.791197.

Billingsley, J., Visala, A. and Dunn, M. 2008. Robotics in agriculture and forestry. In: Siciliano, B. and Khatib, O. (Eds), *Springer Handbook of Robotics*. Berlin, Heidelberg: Springer Berlin Heidelberg, pp. 1065–77.

Brunner, A. and Gizachew, B. 2014. Rapid detection of stand density, tree positions, and tree diameter with a 2D terrestrial laser scanner. *European Journal of Forest Research* 133(5), 819–31. doi:10.1007/s10342-014-0799-1.

Contreras, M. A. and Chung, W. 2007. A computer approach to finding an optimal log landing location and analyzing influencing factors for ground-based timber harvesting. *Canadian Journal of Forest Research* 37(2), 276–92. doi:10.1139/x06-219.

Contreras, M. A. and Chung, W. 2013. Developing a computerized approach for optimizing individual tree removal to efficiently reduce crown fire potential. *Forest Ecology and Management* 289, 219–33. doi:10.1016/j.foreco.2012.09.038.

Cranab. 2015. Forwarder cranes with world leading technology. Available at: www.cranab.com/downloads/Forwarder-Cranes/Cranab-FC-brochure-EN.pdf (accessed on 31 October 2018).

Danfoss. 2015. Robust and efficient in harsh environments. Retrieved from: http://www.danfoss.in/technicalarticles/cf/robust-and-efficient-in-harsh-environments/?ref=17179901637#/ (Retrieved on 31 October 2018).

Eriksson, M. and Lindroos, O. 2014. Productivity of harvesters and forwarders in CTL operations in Northern Sweden based on large follow-up datasets. *International Journal of Forest Engineering* 25(3), 179–200. doi:10.1080/14942119.2014.974309.

Erler, J. 2013. Portalharvester und Flachlandseilkran (Portalharvester and Flat-land cable crane). *Forst & Technik* 9, 18–21.

Ersson, B. T., Bergsten, U. and Lindroos, O. 2011. The cost-efficiency of seedling packaging specifically designed for tree planting machines. *Silva Fennica* 45(3), 379–94. doi:10.14214/sf.108.

Ersson, B. T., Bergsten, U. and Lindroos, O. 2014. Reloading mechanized tree planting devices faster using a seedling tray carousel. *Silva Fennica* 48(2), article id 1064, 14p. doi:10.14214/sf.1064.

Flisberg, P., Forsberg, M. and Rönnqvist, M. 2007. Optimization based planning tools for routing of forwarders at harvest areas. *Canadian Journal of Forest Research* 37(11), 2153–63. doi:10.1139/X07-065.

Giusti, A., Guzzi, J., Ciresan, D. C., He, F. L., Rodríguez, J. P., Fontana, F., Faessler, M., Forster, C., Schmidhuber, J., Caro, G. D., et al. 2016. A machine learning approach to visual perception of forest trails for mobile robots. *IEEE Robotics and Automation Letters* 1(2), 661–7. doi:10.1109/LRA.2015.2509024.

Golob, T. B. 1981. Worker and his tools. Seminar on Occupational Health & Safety and Applied Ergonomics on Highly Mechanised Logging Operations. Ottawa, Canada: Food and Agriculture Organization/Economic Commission for Europe/International Labour Organization.

Grigolato, S., Mologni, O. and Cavalli, R. 2017. GIS applications in forest operations and road network planning: An overview over the last two decades. *Croatian Journal of Forest Engineering* 38(2), 175–86.

Häggström, C. and Lindroos, O. 2016. Human, technology, organization and environment – a human factors perspective on performance in forest harvesting. *International Journal of Forest Engineering* 27(2), 67–78. doi:10.1080/14942119.2016.1170495.

Hallonborg, U. 1997. Ingen man på maskinen – En förarlös vision (No man in the machine – A driverless vision). Arbetsrapport no. 399. Uppsala, Sweden: Skogforsk (in Swedish).

Hansson, A. and Servin, M. 2010. Semi-autonomous shared control of large-scale manipulator arms. *Control Engineering Practice* 18(9), 1069–76. doi:10.1016/j.conengprac.2010.05.015.

Hauglin, M., Hansen, E. H., Næsset, E., Busterud, B. E., Gjevestad, J. G. O. and Gobakken, T. 2017. Accurate single-tree positions from a harvester: A test of two global satellite-based positioning systems. *Scandinavian Journal of Forest Research* 32(8), 774–81. doi:10.1080/02827581.2017.1296967.

Hellström, T., Lärkeryd, P., Nordfjell, T. and Ringdahl, O. 2009. Autonomous forest vehicles: Historic, envisioned, and state-of-the-art. *International Journal of Forest Engineering* 20(1), 31–8. doi:10.1080/14942119.2009.10702573.

Hiab. 2018. HiVision™ – Ground-breaking way to operate a crane. Retrieved from: https://www.hiab.com/en/pages/loglift-jonsered/hivision (Retrieved on 31 October 2018).

Hosseini, A., Lindroos, O. and Wadbro, E. (2019). A holistic optimization framework for forest machine trail network design accounting for multiple objectives and machines. *Canadian Journal of Forest Research 49*, 111–20. doi:10.1139/cjfr-2018-0258.

Hutter, M., Gehring, C., Lauber, A., Gunther, F., Bellicoso, C. D., Tsounis, V., Fankhauser, P., Diethelm, R., Bachmann, S., Bloesch, M., et al. 2017. ANYmal-toward legged robots for harsh environments. *Advanced Robotics* 31(17), 918–31. doi:10.1080/01691864.2017.1378591.

Hyyti, H., Lehtola, V. V. and Visala, A. (2018). Forestry crane posture estimation with a two-dimensional laser scanner. *Journal of Field Robotics* 35(7), 1025–49. doi:10.1002/rob.21793.

John Deere. 2013. Smooth boom control/intelligent boom control. Retrieved from: http://www.deere.com/en_US/docs/forestry/SBC_IBC_FastFact_FNL.pdf (Retrieved on 31 October 2018).

Konrad. 2017. Ground carriage PULLY. Retrieved from: http://www.forsttechnik.at/en/products/ground-carriage-pully.html (Retrieved on 31 October 2018).

Kourtz, P. 1996. Autonomous forestry robots for brushing and thinning in young conifer stands: Early Canadian experiences. In: Gellerstedt, S., Asplund, C. and Wästerlund, I. (Eds), *Robotics with Application to Forestry*. Uppsatser och Resultat nr. 285. Garpenberg, Sweden: Department of Operational Efficiency, SLU. ISSN: 0282-2377.

La Hera, P. 2011. Underactuated mechanical systems: Contributions to trajectory planning, analysis, and control. Doctoral dissertation thesis. Sweden: Umeå University.

La Hera, P. and Ortiz Morales, D. 2015. Model-based development of control systems for forestry cranes. *Journal of Control Science and Engineering* 2015, 15p, article ID 256951.

Leidenfrost, H. T., Tate, T. T., Canning, J. R., Anderson, M. J., Soule, T., Edwards, D. B. and Frenzel, J. F. 2013. Autonomous navigation of forest trails by an industrial-size robot. *Transactions of the ASABE* 56(4), 1273–90.

Lideskog, H. 2018. A methodology for automation of mechanized forest regeneration. Doctoral dissertation thesis. Sweden: Luleå University of Technology.

Lideskog, H., Karlberg, M. and Bergsten, U. 2015. Development of a research vehicle platform to improve productivity and value-extraction in forestry. *Procedia CIRP* 38, 68–73. doi:10.1016/j.procir.2015.07.014.

Lindroos, O. 2012. Evaluation of technical and organizational approaches for directly loading logs in mechanized CTL harvesting. *Forest Science* 58(4), 326–41. doi:10.5849/forsci.11-001.

Lindroos, O. and Cavalli, R. 2016. Cable yarding productivity models: A systematic review over the period 2000–2011. *International Journal of Forest Engineering* 27(2), 79–94. doi:10.1080/14942119.2016.1198633.

Lindroos, O., Ringdahl, O., La Hera, P., Hohnloser, P. and Hellström, T. 2015. Estimating the position of the harvester head – A key step towards the precision forestry of the future? *Croatian Journal of Forest Engineering* 36(2), 147–64.

Lindroos, O., Häggström, C. and La Hera, P. 2017. Drivers of advances in mechanized timber harvesting – A selective review of technological innovation. *Croatian Journal of Forest Engineering* 38(2), 243–58.

Manner, J., Nordfjell, T. and Lindroos, O. 2016a. Automatic load level follow-up of forwarders' fuel and time consumption. *International Journal of Forest Engineering* 27(3), 151–60. doi:10.1080/14942119.2016.1231484.

Manner, J., Palmroth, L., Nordfjell, T. and Lindroos, O. 2016b. Load level forwarding work element analysis based on automatic follow-up data. *Silva Fennica* 50(3). doi:10.14214/sf.1546.

Marra, E., Cambi, M., Fernandez-Lacruz, R., Giannetti, F., Marchi, E. and Nordfjell, T. 2018. Photogrammetric estimation of wheel rut dimensions and soil compaction after increasing numbers of forwarder passes. *Scandinavian Journal of Forest Research* 33(6), 613–20. doi:10.1080/02827581.2018.1427789.

Mathworks. 2016. INCOVA designs intelligent valve-control system for a 20-ton excavator. Retrieved from: https://www.mathworks.com/company/user_stories/incova-design s-intelligent-valve-control-system-for-a-20-ton-excavator.html (Retrieved on 31 October 2018).

Melander, L. and Ritala, R. 2018. Time-of-flight imaging for assessing soil deformations and improving forestry vehicle tracking accuracy. *International Journal of Forest Engineering* 29(2), 63–73. doi:10.1080/14942119.2018.1421341.

Milne, B., Chen, X., Hann, C. and Parker, R. 2013. Robotisation of forestry harvesting in New Zealand – An overview. In: *2013 10th IEEE International Conference on Control and Automation (ICCA)*, pp. 1609–14.

Mohtashami, S., Bergkvist, I., Löfgren, B. and Berg, S. 2012. A GIS approach to analyzing off-road transportation: A case study in Sweden. *Croatian Journal of Forest Engineering* 33(2), 275–84.

Murphy, G. E. 2014. Priority list bucking on a mechanized harvester considering external properties and stiffness of Douglas-fir. *International Journal of Forest Engineering* 25(3), 214–21. doi:10.1080/14942119.2014.973177.

Norouzi, M., Miro, J. V. and Dissanayake, G. 2017. Planning stable and efficient paths for reconfigurable robots on uneven terrain. *Journal of Intelligent and Robotic Systems* 87(2), 291–312. doi:10.1007/s10846-017-0495-8.

Ortiz Morales, D. 2015. Virtual holonomic constraints: From academic to industrial applications. Doctoral dissertation thesis. Sweden: Umeå University.

Ostovar, A., Hellström, T. and Ringdahl, O. 2016. Human detection based on infrared images in forestry environments. In: Campilho, A. and Karray, F. (Eds), *Image Analysis and Recognition. ICIAR 2016. Lecture Notes in Computer Science*, vol. 9730. Cham: Springer, pp. 175–82.

Oveland, I., Hauglin, M., Giannetti, F., Schipper Kjørsvik, N. and Gobakken, T. 2018. Comparing three different ground based laser scanning methods for tree stem detection. *Remote Sensing* 10(4), 538. doi:10.3390/rs10040538.

Parker, R., Karen, B. and Clinton, P. W. 2016. Robotics in forestry. *New Zealand Journal of Forestry* 60(4), 8–14.

Peters, P. A. 1973. Advanced Logging System-II: Balloon logging: A look at current operating systems. *Journal of Forestry* 71(9), 577–9.

Pierzchała, M., Giguère, P. and Astrup, R. 2018a. Mapping forests using an unmanned ground vehicle with 3D LiDAR and graph-SLAM. *Computers and Electronics in Agriculture* 145, 217–25. doi:10.1016/j.compag.2017.12.034.

Pierzchała, M., Kvaal, K., Stampfer, K. and Talbot, B. 2018b. Automatic recognition of work phases in cable yarding supported by sensor fusion. *International Journal of Forest Engineering* 29(1), 12–20. doi:10.1080/14942119.2017.1373502.

Pnevmaticos, S. M. and Mann, S. H. 1972. Dynamic programming in tree bucking. *Forest Products Journal* 22(2), 26–30.

Pohjankukka, J., Riihimäki, H., Nevalainen, P., Pahikkala, T., Ala-Ilomäki, J., Hyvönen, E., Varjo, J. and Heikkonen, J. 2016. Predictability of boreal forest soil-bearing capacity by machine learning. *Journal of Terramechanics* 68, 1–8. doi:10.1016/j.jterra.2016.09.001.

Ponsse 2017. Ponsse active frame. Retrieved from: https://www.ponsse.com/products/forwarders/activeframe (Retrieved on 31 October 2018).

Puliti, S., Talbot, B. and Astrup, R. 2018. Tree-stump detection, segmentation, classification, and measurement using unmanned aerial vehicle (UAV) imagery. *Forests* 9(3), 102. doi:10.3390/f9030102.

Raibert, M., Blankespoor, K., Nelson, G. and Playter, R. 2008. Bigdog, the rough-terrain quadruped robot. *IFAC Proceedings Volumes* 41(2), 10822–5. doi:10.3182/20080706-5-KR-1001.01833.

Ringdahl, O., Lindroos, O., Hellström, T., Bergström, D., Athanassiadis, D. and Nordfjell, T. 2011. Path tracking in forest terrain by an autonomous forwarder. *Scandinavian Journal of Forest Research* 26(4), 350–9. doi:10.1080/02827581.2011.566889.

Ringdahl, O., Hellström, T. and Lindroos, O. 2012a. Potentials of possible machine systems for directly loading logs in cut-to-length harvesting. *Canadian Journal of Forest Research* 42(5), 970–85. doi:10.1139/x2012-036.

Ringdahl, O., Hellström, T., Wästerlund, I. and Lindroos, O. 2012b. Estimating wheel slip for a forest machine using RTK-DGPS. *Journal of Terramechanics* 49(5), 271–9. doi:10.1016/j.jterra.2012.08.003.

Salmivaara, A., Miettinen, M., Finér, L., Launiainen, S., Korpunen, H., Tuominen, S., Heikkonen, J., Nevalainen, P., Sirén, M., Ala-Ilomäki, J., et al. 2018. Wheel rut measurements by forest machine-mounted LiDAR sensors–accuracy and potential for operational applications? *International Journal of Forest Engineering* 29(1), 41–52. doi:10.1080/14942119.2018.1419677.

Silversides, C. R. (Ed.). 1974. *Proceedings of the IUFRO Division 3 Symposium 'Forest Harvesting Mechanization and Automation. Proceedings'*, 25 September – 5 October, Ottawa, Thunder Bay and Sault Ste. Marie, Canada. Ottawa, Ontario, Canada: Canadian Forestry Service, Department of the Environment.

Silversides, C. R. 1997. Broadaxe to flying shear: The mechanization of forest harvesting east of the Rockies. Transformation Series no. 6. Ottawa, Canada: National Museum of Science and Technology.

Smith, G. W. and Harrel, C. 1961. Linear programming in log production. *Forest Products Journal* 11, 8–11.

Søvde, N. E. 2015. Algorithms for estimating the suitability of potential landing sites. *Mathematical and Computational Forestry and Natural Resource Sciences* 7(1), 1.

Suvinen, A. and Saarilahti, M. 2006. Measuring the mobility parameters of forwarders using GPS and CAN bus techniques. *Journal of Terramechanics* 43(2), 237–52. doi:10.1016/j.jterra.2005.12.005.

Talbot, B., Pierzchała, M. and Astrup, R. 2017. Applications of remote and proximal sensing for improved precision in forest operations. *Croatian Journal of Forest Engineering* 38(2), 327–36.

Torresan, C., Berton, A., Carotenuto, F., Di Gennaro, S. F., Gioli, B., Matese, A., Miglietta, F., Vagnoli, C., Zaldei, A. and Wallace, L. 2017. Forestry applications of UAVs in Europe: A review. *International Journal of Remote Sensing* 38(8–10), 2427–47. doi:10.1080/01431161.2016.1252477.

Uusijärvi, R. 1985. Automation and remote controlling of forest machinery. Report from NSR workshop in Sweden, April 1983. Wood Technology Report No. 70. Stockholm, Sweden: Swedish Institute for Wood Technology Research. (In Swedish with English summary).

Vestlund, K. and Hellström, T. 2006. Requirements and system design for a robot performing selective cleaning in young forest stands. *Journal of Terramechanics* 43(4), 505–25. doi:10.1016/j.jterra.2005.07.001.

Vestlund, K., Nordfjell, T. and Eliasson, L. 2005. Comparison of human and computer-based selective cleaning. *Silva Fennica* 39(4), 509–23. doi:10.14214/sf.363.

Vestlund, K., Nordfjell, T., Eliasson, L. and Karlsson, A. 2006. A decision support system for selective cleaning. *Silva Fennica* 40(2), 271–89. doi:10.14214/sf.343.

Visser, R. and Berkett, H. 2015. Effect of terrain steepness on machine slope when harvesting. *International Journal of Forest Engineering* 26(1), 1–9. doi:10.1080/14942119.2015.1033211.

Wells, L. A. 2018. A vision system for automatic dendrometry and forest mapping. PhD thesis. Corvallis, OR: Oregon State University.

Westerberg, S. 2014. Semi-automating forestry machines: Motion planning, system integration, and human-machine interaction. Doctoral dissertation thesis. Sweden: Umeå University.

Westerberg, S. and Shiriaev, A. 2013. Virtual environment-based teleoperation of forestry machines: Designing future interaction methods. *Journal of Human-Robot Interaction* 2(3), 84–110. doi:10.5898/JHRI.2.3.Westerberg.

Chapter 12

Advances in robotic milking

Marcia Endres and Jim Salfer, University of Minnesota, USA

1 Introduction

Robotic milking systems (RMS) are becoming more common worldwide with a rapid growth taking place throughout the United States. It is estimated that there are ~25 000 RMS farms in the world (Rodenburg, 2017). In the same way that mechanization and expansion have improved productivity in the last 20 years, automation will drive dairy industry progress in the future.

The milking process is well suited for robotic technology. It is a repetitive process where the teats are found and cleaned, abnormal milk identified, cows milked and post-dipped. A robotic arm prepares the teats for milking and attaches the milking units. Lasers, cameras or combinations are used to find the teats. Teat cleaning is done by automatic brushes or prepping cups. After milking, teats are automatically sprayed with disinfectant or disinfectant is applied by using the cup. Figure 1 shows an example robotic milking box system with milking cups attached to the cow.

The majority of RMS around the world today are single box systems, with one robotic arm per milking unit. Cows voluntarily visit the robot box for milking. A smaller number of rotary RMS are in operation in Europe, Canada, Australia and the United States, and these systems usually utilize a more conventional way of milking cows, that is cows are moved to the milking parlour in groups by a worker and are batch milked two or three times per day.

One of the advantages of RMS is that they automatically collect large amounts of data on each cow daily including milk production, milk conductivity,

http://dx.doi.org/10.19103/AS.2019.0056.20

Figure 1 Example of a robotic milking system box showing robotic arm and milking cups attached to a cow. Photo courtesy University of Minnesota, St. Paul, USA.

milk colour, bodyweight, activity and rumination among others. Some of the information collected requires investing in additional cow sensors; therefore, not all producers will have those data. If used by the dairy farmer, this information can greatly assist with management of the herd. King and DeVries (2018) reviewed the literature on detection of health disorders using data from RMS and associated technologies and found consistent trends in activity, rumination time and milk production changes prior to diagnosis. Authors suggested that more work is still needed to incorporate validated algorithms and models for more accurate health alerts. In addition, integration of data is needed to support decision making at the farm.

Another advantage of RMS is that robotic milking provides a consistent milking routine with no need for training workers. It is becoming more difficult to find employees that are willing to manually milk cows. These workers are at increased risk of musculoskeletal disorders because of the awkward postures and repetitive motion of the milking process (Douphrate et al., 2013). RMS reduce this risk to dairy farm workers.

There are disadvantages of RMS. Capital investment is higher per cow than other milking systems. The technology is rapidly evolving, so obsolescence of an expensive capital investment is a concern along with expected higher repair and maintenance costs than for conventional milking systems.

Our surveys indicate that dairy farmers had many reasons for installing RMS. The three most often mentioned reasons were time flexibility - which helps improve quality of life; labour management - either the desire to expand without adding employees or the challenge with hiring and managing labour; and human health issues - some farmers would be out of the business because of health issues if they had not installed RMS. Our results agree with other surveys conducted in Europe (Hogeveen and Steeneveld, 2013; Stræte et al., 2017).

2 Barn design considerations

The design of the RMS barn should encourage natural cow movement that results in frequent visits to the milking box and minimizes labour. Cow flow, comfort and easiness of management within the barn are factors that can affect performance of RMS. This can be accomplished by building a new facility where RMS is installed or by retrofitting an older facility. Research has shown no differences in milk production between new and retrofitted RMS barns (Tremblay et al., 2016; Siewert et al., 2018).

2.1 Cow traffic flow

There are two types of cow traffic in RMS systems – free flow and guided flow. In barns with free flow traffic, cows can access all areas of the barn without restriction. In guided flow traffic, one-way gates and selection gates are used to guide cows to milking, feeding and resting areas.

The two types of guided flow traffic are milk first and feed first. In the milk first system, cows leaving the resting area must pass through a pre-selection gate that determines if they are eligible for milking. If a cow meets the requirement to be milked, she is guided to a commitment pen that contains the RMS box. If she is not eligible for milking she is allowed to enter the feed bunk area and can only re-enter the resting area through a one-way gate. Pre-selection gates can also be installed in crossovers away from the RMS box and open only for cows not eligible for milking.

In the feed first system, cow traffic is the reversal of the milk first system. After eating, the cows enter a selection gate that determines if she is eligible for milking. The gate either guides her to the commitment pen for milking or to the resting area. The commitment pen is a gated area next to the RMS box that cows eligible for milking cannot leave until they are milked.

Both free flow and guided flow systems can be successful. The best option is determined by the goals of the producers and their management style. In general, research has shown that there are fewer fetch cows with guided flow systems than free flow systems (Bach et al., 2009; Rodenburg and House, 2007). Fetch cows are those that need to be brought to the RMS by a worker as they did not visit the RMS box voluntarily within a period of time based on milking permission settings. Free flow traffic was associated with greater milk production per cow per day (Tremblay et al., 2016) compared to guided flow. Cows in a guided flow system consumed fewer meals per day but larger meals with longer meal duration when they visited the feed bunk, resulting in no difference in total eating time, eating rate or average daily dry matter intake compared to free flow (Bach et al., 2009).

Capital investment tends to be a little higher in guided flow systems because of the additional pre-selection gates. Cow management complexity can be higher in guided flow systems. In free flow systems cows have access to the RMS, stalls and feed at all times. In guided flow systems, managers must be aware of cows that are confined to the commitment pen for long periods of time and choose not to visit the RMS box. Cows also may choose not to pass through the pre-selection gate on their own on a regular basis. However, feeding management can be more challenging in the free flow system. There are no pre-selection gates to help guide cows to the milking box so ration changes that affect visits are exacerbated.

2.2 Other barn design considerations

In RMS, lactating cows do not go to a parlour to be milked, rather they stay in their home pen at all times and can therefore be in the way of equipment being used to scrape alleys or bed stalls. Rodenburg (2017) recommended that manual scraping of manure is not ideal as it causes more disruption to the cows, and so automated systems are more favourable. In addition, layouts with wide alleys, free traffic and many crossovers will offer more escape routes for cows when equipment enters the barn.

Rodenburg and House (2007) recommended a minimum open space of at least 6 m between the RMS box and the first freestall, 4.2 m wide alleys in the feed lane and 3.3 m wide alleys between two rows of stalls. Rodenburg (2017) suggested that devices such as cow brushes be installed away from the RMS box to entice cows not eligible for milking to move away, and to have a protected exit lane at least one-cow length long at the exit of the RMS box. In addition, the area around the RMS should be free of stray voltage. Rubber flooring should be installed in the RMS box along with cooling fans over the cows as they are milked. It is important to make the RMS box attractive to the cows to increase milking visits.

Fetching cows has to be as simple as possible, with least disruption, and one person should be able to accomplish this task without additional help. Rodenburg (2017) suggests the use of a split entry fetch pen that holds 4–5 cows at a time in free flow systems (Fig. 2). This pen does not have access to water, feed or freestalls, and it is used only for fetched cows to access the RMS box by a free-swinging gate at the entry to the RMS box. Other cows in the pen still have access to the RMS box from the other side of the gate. He suggests that this option provides some protection for more submissive cows. Research in the Netherlands (Heurkens et al., 2016) indicated that the use of a split entry fetch pen could improve milk production per robot.

Figure 2 View from above of a split entry gate that allows fetched cows to enter the RMS box directly from the fetch pen. The RMS box is under the platform to the right of the picture. Photo courtesy University of Minnesota, St. Paul, USA.

A sort pen for cows that need to be handled for breeding, treatment, hoof care and so on is also recommended (Rodenburg, 2017). This pen needs to include resting space and access to feed and water, and the opportunity to return for milking. Housing fresh, lame or injured cows in a bedded pack near the RMS box with robot access could be beneficial (Rodenburg, 2017). Heurkens et al. (2016) found that the presence of a straw bedded pack near the RMS box was associated with greater milk production per RMS. It is also suggested that RMS barns might work better with perimeter feeding.

3 Feeding management

One of the most important factors to success in RMS is how the cows are fed. This section focuses on box robots, not the fully automated rotary parlours where cows are fed similarly to a conventional parlour herd.

Dairy nutritionists have a goal of optimizing milk production and health of dairy cows by balancing a least-cost ration that meets cows' nutrient requirements. In most conventional confinement herds, this goal is accomplished by feeding a total mixed ration (TMR) where all the ingredients are mixed together and delivered to the cows. For RMS herds, a partial mixed ration (PMR), containing all of the forage and a portion of the concentrate, is offered in the feed bunk. An additional amount of concentrate is fed in the RMS box, but this amount varies according to the cow's stage of lactation and/or milk production.

3.1 What entices cows to visit the RMS box?

The major motivating factor to attract cows to consistently visit the RMS box is the pelleted concentrate that is offered in the box, not the fact that cows 'feel' a need to be milked at that time (Prescott et al., 1998). However, cows' visits to the RMS box are not only dependent on the PMR delivered in the feed bunk and pellets offered in the RMS, but also on feeding management, cow comfort, cow health and social interactions among cows. Heat stress can result in reduced feed intake and decreased number of visits to the RMS box. Severely lame cows are also less likely to visit the RMS box voluntarily (Bach et al., 2009). These factors need to be considered in a feeding programme for RMS herds.

A recent study in Canada (Hare et al., 2018), conducted using a feed first guided flow RMS, indicated that providing a high-energy PMR and limiting the quantity of concentrate pellet in the RMS box to approximately 0.5 kg/cow/day might improve overall daily dry matter intake, increase PMR eating rate and meal size, and reduce sorting of the PMR as compared to feeding a low-energy PMR along with 5 kg/cow/day of concentrate pellet in the RMS box. They observed that for every 1 kg increase in concentrate offered in the RMS, there was a 1.58 kg reduction in PMR intake. They did not observe significant differences in milk production between treatments. In contrast, the amount of concentrate offered in the RMS was positively associated with milk production per cow and per RMS in a study with free flow RMS farms (Siewert et al., 2018).

Bach and Cabrera (2017) suggested that concentrate meal sizes should be limited to about 1.5 kg or less per visit so that cows consume all the feed that is allocated to them at each visit, and the total daily concentrate offered in the RMS be limited to approximately 4 kg/cow. More research is needed to conclusively determine amounts of concentrate to be fed in the RMS and we suggest that this will be highly dependent on the type of cow traffic flow.

3.2 Composition of the RMS pellet

Halachmi et al. (2006) found that both pellets high in starch (high inclusion of ground barley, corn, sorghum and wheat bran) and pellets high in digestible neutral detergent fibre (NDF, high inclusion of soy hulls, corn gluten feed and soybean meal) could be used successfully to attract cows to the RMS box. The two pellets resulted in similar daily milk visits, milk production and fat-corrected milk production. However, concentrate allowance was kept low. In a follow-up study (Halachmi et al., 2009) comparing a high starch pellet (53% starchy grains) to a low starch pellet (25% starchy grains), they found no differences in milking frequency or milk components; however, milk production average was greater for cows fed the low starch pellet.

Miron et al. (2004) reported that concentrates high in starch resulted in greater milk protein per cent whereas concentrates high in digestible fibre resulted in greater milk fat per cent. Madsen et al. (2010) compared feeding up to 5 kg/day of six different pelleted feeds - barley-oat, maize, wheat, high fat or dried grass - compared to a control pellet. The barley-oat pellet had higher visits than the control whereas the maize- and wheat-based pellets had a similar number of visits to the control. However, the high fat and dried grass pellets had lower visits than the control pellet. Results of these studies may indicate that palatability and performance can be maintained with various ingredient compositions of the pelleted concentrate.

3.3 Considerations for RMS in grazing herds

When RMS is used in grazing herds, there is an additional challenge of enticing cows to leave the pasture and voluntarily attend the RMS box. In pasture-based systems there appears to be a relatively large percentage of cows with long milking intervals (defined as greater than 16 h). Lyons et al. (2013a) found 47% and 38% of milking intervals exceeded the 16-h threshold in groups of cows fed a PMR and concentrate pre- and post-milking, respectively. Cows fed pre-milking returned from pasture to the milking barn sooner (11.9 h) than cows fed post-milking (13.3 h); however, the cows fed pre-milking spent more time in the feeding and waiting areas before entering the RMS platform (voluntary rotary RMS), resulting in a decreased average milking frequency compared to those fed post-milking (1.6 vs. 1.7 milkings per day for groups fed pre- and post-milking, respectively). It is important to note that although there were differences in cow behaviour, no differences were found in daily milk yield between the two feeding management systems. In another study in Australia, Scott et al. (2015) found that feeding a supplement post-milking reduced (21% shorter) the voluntary waiting time for cows to be milked compared to feeding them a supplement pre-milking.

Spörndly and Wredle (2004) showed that a longer distance (260 m) to pasture from the RMS box resulted in reduced daily milk yield (26.4 vs. 29.1 kg) and milking frequency (2.3 vs. 2.5 milkings/day) than a shorter distance (50 m). Lyons et al. (2013b) also reported longer milking intervals when the distance from the RMS box was greater than 500 m.

Research has shown that as the sward height of the pasture decreases, the number of milkings and visits per cow per day to the RMS box increases (Ketelaar-de Lauwere et al., 2000; Lyons et al., 2013b). Ketelaar-de Lauwere et al. (2000) also observed that cows returned from pasture to the RMS area in groups, resulting in peak periods of RMS visits and periods where the RMS was unvisited for several hours.

Lyons et al. (2014) reviewed the literature on milking frequency in pasture-based RMS. They suggested that in general, cows in pasture systems have lower milking frequency than what is observed in confinement RMS. They suggested more research is needed to investigate location and frequency of feed incentives that could potentially increase cow traffic to the RMS box.

4 Milk quality and udder health

One area that can be challenging to manage in RMS is udder health. Epidemiological studies have indicated that udder health was compromised after introduction of RMS (reviewed by Hovinen and Pyörälä, 2011); however, those studies were conducted a decade ago or earlier, and RMS technology has changed. A recent survey in Canada showed that change in bulk tank somatic cell count (SCC) varied across farms when changing from conventional milking to RMS, with 43%, 37% and 20% of respondents indicating a decrease, no change or increase, respectively. The geometric mean SCC was 180 000 cells/mL (Tse et al., 2018). However, other more recent research showed that SCC was higher on farms with RMS (194 000 cells/mL) compared to conventional milking systems (161 000 cells/mL) (Shishanthi, 2017).

Producers depend on the RMS detection tools to identify mastitis cases and divert the milk from the bulk tank since workers are not evaluating the foremilk in the parlour like in a conventional system. A machine prepares the udder for milking, which is more difficult to do effectively with dirty cows. One of the advantages in RMS is that quarters are individually milked and detached which reduces overmilking. There is also less risk for antibiotic contamination with a treated cow, as long as her ID is entered in the computer correctly and the milk is discarded.

4.1 Identification of mastitis cases

Electrical conductivity (EC) of the milk is one of the parameters used to help detect mastitis cases in RMS. Previous review of the literature (Hovinen and Pyörälä, 2011) indicated that EC is an inadequate method to detect subclinical or clinical mastitis, and abnormal milk in general. In a field study, sensitivity (Se) of EC for detecting mastitic quarters (single sample, SCC > 200 000 cells/mL) ranged from 5% to 18%, depending on the RMS. When mastitis was defined as SCC >1 000 000 cells/mL, Se ranged from 9% to 43%. In a small number of clinical cases of mastitis, Se of EC was comparable to the average Se of workers to detect abnormal milk by visual observation (70–80%).

However, most RMS will include additional measurements. van der Tol et al. (2012) reported a clinical mastitis (CM) detection model including EC, milk colour, dead milk time and milk yield proportion per quarter. Combined with defined variables (a CM time window of five days and including three previous milkings), the model achieved a Se and specificity (Sp) of 88.9% and 99.3%, respectively.

4.2 Challenges for RMS

One current weakness with RMS (depending on the system) is teat preparation for milking and post-dipping. Hovinen and Pyörälä (2011) summarized that 8% of teat cleanings per cow failed due to machine problems and 4% because of cow-related problems, including kick-offs. One of the studies mentioned in their review that only 67% of the cleanings were technically successful, that is all four teats were brushed. In the best performing farm, over 95% of the teat cleanings were technically successful. Reasons for most of the failed teat cleanings were undetermined but of the known causes, a device failure in one herd and restless behaviour of the cows in several herds were associated with most of the totally unsuccessful teat cleanings, whereas abnormal udder and teat structure were associated with most of the partly unsuccessful teat cleanings.

Another challenge relates to identification of CM cases. Although additional metrics have improved detection in recent years, some cows might still be missed. Treating cows is also more difficult since they no longer go to a parlour to be milked. Contagious organisms can be a real challenge and contagious mastitis can more easily spread. In addition, infected cows generally cannot be grouped separately.

4.3 Factors influencing udder health in RMS

It is important that cows visit the RMS box on a regular basis. If cows have a very long milking interval, leaking of milk greatly increases and these cows are at higher risk for mastitis. Penry et al. (2017) showed an increased risk of quarter CM with increasing milking interval in a case-control study using data from a 1549-cow farm with 20 RMS boxes.

Bulk tank milk should be cultured regularly. If contagious organisms are present in the herd there can be some special add-on equipment (e.g. steam cleaners) that may help. Post-dip coverage should be monitored and a dip that has the proper kill time and emollient for the herd's bedding and barn conditions should be used.

Keeping the barn and stalls clean is very critical; the RMS cannot distinguish between dirty and clean udders, therefore it is important that cows enter the RMS unit with a lower bacteria load. A study at the University of Wisconsin showed that teat skin bacterial counts were significantly greater for swabs obtained from teats of cows with dirtier udders compared with swabs obtained from cows with cleaner udders (Guarin et al., 2017).

Dohmen et al. (2010) showed that the annual average herd SCC in RMS herds was positively associated with the proportion of cows with dirty teats before milking and the proportion of cows with dirty legs. In addition, the annual average percentage of new cows with high SCC was positively associated with the proportion of cows with dirty teats before milking. At the cow level, hygiene scores of the udder, thighs and legs were positively associated with SCC.

Cleanliness of milking units, robot and area around the robot is also critical. In addition, it is important to fine-tune RMS settings for optimum performance on a routine basis and optimize/adjust the machine for the barn and bedding type. Default values might not be adequate. Working with service providers helps in this process.

Prep time prior to milking needs to be monitored and optimized to achieve ideal milk let down. Most RMS provide information on the time from unit attachment to milk flow. Settings should be adjusted if this is longer than optimum (>15 s). Post dip sprayers (not all RMS use them) also need to be adjusted so they actually get appropriate amounts of dip on the teats.

Other items to keep in mind are adjusting the pre-stimulation time (to result in rapid milk let down) and teat drying time, plus adjusting for breed, size of teat and shape of the udder. If the pen is mostly primiparous cows, the settings will be more aggressive because they have smaller teats. Hyperkeratosis should be monitored and machine settings adjusted accordingly. A system to find failed cows rapidly should be developed.

5 Field observations

5.1 Feeding cows in RMS

We observed that at start-up of a new RMS, nutritionists and farmers focused on developing a pellet formulation that encouraged RMS box visits. Pellets that are made from high-quality, palatable ingredients and hard, with high sheer force and no fines, promote increased visits and more rapid feed consumption. The RMS box concentrate pellet is designed to complement the farms' forages and other ingredients in the PMR. For example, if the PMR is high in corn silage and thus high in starch, a pellet with highly digestible NDF from by-products might be considered to minimize the risk of sub-acute ruminal acidosis.

Once producers had a pellet(s) that worked well, other factors became more important. Many producers commented that even minor changes in the PMR moisture, consistency of the mix (e.g. particle size) and changes in forage quality affected visits to the RMS box. A drop in RMS visits will result in a decrease in milk production and an increase in the number of fetch cows. The increase in fetch cows may disrupt other cow behaviours, resulting in even greater decreases in visits and milk production. It is crucial to have consistent feeding in order to maintain high milk production and minimize the number of fetch cows.

In a study we conducted in RMS herds in Minnesota and Wisconsin we found that the amount of pellets offered through the RMS box in free flow system farms averaged 5 kg/cow/day and ranged from 0.9 to 11.3 kg/cow/day. The PMR was balanced for a milk production of 4.5–13.5 kg less than the herd's bulk tank average milk production.

Lead feeding was generally used in early lactation. Until about 28 days in milk, cows were fed for 35–45 kg of milk production per day. From about 28 days in milk through peak lactation, cows continued to be fed nutrients that supported 35–45 kg of milk per day or for actual milk production, whichever was higher. After peak lactation, cows were fed for actual milk production and regaining body condition. Primiparous cows were frequently fed additional robot concentrate to support growth and because of their lower PMR intake.

For cows close to dry off, producers used a feed table for late lactation cows that decreased RMS pellet feed in order to reduce milk production before dry off, especially farms with high producing late lactation cows. One challenge of free flow systems is that late lactation cows can become fetch cows. A key to preventing this is to have an excellent reproductive programme that maintains high milk production through the end of lactation.

We observed that milk first guided flow traffic farms used a different feeding approach than free flow. The amount of feed offered in the RMS box was minimal and only used to entice cows to visit the RMS box. A higher percentage of the cow's feed intake was delivered through the PMR. One main reason farmers installed guided flow RMS was because of the desire to feed less of the more expensive pelleted feed in the RMS box.

Farmers with milk first guided flow systems were feeding from 0.9 to 5.5 kg of pellets per cow per day. The average amount fed across all herds was approximately 3.5 kg/cow/day. Most herds were fed 0.7–1.4 kg of pellets per milking visit. Because of earlier lactation, higher producing cows are guided to the RMS box more frequently, and they received more RMS pelleted concentrate than late lactation cows.

The PMR in guided flow systems tended to be slightly higher in energy and lower in neutral detergent fibre than the PMR in free flow systems. For guided flow herds, the PMR was balanced for 4–9 kg less milk production than the average of the herd. This difference should be expected between free flow and guided flow systems. Using a high-energy density PMR in free flow barns may

Table 1 Range of select partially mixed ration nutrient values on two surveys of RMS farms

Item	University of MN survey	Ontario survey[a]
Net energy for lactation (Mcal/lb)	0.60–0.78	0.63–0.81
Neutral detergent fibre (% of DM)	28–40	30–50
Crude protein (% of DM)	12.0–17.7	13–18

[a] Tom Wright, Ontario Ministry of Agriculture, *pers. comm.*

lead to more fetch cows or decreased milking frequency, resulting in less milk production per cow, whereas in guided flow barns, cows are guided to the RMS box using selection gates.

Farmer comments and our observations indicate that the milk first system is superior with the US style of dairying where economics demand high production. In feed first systems, cows tend to stand in the feed alley or commitment pen ruminating after eating the PMR without entering the selection gate or visiting the RMS box.

Table 1 summarizes key PMR nutrient concentrations from one of our field studies on RMS farms in Minnesota and Wisconsin and a 2013 Ontario survey (T. Wright, Ontario Ministry of Agriculture, *pers. comm.*). Preliminary results of an ongoing field study we are conducting on RMS farms in Minnesota and Wisconsin show similar concentrations of NDF, crude protein (CP) and net energy (NE_L) as shown in Table 1. All studies included both free and guided flow farms (approximately 30% guided flow in the Minnesota study).

Wright also evaluated the PMR particle size using the Penn State Particle Separator and reported a higher percentage of particles on the top sieve and a lower percentage on the bottom pan than those recommended for a TMR (average 13.1% on the top sieve). This is expected considering some of the concentrate is fed in the RMS box separate from the PMR.

One potential advantage of RMS is the opportunity to feed each cow closer to her nutrient requirements by providing nutrients through a combination of the PMR and RMS box pellet. Even though RMS allows for feeding more than one concentrate feed in the RMS box, many producers in our previous field studies only used one feed. Our ongoing field study indicates that some producers are now using more than one feed to better target cows' nutrient requirements. Feeding a combination of concentrates in the RMS box at different proportions and amounts according to milk yield, body weight, stage of lactation and potentially milk components may maximize returns from RMS.

5.2 Milk quality observations

We have observed that typically SCC will increase for the first year after transitioning to RMS then might decrease. Conventional herds with poor training of workers and parlour technique commonly have a reduction in SCC after transition to RMS, whereas herds with excellent milk quality and cow prep technique might stay the same or have increased SCC.

5.3 Milk production per RMS per day

To achieve optimum efficiency in RMS we suggest that it is important to focus on milk per RMS per day, especially in the US market where profit margins are tight. This can be accomplished by maximizing the amount of milk per minute of box time and optimizing the per cent of time the robot is idle. There is a wide range among farms in these parameters. Our research showed an average milking time of 5.5 min per milking, with a range of 4.5–6.5 min. The per cent time the robots were idle averaged 19%. Most manufacturers recommend a goal of about 10% idle time to allow for maintenance and optimize visits per cow.

Cow, management and machine factors that affect milk per minute of box time include:

- Prep and attach time – this is affected by cow cooperation, teat placement, udder balance, udder singeing and the ability of the laser and/or cameras to find the teats
- Ideal udder conformation – RMS have difficulty attaching to udders with crossed rear teats, too deep and with severe reverse tilt
- Cow milking speed
- Milking permission settings – not allowing cows to milk until predicted milk production is higher will increase milk per minute of box time. However, this may decrease milk per cow per day and visits per cow per day. More research is needed to determine optimal milking permission settings
- Machine settings and maintenance – clean lasers and cameras and prep settings can minimize prep and attach time
- Optimum free time – suggested free time is 10%, but some farms have excellent visits and milk per cow with less free time
- Optimum milking refusals per cow. Refusals are visits to the RMS where no milking attachment was attempted because of milking permission settings

Factors positively associated with milk production per RMS in one of our studies with 32 free flow farms in Minnesota and Wisconsin included number of milking visits to the RMS box per day, cow milking speed, number of cows per RMS, average age of the cows in the herd and daily amount of robot feed offered in the RMS per cow (Siewert et al., 2018). Conversely, factors that were negatively associated with milk per RMS included residual feed (feed programmed to be offered to the cow, but not offered), treatment time (the time spent preparing the udder before milking and applying a teat disinfectant after milking), failures (visits where the cow had permission to be milked but the milking unit could not attach or she did not complete the milking) and refusals (cows visited the milking box, but milking permissions did not allow her to be milked at that time).

The focus is to balance milk per cow per visit, milk per cow per day and number of cows per RMS. To optimize RMS performance long term, cows with slow milking speed, and slow prep and attach time should be culled. This will make RMS much more efficient.

Another factor that might be influencing RMS efficiency is training of primiparous cows. One of our studies with 33 free flow RMS farms showed that primiparous cows had a lower milking frequency than multiparous (second and greater parity) cows in early lactation along with a greater frequency of failures. As previously mentioned, failures are characterized as visits to the RMS box where the cow should have been milked based on permission settings; however, due to equipment failures or the inability of the RMS arm to attach or reattach the milking unit due to conformation or cooperation issues with the cow, a milking visit was not successfully completed.

Better training of fresh cows to use the RMS, especially primiparous cows, could potentially improve productivity of RMS farms. Most farms fetch their fresh cows for a few days in the beginning of lactation to ensure they are visiting the RMS. The use of automated concentrate feeders in the pre-fresh pen is becoming more common. This helps cows to become more acquainted with an automated system and more likely to voluntarily visit the RMS in early lactation.

We conducted a case study on a free flow RMS farm that trained their first-calf heifers during the pre-fresh period by housing them in a bedded pack pen within the RMS barn and physically bringing them through the RMS box twice a day where they were offered 0.5 kg of concentrate pellet per visit. Our analysis comparing milking frequency of primiparous cows for the year before they started their training protocol and the year after training showed a difference of approximately 1.2 more milking visits per day in early lactation for trained cows (Fig. 3). This change in milking frequency was probably one of the factors contributing to an increase of approximately 3 kg of peak milk production for the trained primiparous cows. It is important to point out that this farm had enough free time available in the RMS to do the training. This approach might not be feasible if the RMS is being used to full capacity (suggested by manufacturers to be 60–65 cows per RMS box in free flow RMS).

5.4 Economic considerations

The RMS is hi-tech and expensive; this sophisticated piece of equipment requires money to maintain and repair. If producers learn how to fix little things,

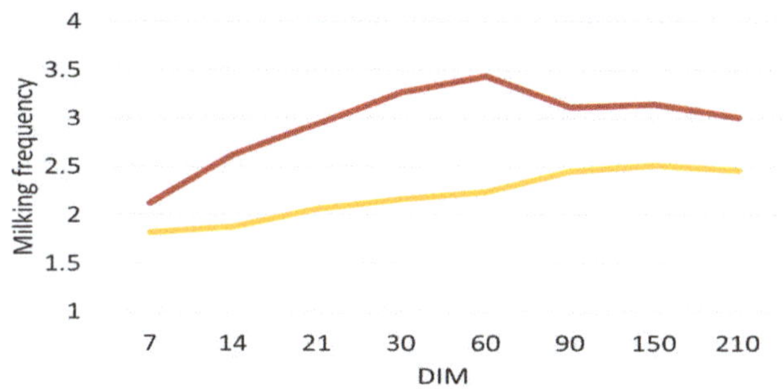

Figure 3 Milking frequency (milking visits to the RMS/day) by days in milk (DIM) for primiparous cows in a case study RMS herd before and after training to the RMS. Trained cows = red; non-trained cows = yellow.

it will help make RMS more affordable in the long run and reduce the number of failures and problems that can affect robot efficiency. It is not possible to just go to the local hardware store to purchase all the RMS parts needed.

Even though one of the motivating factors of most producers installing robots is to improve quality of life, it is important to develop a realistic cash flow and carefully think about how much capital can be tied up in the milking system. The University of Minnesota developed a partial budget online tool with corresponding spreadsheets examining the economics of RMS compared to a conventional milking system. It is available at: http://z.umn.edu/RobotParlor. This tool can allow for a quick assessment on how RMS may affect profitability. The main management factors affecting whether RMS are more profitable than parlours are increased milk production per cow, labour savings, and labour wages and wage inflation. Another major factor is years of economically useful life (Salfer et al., 2017).

6 Summary and future trends

The milking process fits well with robotic technology. This technology continues to improve and more options will be available in the future. Labour availability in many parts of the world will most likely continue to drive the adoption of RMS. We suggest that taking a whole system approach in design and management will maximize RMS performance and efficiency.

Feeding cows in RMS requires adjustments on ration formulation to address the need to entice cows to the RMS box. Many factors affect attendance to the RMS and influence milk production. Along with balancing the PMR and concentrate pellet for the targeted milk production goal of the farm, factors related to feeding management, cow comfort, cow health and transition cow programmes also play a major role. Offering multiple feeds in the RMS box (both amount and composition) to more closely match the nutrient needs of individual cows is an area that has not yet been extensively implemented and could potentially be beneficial to the success of RMS.

The RMS will continue to improve and provide additional precision individual cow measurements that help improve herd performance and animal health. Udder health can be more challenging to manage in RMS but many RMS farms have low SCC and mastitis incidence. The knowledge about RMS continues to grow and producers' satisfaction with the system has continued to improve in recent years. It takes a team approach including producers, service providers, nutritionists, veterinarians and other advisors to best optimize RMS utilization.

7 Where to look for further information

Several reviews of the literature (udder health: Hovinen and Pyörälä, 2011; cow behaviour: Jacobs and Siegford, 2012; barn design: Rodenburg, 2017; feeding strategies: Bach and Cabrera, 2017; economics: Salfer et al., 2017; disease detection: King and DeVries, 2018) provide additional information on RMS research (see References section for full details).

Centres of expertise or additional documents include:

- Future Dairy Australia (http://futuredairy.com.au/)
- Precision Dairy Farming – Australia NSW Government (https://www.dpi.nsw.gov.au/animals-and-livestock/dairy/robotic-milking-systems)
- International Precision Dairy Farming Association (http://www.precisiondairyfarming.com/)
- Robotic Milking – New Zealand (https://www.sciencelearn.org.nz/resources/2089-robotic-milking)
- Robotic Technologies in UK Dairy Farming (https://orca.cf.ac.uk/40232/1/Robotic%20milking%20project%20brochure.pdf)
- Automatic Milking Systems – CoBank (https://www.cobank.com/-/media/files/ked/dairy/automatic-milking-systems-report-mar-2018.pdf)

8 References

Bach, A. and V. Cabrera. 2017. Robotic milking: Feeding strategies and economic returns. *J. Dairy Sci.* 100:7720–8.

Bach, A., M. Devant, C. Iglesias and A. Ferret. 2009. Forced traffic in automatic milking systems effectively reduces the need to get cows, but alters feeding behavior and does not improve milk yield of dairy cattle. *J. Dairy Sci.* 92:1272–80.

Dohmen, W., F. Neijenhuis and H. Hogeveen. 2010. Relationship between udder health and hygiene on farms with an automatic milking system. *J. Dairy Sci.* 93:4019–33.

Douphrate, D. I., C. Lunner Kolstrup, M. W. Nonnenmann, M. Jakob and S. Pinzke. 2013. Ergonomics in modern dairy practice: A review of current issues and research needs. *J. Agromedicine.* 18:198–209.

Guarín, J. F., C. Baumberger and P. L. Ruegg. 2017. Anatomical characteristics of teats and premilking bacterial counts of teat skin swabs of primiparous cows exposed to different types of bedding. *J. Dairy Sci.* 100:1436–44.

Halachmi, I., E. Shoshani, R. Solomon, E. Maltz and J. Miron. 2006. Feeding of pellets rich in digestible neutral detergent fiber to lactating cows in an automatic milking system. *J. Dairy Sci.* 89:3241-2.

Halachmi, I., E. Shoshani, R. Solomon, E. Maltz and J. Miron. 2009. Feeding soyhulls to high yielding dairy cows increased milk production but not milking frequency, in an automatic milking system. *J. Dairy Sci.* 92:2317–25.

Hare, K., T. J. DeVries, K. S. Schwartkopf-Genswein and G. B. Penner. 2018. Does the location of concentrate provision affect voluntary visits, and milk and milk component yield for cows in an automated milking system? *Can. J. Anim. Sci.* 98:399–404.

Heurkens, D., C. Kamphuis and A. J. van der Kamp. 2016. Effects of free-stall barn layout on efficiency of Dutch dairy farms with an automatic milking system. *Proceedings of the Precision Dairy Farming Conference, Leeuwarden, The Netherlands*. Wageningen Academic Publishing, Wageningen, The Netherlands, pp. 157–62.

Hogeveen, H. and W. Steeneveld. 2013. Integrating it all: Making it work and pay at the farm. *Proceedings of the Precision Dairy Conference and Expo*, pp. 113–21. https://www.ansci.umn.edu/sites/ansci.umn.edu/files/2013_precision_dairy_proceedings.pdf

Hovinen, M. and S. Pyörälä. 2011. Invited review: Udder health of dairy cows in automatic milking. *J. Dairy Sci.* 94:547–62.

Jacobs, J. A. and J. M. Siegford. 2012. The impact of automatic milking systems on dairy cow management, behavior, health, and welfare. *J. Dairy Sci.* 95:2227–47.

Ketelaar-de Lauwere, C. C., A. H. Ipema, C. Lokhorst, J. H. M. Metz, J. P. T. M. Noordhuizen, W. G. P. Schouten and A. C. Smits. 2000. Effect of sward height and distance between pasture and barn on cows' visits to an automatic milking system and other behaviour. *Livest. Prod.* 64:131–42.

King, M. T. M. and T. J. DeVries. 2018. Detecting health disorders using data from automatic milking systems and associated technologies. *J. Dairy Sci.* 101:8605–14.

Lyons, N. A., K. L. Kerrisk and S. C. Garcia. 2013a. Effect of pre- versus postmilking supplementation on traffic and performance of cows milked in a pasture-based automatic milking system. *J. Dairy Sci.* 96:4397–405.

Lyons, N. A., K. L. Kerrisk, N. K. Dhand and S. C. Garcia. 2013b. Factors associated with extended milking intervals in a pasture-based automatic milking system. *Livest. Sci.* 158:179–88.

Lyons, N. A., K. L. Kerrisk and S. C. Garcia. 2014. Milking frequency management in pasture-based automatic milking systems: A review. *Livest. Sci.* 159:102–16.

Madsen, J., M. R. Weisbjerg and T. Hvelplund. 2010. Concentrate composition for automatic milking systems – effect on milking frequency. *Livest. Sci.* 127:45–60.

Miron, J., M. Nikbachat, A. Zenou and D. Ben-Ghedalia. 2004. Lactation performance and feeding behavior of dairy cows supplemented via automatic feeders with soy hulls or barley based pellets. *J. Dairy Sci.* 87:3808–15.

Penry, J. F., P. M. Crump, P. L. Ruegg and D. J. Reinemann. 2017. Short communication: Cow- and quarter-level milking indicators and their associations with clinical mastitis in an automatic milking system. *J. Dairy Sci.* 100:9267–72.

Prescott, N. B., T. T. Mottram and A. J. F. Webster. 1998. Relative motivations of dairy cows to be milked or fed in a Y-maze and an automatic milking system. *Appl. Anim. Behav. Sci.* 57:23–33.

Rodenburg, J. 2017. Robotic milking: Technology, farm design, and effects on work flow. *J. Dairy Sci.* 100:7729–38.

Rodenburg, J. and K. House. 2007. Field observations on barn layout and design for robotic milking. *Proceedings of the Sixth International Dairy Housing Conference*. ASABE publication number 701P057e.

Salfer, J. A, K. Minegishi, W. Lazarus, E. Berning and M. I. Endres. 2017. Finances and returns for robotic dairies. *J. Dairy Sci.* 100:7739–49.

Scott, V. E., K. L. Kerrisk, P. C. Thomson, N. A. Lyons and S. C. Garcia. 2015. Voluntary cow traffic and behaviour in the premilking yard of a pasture-based automatic milking system with a feed supplementation regime. *Livest. Sci.* 171:52–63.

Shishanthi, J. 2017. Plasmin, plasminogen, protein and somatic cells variation of bulk milk. Second cycle, A2E. Uppsala: SLU, Department of Molecular Sciences. http://stud.epsilon.slu.se

Siewert, J. M., J. A. Salfer and M. I. Endres. 2018. Factors associated with productivity on automatic milking system dairy farms in the Upper Midwest United States. *J. Dairy Sci.* 101:8327–34.

Spörndly, E. and E. Wredle. 2004. Automatic milking and grazing – Effects of distance to pasture and level of supplements on milk yield and cow behavior. *J. Dairy Sci.* 87:1702–12.

Stræte, E. P., J. Vik and B. Gunnar Hansen. 2017. The social robot: A study of the social and political aspects of automatic milking systems. *Proceedings in System Dynamics and Innovation in Food Networks 2017*, pp. 220–33. http://dx.doi.org/10.18461/pfsd.2017.1722

Tremblay, M., J. P. Hess, B. M. Christenson, K. K. McIntyre, B. Smink, A. J. van der Kamp, L. G. de Jong and D. Döpfer. 2016. Factors associated with increased milk production for automatic milking systems. *J. Dairy Sci.* 99:3824–37.

Tse, C., H. W. Barkema, T. J. DeVries, J. Rushen and E. A. Pajor. 2018. Impact of automatic milking systems on dairy cattle producers' reports of milking labour management, milk production and milk quality. *Animal* 12:2649–56. doi:10.1017/S1751731118000654.

van der Tol, R., R. X. Song and P. N. Kool. 2012. Validation of a clinical mastitis indicator. *Proceedings of the National Mastitis Council, 51st Annual Meeting*, pp. 233–4.

Chapter 13

Advances in automating meat processing operations

Ai-Ping Hu, Georgia Tech Research Institute, USA

1 Introduction

Robotics has revolutionized mass manufacturing, ranging from the automotive and electronics industries to the harvesting of commodity crops (wheat, soybean and corn). The common feature shared by these different application domains is the presence of a very structured environment. Within a factory, pains are taken to precisely fixture parts (such as a car chassis or a printed circuit board) so that they are presented to the robot in a predictable and repeatable way. Likewise, commodity farms are laid out in acres and acres of regular rows. By contrast, comparatively unstructured specialty farms (fruits and nuts) have largely not yet benefited from automation. The technology gap is analogous to the one between mass manufacturing and the production of handmade goods. The former, taking place in a standardized environment, handles uniform products while the latter deals with a 'lot size of one'. The former is relatively easy to automate while the latter, which has many adept human practitioners, has proven extremely difficult to automate with robots.

Meat processing falls into this latter challenging category, as it deals with deformable biological products that lack uniformity: no two fish, sides of beef or chicken carcasses are exactly alike. Walk into any meat processing plant and the majority of the line workers you will encounter are engaged in some form of deboning, portioning or trimming of non-uniform product with a knife.

http://dx.doi.org/10.19103/AS.2019.0056.21

This chapter selectively surveys the advances in robotic automation of the processing of fish, beef and pork, as well as poultry (for which the author will present a case study from his own research).

In what follows, a distinction is made between robotic, or *flexible*, automation (our topic) and what is known as *fixed automation*. In a fixed automation environment, the sequence of processing operations is determined by the configuration of the equipment. The 'programmed' sequence of operations is embodied in mechanical cams, gears and other physical mechanisms that are not easily changed. Fixed automation is inherently not an optimal approach to follow when dealing with unstructured environments and with variable, non-uniform products. By contrast, robotic automation is comparatively versatile and can adapt to a myriad of tasks without requiring hardware changes.

To categorize the surveyed work, each is tagged with its *manipulation* method, *perception* method and whether it takes a *static* or *dynamic* approach. Manipulation describes the physical interaction with the meat: deboning (cutting with respect to bone structure), portioning (of fillets, while applying some optimality criteria), harvesting (parts extraction/transfer) and trimming (removal of fat or viscera). In each of these operations, the de facto tool is a knife. Perception describes the sensing modality used (e.g. machine vision). Our distinction between a static approach and a dynamic one is illustrated in Fig. 1.

The static approach proceeds sequentially in an open-loop fashion: the meat is examined (once) and then operated upon. By contrast, a dynamic approach effects a feedback loop so that the meat is continuously sensed in real-time and the physical interaction is adjusted commensurately.

2 Fish

Reference	Manipulation	Perception	Static/dynamic
Gamage et al. (1993)	Deboning	2D camera	Static
De Silva and Wickramarachchi (1997)	Deboning	2D camera	Static
Omar and De Silva (2000, 2001)	Portioning	2D camera, structured laser	Static
Bar et al. (2016)	Trimming	3D line scan camera	Static

For fish processing, the main manipulation tasks are head and tail removal (deboning), portioning of fillets and trimming. Because of their relatively low-profile morphology, from an automation standpoint fish have been effectively treated as nearly two-dimensional (2D, i.e. flat), conveyed horizontally, and sensed and manipulated from above, such as what takes place on a canning line.

Figure 1 Illustration of static (left) and dynamic (right) approach to automating meat processing operations.

Gamage et al. (1993) applied computer vision pattern recognition techniques to identify fish head features, such as the eye, nose, flap and gill (Fig. 2, left). Such features are then used as references for positioning and orienting cutters executing head removal. As detailed by Gamage et al., the cutters may not necessarily intersect with the vision-identified features themselves. Rather, an offline study was first performed with a quantity of roughly 500 salmon heads to statically correlate feature dimensions, such as eye-to-gill distance, eye-to-edge-of-fin distance and so on, to cutter location using multiple regression models. The fish heads used in the research were obtained from a local fish processing plant and consultation with fish processing experts was used to determine the optimal cutter location for maximum meat recovery as: 'the point of intersection of the line along the back of the roof of the mouth and the centre plane of the head'. This optimal cutter location is not visible from the outside of the fish and is essentially being statistically predicted based on identifiable head features. The accuracy of the method was assessed to be within 2.5 mm of known optimal via manual butchering, an improvement over existing fixed automation cutters calibrated using average settings. De Silva and Wickramarachchi (1997) describe the research and development of a flexible automation solution for fish head cutting that eliminates many of the drawbacks of a fixed automation system. The described system uses a CCD

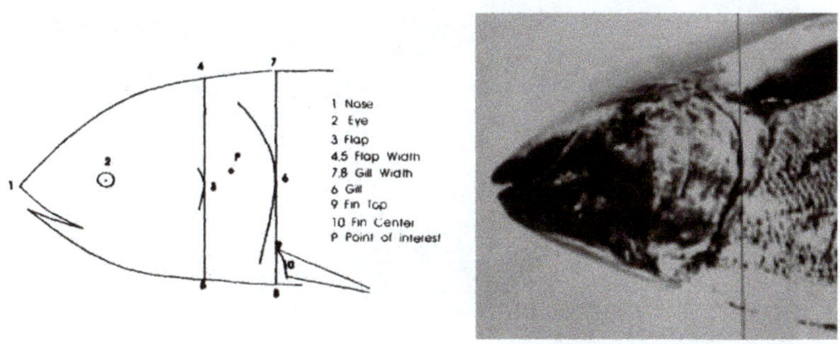

Figure 2 Left: anatomical features of a fish head. Source: Adapted from Gamage et al. (1993). Right: detected gill and corresponding cutter position. Source: Adapted from De Silva and Wickramarachchi (1997).

camera to detect the fish gill, the location of which has been associated with the desired cutter position (Fig. 2, right). The cutter position is adjusted by a servomotor. This added customizability achieved by the flexible automation solution led to an average 3.4% in improved yield compared to existing fixed automation.

Omar and De Silva (2000, 2001) studied optimal fish fillet portioning for can-filling. The goal of the optimization algorithm is to make each portion as close as possible to the target weight on a can's label, within a tolerance. For this problem, a 2D camera was used along with structured laser light to obtain a volume measurement of the fillets, from which weight could be estimated. A numerically tractable optimization algorithm was formulated based on overlapping adjacent fillets (to reduce the number of constraints) and casting the task as a nonlinear integer programming problem. The can-filling portions could be achieved by a straight-blade cutter with variable spacing. Simulation results based on real data, as well as a schematic for a physical embodiment, were presented.

Bar et al. (2016) studied automated post-trimming of fish fillets, performing the final removal of unwanted fat, bone remnants and other defects left behind by current commercial trimming machines. A prototype implementation consisting of a three-dimensional (3D) line scan camera, a 6 degree-of-freedom (DOF) robotic arm and a flexible knife end effector was described. The 3D line scan camera obtains intensity and depth information for the fillet, from which key feature points could be extracted using computer vision methods (Fig. 3). Through a combination of experimentation and empirical observation of human practitioners, Bar et al. designed spatial knife trimming paths (including one characterized by a sawing motion) for the tail trim operation that was the focus of their work. The knife paths were mathematically parameterized by the identified key feature points, so that the paths could be scaled, rotated and

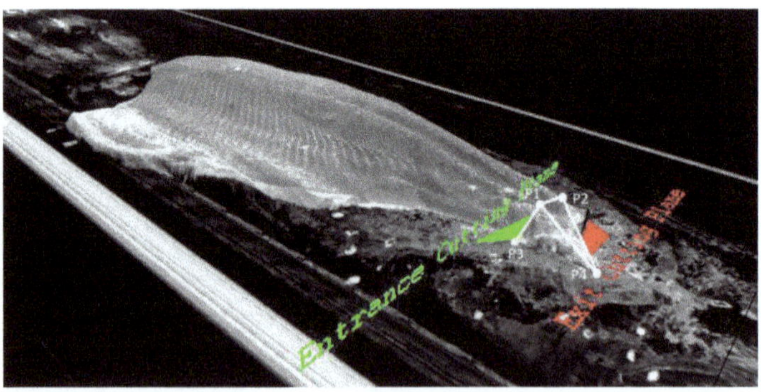

Figure 3 3D line scan of fish fillet on conveyor for post-trimming. Source: Adapted from Bar et al. (2016). P1 indicates the thickest point on the fillet between fitted line segment from P2 to P3 and P4 is the tip of the tail. Points P2, P3 and P4 lie in the conveyor plane.

otherwise transformed and customized for each unique fillet. Once computed, the robotic arm executed the trimming path in an open-loop fashion.

3 Beef and pork

Reference	Manipulation	Perception	Static/dynamic
Khodabandehloo (1989)	Deboning	CCD camera, force sensor	Dynamic
Wadie et al. (1995)	Deboning	Camera	Static
Guire et al. (2010)	Deboning	Camera, structured lighting, force sensor	Dynamic
Long et al. (2013, 2014)	Portioning	Camera, force sensor	Dynamic
Subrin et al. (2014)	Deboning	Force sensor	Dynamic

The flexible automation approach combining *customized perception* plus *adaptable manipulation* that characterized fish processing (above) is also applicable to beef, pork and poultry. Khodabandehloo (1989) articulated the challenges inherent in automating meat processing (non-uniform nature, wide range in size and weight) and described the new technologies that would be needed (grippers/mechanical hands for food products, vision and tactile sensors, expert knowledge transfer (and active learning) from human practitioners to machines). The work laid out the basic perception plus manipulation paradigm for robotic meat cutting (see Fig. 4, left). Force feedback (how much mechanical effort is being expended while cutting and also: Can a bone be used as an opposing-force cutting guide?) was inserted as a form of continuous sensing for dynamic knife adjustment, as was a capability for the overall system to learn from its historical database of imaged carcasses and adjust its decision processing for robot control.

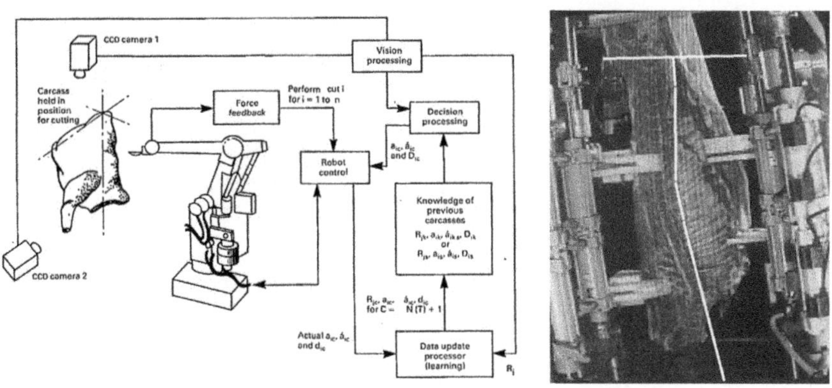

Figure 4 Left: robotic meat cutting general configuration. Source: Adapted from Khodabandehloo (1989). Right: vision system image showing computed knife path overlaid onto pig carcass. Source: Adapted from Wadie et al. (1995).

Wadie et al. (1995) developed an automated system for pig carcass cutting and evisceration that uses a camera and a custom robotic cutter. Figure 4 (right) shows sample cutting paths that were computed using identifiable anatomical features such as the centre of the last lumbar vertebra, the edge of the belly, the minimum of the spine curve at the neck and the centre of the fifteenth vertebra from the last lumbar vertebra. (Note that the derivation of a cutting path that is computed based on a correlation with perceived anatomical features is similar to automated fish deboning and trimming.) In cases where the vertebrae are obscured by fat or skin, the raw vision image is compared to an idealized spine model to fill in missing data.

Guire et al. (2010) considered the automated quartering of beef carcasses with a 6-DOF robotic arm with a knife end effector, which involves performing the cut ABCD, known as the 'Z-cut', on a beef carcass that has been halved (Fig. 5, left). Cut AB is performed along the thirteenth rib, cut BC is between the fifth and seventh ribs (depending on season and market demand for particular cuts) and must remain parallel to the spine, and cut CD severs the spinal column. A known structured light pattern is projected onto the halved carcass (Fig. 5, right) to facilitate image processing to locate the relevant anatomical features from which a nominal path ABCD may be computed for the specific carcass half. Guire et al. then used a force sensor integrated with the knife to use rib bones as cutting guides by monitoring lateral forces on the knife blade. Different blade angles and speeds were experimented with, to identify cutting paths with minimal energy requirements and producing a desired texture quality.

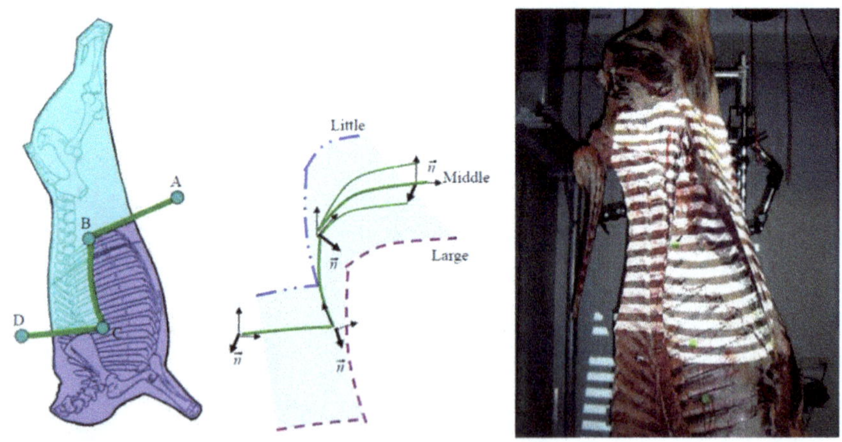

Figure 5 Left: nominal Z-cut (denoted by ABCD) for quartering of a halved beef carcass. Centre: natural carcass size variation (little, middle, large) will determine the exact shape of the cut. Right: structured light pattern projected onto beef carcass. Source: Adapted from Guire et al. (2010).

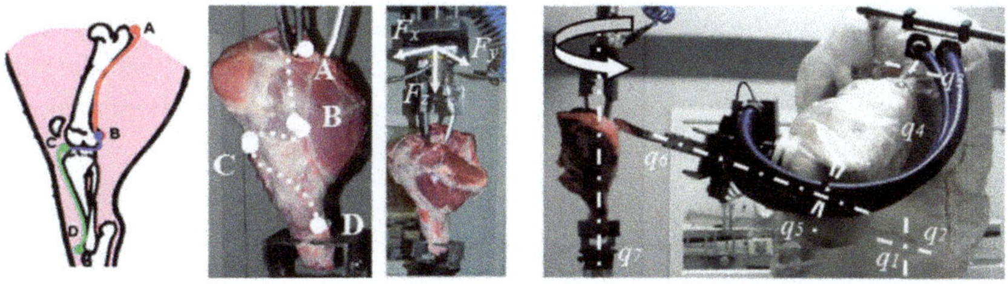

Figure 6 Ham deboning: desired cutting path (schematic), overlaid onto real ham, ham support equipped with a force sensor on the upper clamp and robotic cell showing the seven total degrees of freedom. Source: Adapted from Subrin et al. (2014).

Subrin et al. (2014) used (only) a force sensor for feedback during ham deboning automated with a 7-DOF robotic system wielding a knife end effector (Fig. 6). The ham is mounted onto a rotating fixture and operated upon by a 6-DOF robotic arm. A force sensor is located on the rotating fixture. The nominal desired knife path ABCD for the ham deboning operation is shown in Fig. 6, obtained by studying human arm gestures while performing the task. The kinematics (i.e. motion geometry) of the human arm is redundant, which informed the use of the rotating fixture that contributes an additional degree of freedom. Similar to Guire et al. (2010), this implementation seeks to use bone as a guide for the force-controlled knife. However, unlike other approaches reviewed, no visual sensing is used at all.

Long et al. (2013, 2014) considered the problem of robotic portioning of beef slabs, specifically: separating two muscles joined by a non-homogeneous, tendon-like region known as the aponeurosis. Two robotic arms working together are used for the operation: one equipped with a knife and a second to pull one side of the beef slab (see Fig. 7). The desired behaviour is a gradual opening of the 'cutting valley' as the cut depth increases. Vision and force

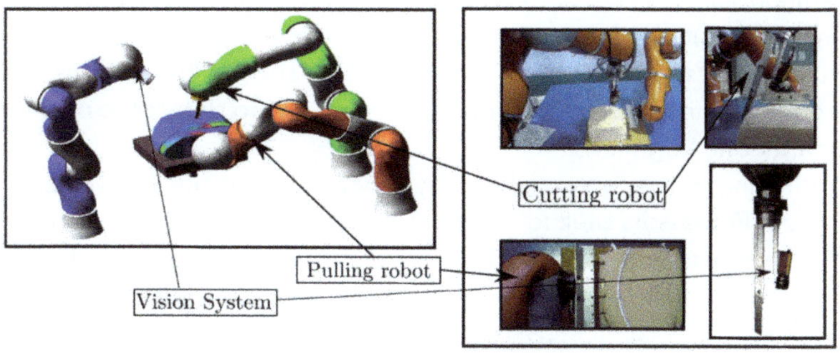

Figure 7 Comparison of simulation and experimental meat slab cutting cell. Source: Adapted from Long et al. (2014).

sensors are used to monitor real-time object deformation, so that the knife path can be continually updated. Long et al. modelled the operation (including the beef slab) in a computer simulator and then performed experiments using a foam stand-in for the beef slab.

4 Poultry portioning and harvesting

Reference	Manipulation	Perception	Static/dynamic
Daley et al. (1993)	Deboning	2D camera	Static
Heck (2006)	Portioning	2D camera, laser line scanner	Static
Barbut (2010, 2015)	Deboning, harvesting	2D camera	Static
Misimi et al. (2016)	Harvesting	RGB-D camera	Static

The majority of the line workers in a poultry processing plant are engaged in some form of handling, deboning or trimming of natural product with a knife. A handful of fixed automation solutions currently exist, though they are no match for human deboners in terms of their ability to extract the maximum amount of yield in meat. Furthermore, poultry processing plants are chronically plagued by labour turnover – the line jobs can be dull, dirty and dangerous. An automation solution that can spare people from this work environment, while performing at parity with humans, is highly desirable. As has been shown, the non-uniform and deformable nature of bio-materials makes processing them a challenging operation to automate.

Daley et al. (1993) described early challenges in applying flexible automation to poultry processing, considering gripper design for variable natural product, programming and cycle time optimization and robot work cell design. Early work in applying 2D greyscale and colour imaging to poultry inspection, quality control and robot guidance is also considered. Barbut (2010, 2015) summarizes the current state of the art in automation of poultry processing plants, covering operations such as stunning, scalding/plucking, evisceration and chilling. Also covered are the commercially available fixed automation solutions for poultry leg meat deboning, innovations in using a flexible carcass carrier ('cone') that can rotate about different axes and the implementation of sensing solutions (such as X-ray) to detect the presence of unwanted bone fragments.

Portioning in the context of poultry processing principally involves breast meat fillets. Processors seek to maximize profits by efficiently using each piece of deboned meat (refer to Fig. 8): one breast fillet can be divided into bun-sized portions (most valuable), nuggets, strips and trim (least valuable). Heck (2006) described a commercial portioning system that uses high-speed image processing (2D camera and a laser line scanner, which together can estimate volume and therefore weight) and automated focused water jets for processing

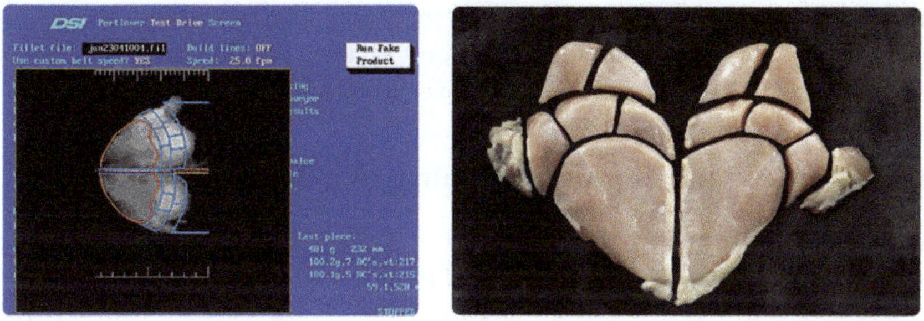

Figure 8 Left: commercial template software showing poultry fillet portioning into two bun-sized portions and several nuggets. Right: example end product. Source: Adapted from Heck (2006).

breast fillets. The portioning system images each fillet, determines an optimal cutting pattern and controls the water jets to make the desired cuts. The conveyor moving the fillets can have a speed of up to 85 feet/min and the system can process 180–200 fillets/min for simpler portioning configurations and 60–70 for more complex cuts.

Misimi et al. (2016) have developed a prototype robotic poultry harvesting system for removing tender fillets from a bird front-half (the final operation in front-half harvesting). The prototype uses an RGB-D (i.e. colour plus depth) camera as the perception sensor and a custom, 3D-printed beak-shaped gripper with integrated supporting plate, for grasping and holding tender fillets and simultaneously scraping the carcass while pulling the tenders (refer to Fig. 9). The colour image and depth data from the RGB-D camera is processed to locate the 3D coordinates of the grasping point that is the initial contact

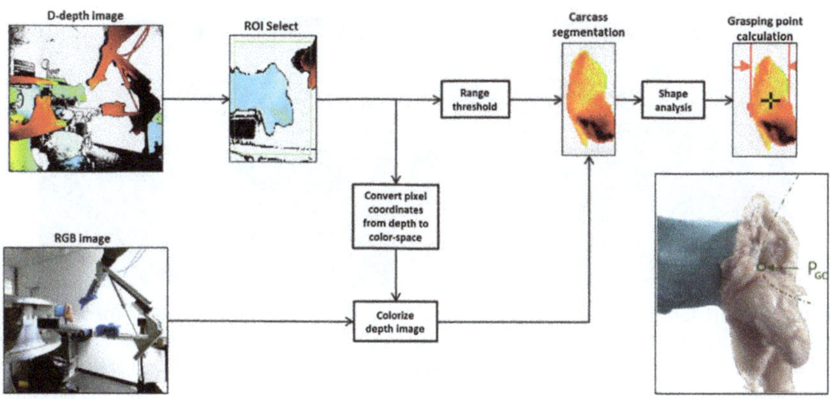

Figure 9 Sequence of RGB-D camera image processing operations leading to computation of the grasping point P_{GC}. The GRIBBOT beak may be seen in the inset 'RGB Image'. Source: Adapted from Misimi et al. (2016).

point of the gripper with the bird carcass for tender fillet harvesting. The scrape and pull motions are computed based on the anatomical boundary between the meat and the carcass rib cage, and executed in an open-loop fashion.

5 Case study: intelligent deboning of poultry

Second and further processing operations constitute the largest use of on-the-line labour in today's poultry processing plants. Deboning is a significant part of these operations. This section describes continuing work in the intelligent automation of bird deboning, a task that is conceptualized into three parts: characterizing the non-uniform bird product (using statistical studies and image processing), obtaining a nominal cutting path (using image features correlated to internal bird structures and robot kinematics) and correcting for deviations from the nominal cutting path (using force control). Results from deboning experiments using a prototype developed to perform bird shoulder cuts are described.

Deboning in the context of beef and pork has relatively coarse requirements. For example, cutting through hard bone is permissible in those industries, in contrast to the poultry industry, wherein the generation of bone chips is deemed a safety hazard to the consumer. Researchers at the Georgia Tech Research Institute's Food Processing Technology Division (FPTD) have been engaged in automating poultry deboning (Zhou et al., 2007), in particular, severing the tendons and joints on bird front-halves (without cutting into bone) in preparation for the removal of the wings and breast meat. FPTD has designed and fabricated a prototype robot deboning system, shown in Fig. 10, with the capability to adapt to individual birds. The 2-DOF test bed cutting robot (its

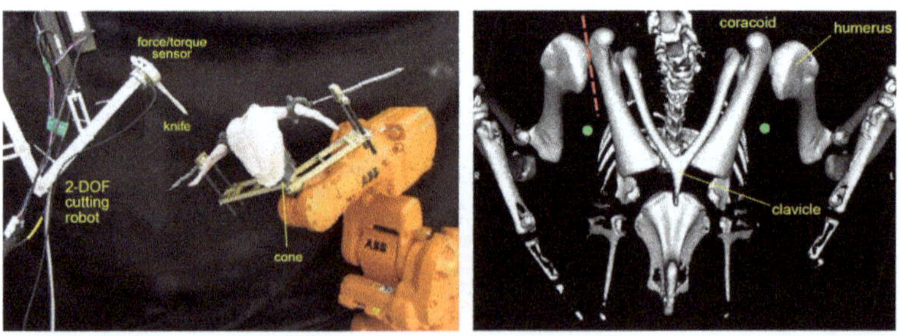

Figure 10 Left: Georgia Tech Research Institute's prototype embodiment of the Intelligent Deboning System. The 2-DOF robot manipulates the knife to cut a bird being presented by the 6-DOF robotic arm. The bird sits on a cone that is rigidly attached to the wrist of the robotic arm. Right: MRI scan of a chicken showing a frontal view of the bird skeleton. Three bones (the coracoid, clavicle and humerus) come together to form the shoulder joint. The green dots indicate points of interest on the bird's exterior. The dashed red line denotes the shoulder joint cut.

kinematics have been described in Zhou et al., 2009) is able to manipulate a knife blade within a suitable motion envelope. A 3D machine vision system (not shown) views each bird individually, in order to extract the location of feature points on the exterior of the bird carcass. A 6-DOF articulated robotic arm manufactured by ABB is used to properly position and orient each cone-fixtured bird with respect to the test bed cutting robot, in facilitation of respective cuts. Early stage studies on bird fixturing and different modes of wing manipulation have been performed to effect the optimal presentation of the bird front-half to the cutting robot blade. Once in place, the bird and cone are stationary during the cut. The cutting robot wrist is instrumented with a force/torque sensor. This section details FPTD's approaches using this test bed hardware for cutting the bird shoulder joint.

Daley et al. (1999) initially considered the problem of natural product deboning for poultry, recognizing that the ability to adapt to varying bird sizes and shapes would be key to a viable automation solution. Two methods were described to obtain knife cutting paths during deboning, one based on X-ray images of internal bird structures (correlated with visual images) and the second based on motion capture of a person performing a deboning task with a knife. That early work has been translated into FPTD's more recent efforts, where the deboning task has been conceived of being composed of three steps that emulate how a human deboner on a poultry line operates, by:

1 characterizing the non-uniform product (by statistically correlating external bird features to internal structure);
2 obtaining nominal cutting trajectories (by using 3D image processing to detect external landmarks); and
3 correcting for deviations from a nominal knife path (by applying real-time force control).

The implementation of these steps (in hardware and software) is described below where (only) the shoulder joint cut will be considered. Figure 10 shows an MRI scan of a chicken skeleton. The connection between the humerus and the carcass includes ligaments, tendons, meat and skin, which form the ball-and-socket shoulder joint. This connection must be severed in order to separate the humerus from the coracoid and clavicle (refer to the dashed red line in Fig. 10).

5.1 Characterizing the non-uniform product

It turns out that the relative bodily proportions of chickens, like human beings (and fish and cows and pigs), adhere to a quantifiable standard. Given known measurements of key points (e.g. the mid-line of body to the tip of an outstretched appendage), the dimensions of other parts of the body may be deduced. Based on statistical studies performed by FPTD on a quantity of birds, it is possible to deduce the location of the shoulder joint based on just a few key measurements on the exterior of the bird. The approximate locations of these points appear in Fig. 10; the actual points lie on the surface of the bird carcass, which is not visible in the MRI scan shown.

External bird feature measurements are obtained by placing the bird on a cone that is rigidly mounted onto the wrist of the 6-DOF articulated robotic arm (Fig. 10) and then inserting the bird/cone into a machine vision station. The machine vision station contains two pairs of stereo cameras (four cameras total). Each pair looks at one side (left or right) of the bird carcass. The stereo pair allows determination of the 3D position of any given point viewed in the scene. Novel image processing techniques developed at FPTD are used to find the feature points of interest. Statistical studies have been performed to correlate the spatial coordinates of these feature points to bird size and to joint/tendon locations. This has been done by using measurements taken at 64 different points on the bird's anatomy. In robot implementations of the cut, predictions of shoulder joint locations were found to be accurate to within 3 mm.

Each stereo camera pair is first calibrated to (i.e. points within its field of view were referenced with respect to) a checkerboard pattern that has been rigidly affixed to the articulated robotic arm's wrist and inserted into the machine vision station. The kinematic relationship between the checkerboard and the articulated robotic arm is precisely known. The 3D coordinates of a common point viewed by both pairs of stereo cameras is solved for using the least-squares algorithm described in Arun et al. (1987). A coordinate transformation may then be used to relate these 3D points (known with respect to the checkerboard) to the cone (which may be regarded as an end effector with its own coordinate frame, shown in Fig. 11). The transformation between the base of the articulated robotic arm and the base of the 2-DOF cutting test bed robot is known. Therefore, cutting paths that are defined with respect to the cone frame may be easily transformed to the 2-DOF cutting test bed robot, and thereby used to define trajectories for the knife. The right side of Fig. 11 shows schematically the relationship between the cone, the knife and the bird shoulder during a shoulder joint cut.

5.2 Obtaining nominal cutting paths

The ideal cutting path for the automated bird deboning process depends directly on the relationship between the internal bone/tendon structure and the external features on the bird. A deboner's past experiences handling the birds by (a combination of) touch and sight result in a learned muscle memory. Based on this sensory input, a deboner will conceive of imaginary lines on the bird carcass, along which her knife's motion will trace a path.

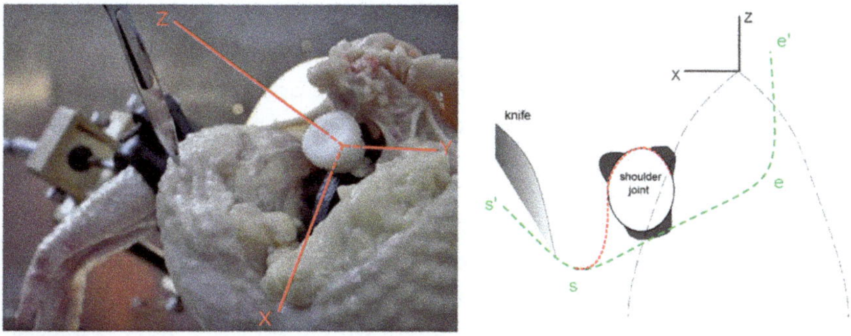

Figure 11 Left: cone coordinate frame. The cone is rigidly attached to the wrist of the 6-DOF robotic arm. Right: shoulder joint cross section (not drawn to scale). Nominal planar trapezoidal shoulder joint cutting path is shown as the green dashed line. The approximate cutting path in the event that bone is contacted is shown as the red dashed line.

Referring to the right side of Fig. 11, the ideal shoulder joint cut takes place when the plane of the knife blade coincides with the spacing between the humerus and the rest of the bird carcass. In this case, the knife is able to slip through the joint interface, severing the tendons (and other pliable biomaterial), without contacting hard bone. We note that the joint interface width, within which no bone contact will take place, is approximately 5 mm. This width is somewhat dependent on the manner in which the wing is manipulated during a cut.

FPTD has implemented fairly simple planar trapezoidal cutting paths (Fig. 11): knife insertion (s′ s), slice across (s e) and knife extraction (e e′). The bird/cone is moved into place within the workspace of the 2-DOF cutting robot such that the knife blade is parallel to the shoulder joint interface (i.e. the gap between the humerus and the coracoid/clavicle). The deduced size of the bird will govern the depth of the blade insertion as well as the slice length. This simple planar cutting path has sufficed for the nominal shoulder joint cut. Experimental results are described below for shoulder joint cuts using this nominal knife path.

5.3 Correcting for deviations from a nominal knife path

Simply following the nominal predicted cutting trajectory in an open-loop fashion will rarely be robust enough to account for natural variations in the bird front-halves. Even within the fairly small 3 mm errors FPTD has achieved in shoulder joint prediction, there have been instances where the robot has inadvertently cut through bone. In such an instance, a bone chip is generated that is unacceptable to the poultry industry because it poses a choking hazard to the consumer. To correct for deviations from nominal, FPTD has implemented a form of force control on the existing cutting robot.

During a cut, if contact with bone is detected, position control of the blade transitions to force control. The latter makes use of the force/torque sensor affixed to the blade handle located at the 'wrist' of the cutting robot end effector (Fig. 10). The sensor is used to measure forces along three orthogonal axes. A modified version of the dynamic hybrid position/force control method developed by Yoshikawa and Sudou (1993) has been implemented on the 2-DOF cutting robot. The method assumes that the profile of the bone cross section – which lies in a plane – is not known a priori. The method uses the orthogonal force sensor readings, properly transformed from the wrist to the blade edge, to estimate, in real-time, the shape of the bone (characterized by unit vectors tangent and normal to the bone surface). Given the deduced shape profile, feedback control based on both cutting robot joint encoder and force/torque sensor readings is then used to command the blade's motion to cut (trace) around the bone, as depicted by the red dashed line on the right side of Fig. 11. When bone is encountered by the blade (i.e. a deviation from the nominal cutting path is encountered), the force/torque sensor effectively aids in closing the feedback control loop. This is similar to what a deboner on the line would do with his/her knife.

The results from two shoulder cut experiments on fresh (i.e. never frozen) birds are reported here. The first experiment is for a 'clean' cut through the shoulder joint, that is a cut through meat and tendon, with no bone contact. The second experiment was devised to intentionally contact the bird's humerus bone. During this second experiment, force control was activated after a threshold force was exceeded.

The procedure described above in this section has been used for each experiment: mount bird onto cone and secure wings, place cone/bird inside machine vision station (at a predefined pose), capture stereo images of the bird, apply tendon prediction image processing algorithm, transform predicted shoulder point from cone to the 6-DOF articulated robotic arm base frame, transform predicted shoulder point from 6-DOF robotic arm to cutting robot base frame, manipulate cone/bird into workspace of 2-DOF cutting robot and perform shoulder joint cut (transitioning to force control if bone is encountered during the cut).

Figure 12 shows the result for the clean shoulder cut. The trapezoidal knife path is shown superimposed onto the photograph of the cut bird. In this case, the knife blade has cut through the shoulder joint interface between the humerus and the coracoid bones, severing the tendons forming the shoulder girdle. The forces experienced by the knife during the cut (in the direction normal to the blade's cutting edge) is shown in the labelled plot in Fig. 12, with four points corresponding to the knife's trapezoidal path indicated. The knife

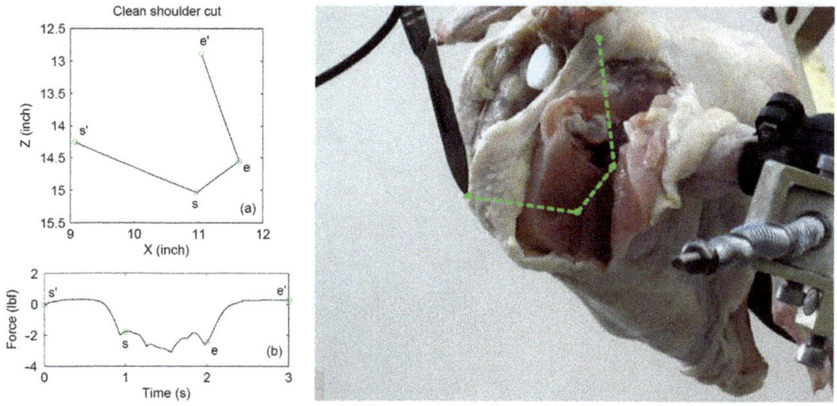

Figure 12 Clean shoulder cut, with no bone contact. Left: (top) the path followed by the knife, plotted with respect to the 2-DOF cutting robot's base frame and (bottom) the corresponding force sensor measurements (in the direction normal to the blade cutting edge) during the cut. Right: photograph of a bird showing a clean shoulder joint cut. The trapezoidal cut path is superimposed onto the photograph. Note that in the image, additional meat has been cut away from the frontal bird breast to better show the characteristics of the cut.

is first inserted and pierces the bird's breast meat along the s' s segment of the cutting path. (Note that the magnitude of the force on the blade increases after meat contact.) After insertion, the segment of the cut along s e commands the blade to draw across and through the shoulder joint. The variations in force (see the local peaks) in this portion of the cut correspond to the blade encountering and severing the various tendons of the shoulder joint. The force magnitude does not exceed approximately 3 lbf for a clean shoulder cut - this is a consistent feature in multiple experiments that have been run. The force profile during withdrawal of the knife (segment e e') is an approximate mirror of the force profile during knife insertion.

Figure 13 shows the results for a shoulder cut during which bone contact (with the humerus) was made by the knife blade. The bird's placement relative to the 2-DOF cutting robot was intentionally offset from the pose determined by the tendon prediction algorithm in order to assure bone contact. The nominal trapezoidal path was run for this experiment, with force control transitioned to upon exceeding a contact force threshold of 5 lbf. The subsequent path followed by the knife blade is plotted in Fig. 13. In this experiment, the desired contact force to be maintained by the force controller was set to 3 lbf. Upon transition, the controller causes the contact force to decrease from the 5 lbf threshold to the desired contact force (Fig. 13).

The approximate bone location during the experiment is indicated in the upper-left plot of Fig. 13. Because the humerus moves as the knife advances

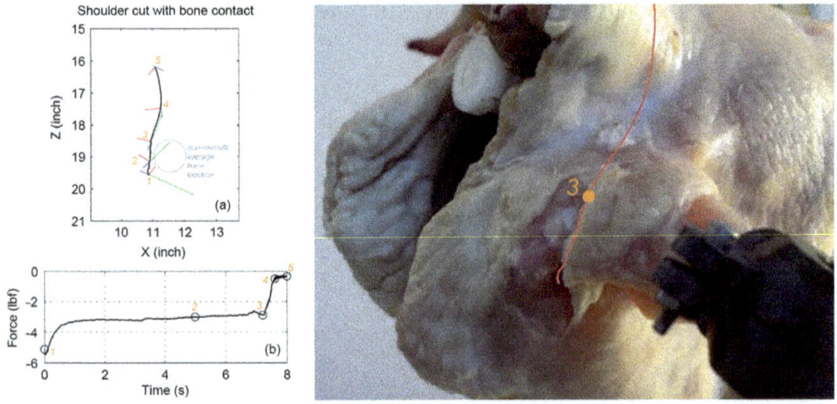

Figure 13 Shoulder cut during which bone was contacted. Left: (top) the path followed by the knife subsequent to bone contact and activation of force control, plotted with respect to the 2-DOF cutting robot's base frame and (bottom) the corresponding contact force on the knife blade. Right: corresponding photograph of shoulder joint cut in which the bird humerus bone was contacted and force control applied. An approximation of the knife path is shown in red, with the onset point of the blade retracting off the bone surface indicated.

during this cutting experiment, an average position is shown. The force control path overlaid onto the photograph of the bird in Fig. 13 depicts an approximation of the knife's motion with respect to the humerus during the force control experiment. At point labelled '3', the knife begins to lose contact with the bone surface. An inspection of the upper-left plot of Fig. 13, in which the computed normal (blue) and tangential (red) unit vectors and the contact force vector (green) are shown at a few key points, provides an explanation. Recall that the force/torque sensor the knife is affixed to is used to deduce in real-time the contour of an unknown surface (in this case, the cross section of a humerus bone). In the present case of cutting through a bird shoulder, the knife is interacting with the bird along multiple points along its blade. The effect is that the force components can be difficult to resolve. In addition, a kinematic limitation is imposed on the orientation of the knife by the 2-DOF cutting robot. It is likely desirable to maintain the same general portion of the blade cutting edge in contact with the bone (by constantly changing the angle of the knife). This is not possible under the described cutting robot configuration. The result is that the blade will come off the bone surface near the apex of the bone cross section, due to an inability to accurately determine the surface normal and tangent vectors. However, the preliminary force control results presented here are promising due to the ability to accurately maintain a desired contact force. In particular, the goal of not generating bone chips is realizable by setting the desired contact force to a suitably low value.

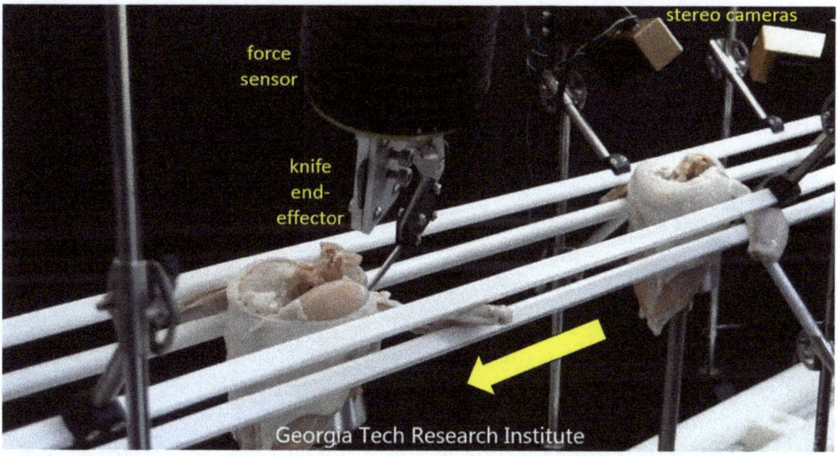

Figure 14 Updated embodiment of the Georgia Tech Research Institute's prototype bird front-half shoulder deboning system, integrated onto a standard moving cone line. The bird wings are passively constrained by the rails. Two stationary stereo cameras are used to image both sides of the unique bird as it passes within their field of view. A force sensor is integrated into the wrist of the 6-DOF robotic arm, preceding the knife end effector.

This section presented a case study describing the intelligent automation of bird shoulder joint cutting according to a *dynamic* conceptual framework ((1) characterizing the non-uniform bird product, (2) obtaining a nominal cutting trajectory and (3) correcting for deviations from the nominal cutting path) that has been implemented on the test bed hardware and software. The results from experiments demonstrate the approach to be effective. The researchers at FPTD have extrapolated the early prototype Intelligent Deboning System described in this section to the one shown in Fig. 14: (1) the bird front-half carcasses are now moving on a standard cone conveyor line and (2) the knife end effector is affixed to the wrist of a 6-DOF robotic arm that is equipped with an integrated force sensor. The force sensor is used to measure the real-time forces encountered by the knife blade and to modify the knife path to avoid cutting through bone. Preliminary experiments show promising results in the dynamic performance of bird front-half shoulder deboning on a moving cone line.

6 Conclusion and future trends

The robotic automation of meat processing operations (including fish, pork, beef and poultry) has two component considerations: *perception* plus *manipulation*. In the work surveyed in this chapter, perception modalities included 2D cameras, structured lighting, laser line scanners, RGB-D (colour plus depth) cameras and force sensors. Manipulation describes the physical interaction with the meat: deboning (cutting with respect to bone structure), portioning (of fillets, while applying some optimality criteria), harvesting (parts extraction/transfer) and trimming (removal of fat or viscera). The visual perception methods in this chapter may be catalogued into one of two varieties: (1) measure features on the exterior of the animal and predict, through prior physiological correlation, relevant interior cutting locations such as internal joints (processing of fish and poultry fall into this method) and (2) measure visible features and superimpose cutting paths onto them (pork and beef processing use this method). Furthermore, the automation approach of the work surveyed may be characterized as either *static* (meat is sensed once, followed sequentially by open-loop manipulation) or *dynamic* (a feedback loop is effected so that the meat is continuously sensed and manipulation is adjusted commensurately). In the dynamic cases, a force sensor was often used for the continuous feedback.

Two important current trends in robotics will have an impact on future automation of meat processing. The first is the application of *deep learning* methods (see, e.g. Goodfellow et al., 2016) to more accurately characterize the non-uniform biological product encountered in the meat industry. Deep learning has already begun to make an impact in automated medical imaging for detecting abnormalities and applied to the control of self-driving cars, among numerous additional applications. The second concerns the arrival on the market of a new breed of so-called *collaborative robots* that are rated to work alongside people without intermediary physical guarding. These collaborative robots are constructed with compliant joints, soft surfaces, reduced pinch points and equipped with force/torque sensors at their joints (to detect unexpected collisions). Their use in the meat processing industry opens up possibilities of achieving synergistic results in tasks that currently remain impossible to fully automate.

7 Where to look for further information

7.1 Further reading

- Barbut, S. 2014. Review: Automation and meat quality-global challenges. *Meat Science* 96(1), 335–45. doi:10.1016/j.meatsci.2013.07.002.
- Sanchez, J., Corrales, J.-A., Bouzgarrou, B.-C. and Mezouar, Y. 2018. Robotic manipulation and sensing of deformable objects in domestic and industrial applications: A survey. *International Journal of Robotics Research* 37(7), 688–716. doi:10.1177/0278364918779698.

7.2 Key journals/conferences

- *Journal of Food Engineering*
- *Industrial Robot: An International Journal*

- *Computers and Electronics in Agriculture*
- *Meat Science*
- Food Automation and Manufacturing Conference and Expo
- International Food Automation Networking Conference
- International Production and Processing Expo

8 References

Arun, K. S., Huang, T. S. and Blostein, S. D. 1987. Least-squares fitting of two 3-D point sets. *IEEE Transactions on Pattern Analysis and Machine Intelligence* 9(5), 698–700. doi:10.1109/TPAMI.1987.4767965.

Bar, E., Mathiassen, J. R., Eilertsen, A., Mugaas, T., Misimi, E., Linnerud, Å. S., Salomonsen, C. and Westavik, H. 2016. Towards robotic post-trimming of salmon fillets. *Industrial Robot: An International Journal* 43(4), 421–8. doi:10.1108/IR-11-2015-0205.

Barbut, S. 2010. Past and future of poultry meat harvesting technologies. *World's Poultry Science Journal* 66(3), 399–410. doi:10.1017/S0043933910000498.

Barbut, S. 2015. Developments in turkey meat harvesting technologies. *World's Poultry Science Journal* 71(1), 59–70. doi:10.1017/S0043933915000069.

Daley, W. D. R., Wyvill, J. C., Thompson, J. C., Holcombe, W. D. and McMurray, G. V. 1993. Robotics and the poultry processing industry. In: Khodabandehloo, K. (Ed.), *Robotics in Meat, Fish and Poultry Processing*. Boston, MA: Springer, pp. 48–69.

Daley, W., He, T., Lee, K.-M. and Sandlin, M. 1999. Modeling of the natural product deboning process using biological and human models. In: *Proceedings of 1999 IEEE/ASME International Conference on Advanced Intelligent Mechatronics*. IEEE, pp. 49–54.

De Silva, C. W. and Wickramarachchi, N. 1997. An innovative machine for automated cutting of fish. *IEEE/ASME Transactions on Mechatronics* 2(2), 86–98. doi:10.1109/3516.588627.

Gamage, L. B., De Silva, C. W. and Gosine, R. G. 1993. Statistical pattern recognition for cutter positioning in automated fish processing. In: *Proceedings of IEEE Pacific Rim Conference on Communications, Computers and Signal Processing* (vol. 2). IEEE, pp. 786–9.

Goodfellow, I., Bengio, Y. and Courville, A. 2016. *Deep Learning* (vol. 1). Cambridge, MA: MIT Press.

Guire, G., Sabourin, L., Gogu, G. and Lemoine, E. 2010. Robotic cell for beef carcass primal cutting and pork ham boning in meat industry. *Industrial Robot: An International Journal* 37(6), 532–41. doi:10.1108/01439911011081687.

Heck, B. 2006. Automated chicken processing: Machine vision and water-jet cutting for optimized performance. *IEEE Control Systems* 26(3), 17–9. doi:10.1109/MCS.2006.1636305.

Khodabandehloo, K. 1989. Getting down to the bare bones. *Industrial Robot: An International Journal* 16(3), 160–5. doi:10.1108/eb005062.

Long, P., Khalil, W. and Martinet, P. 2013. Modeling & control of a meat-cutting robotic cell. In: *16th International Conference on Advanced Robotics (ICAR)*. IEEE, pp. 1–6.

Long, P., Khalil, W. and Martinet, P. 2014. Robotic deformable object cutting: From simulation to experimental validation. In: *European Workshop on Deformable Object Manipulation (EWDOM)*.

Misimi, E., Øye, E. R., Eilertsen, A., Mathiassen, J. R., Åsebø, O. B., Gjerstad, T., Buljo, J. and Skotheim, Ø. 2016. GRIBBOT–Robotic 3D vision-guided harvesting of chicken fillets. *Computers and Electronics in Agriculture* 121, 84–100. doi:10.1016/j.compag.2015.11.021.

Omar, F. K. and De Silva, C. W. 2000. Optimal portion control of natural objects with application in automated cannery processing of fish. *Journal of Food Engineering* 46(1), 31–41. doi:10.1016/S0260-8774(00)00068-6.

Omar, F. K. and De Silva, C. W. 2001. Optimal portion control using variable cutter-blade spacing in can-filling. *Packaging Technology and Science* 14(4), 137–48. doi:10.1002/pts.543.

Subrin, K., Sabourin, L., Stephan, F., Gogu, G., Alric, M. and Mezouar, Y. 2014. Analysis of the human arm gesture for optimizing cutting process in ham deboning with a redundant robotic cell. *Industrial Robot: An International Journal* 41(2), 190–202. doi:10.1108/IR-04-2013-346.

Wadie, I., Maddock, N., Purnell, G., Khodabandehloo, K., Crooks, A., Shacklock, A. and West, D. 1995. Robots for the meat industry. *Industrial Robot: An International Journal* 22(5), 22–4. doi:10.1108/01439919510104111.

Yoshikawa, T. and Sudou, A. 1993. Dynamic hybrid position/force control of robot manipulators-on-line estimation of unknown constraint. *IEEE Transactions on Robotics and Automation* 9(2), 220–6. doi:10.1109/70.238286.

Zhou, D., Holmes, J., Holcombe, W. and McMurray, G. 2007. Automation of the bird shoulder joint deboning. In: *2007 IEEE/ASME International Conference on Advanced Intelligent Mechatronics*. IEEE, pp. 1–6.

Zhou, D., Daley, W. and McMurray, G. 2009. Kinematics and verification of a deboning device. In: *International Conference on Mechatronics and Automation*. IEEE, pp. 2143–8.

Index

www.ingramcontent.com/pod-product-compliance
Ingram Content Group UK Ltd.
Pitfield, Milton Keynes, MK11 3LW, UK
UKHW021826100926
8556IPUK00003B/181

9 781786 762726